APPLIED STATISTICAL SCIENCE

RESEARCH IN APPLIED STATISTICAL SCIENCE

APPLIED STATISTICAL SCIENCE

Additional books in this series can be found on Nova's website under the Series tab.

Additional e-books in this series can be found on Nova's website under the e-book tab.

APPLIED STATISTICAL SCIENCE

RESEARCH IN APPLIED STATISTICAL SCIENCE

MOHAMMAD AHSANULLAH
EDITOR

nova publishers
New York

LIBRARY OF CONGRESS CATALOGING-IN-PUBLICATION DATA

Research in applied statistical science / [edited by] Mohammad Ahsanullah.
pages cm. -- (Applied statistical science)
Includes bibliographical references and index.
ISBN 978-1-63321-818-5 (hardcover)
1. Statistics. I. Ahsanullah, Mohammad, editor.
QA276.18.R47 2014
519.5--dc23

2014031862

Published by Nova Science Publishers, Inc. † New York

CONTENTS

Preface **vii**

Chapter 1 EDF Goodness-of-Fit Tests for Testing the Distributional Assumptions in One-Way Regression Model **1**
Dhanuja Kasturiratna, Truc T. Nguyen and Arjun K. Gupta

Chapter 2 Sample Size Determination for between Group Comparisons in Mixed-Effects Logistic Regression Models for Analysis of Longitudinal Data **11**
Dulal K. Bhaumik, Subhash Aryal, Anup Amatya and Kush Kapur

Chapter 3 Mathematical Modeling of Mortality for Countries with Limited and Defective Data **23**
Jyoti U. Devkota and Radhey S. Singh

Chapter 4 On Characterizations of the Exponential Distributions **37**
Mohammad Ahsanullah, Alexandre Berred and Valery Nevzorov

Chapter 5 Estimation of $P[Y<X]$ for Generalized Logistic Distribution **45**
H. M. Okasha

Chapter 6 A Class of Saturated Binary *S*-Optimal Row-Column Designs **59**
Theophilus Ogunyemi

Chapter 7 Multiply and Partially Confounded Factorial BIB Designs **75**
M. Shamsuddin and M. Albassam

Chapter 8 An Extended Version of Hyper Poisson Distribution and Some of Its Applications **83**
C. Satheesh Kumar and B. Unnikrishnan Nair

Chapter 9 Repeated Measurements Designs for Comparing Two Disjoint Sets of Formulations in Bioequivalence Trials **93**
M. N. Sonawane, Cini Varghese and Seema Jaggi

Chapter 10	Role of Categorical Variables in Multicollinearity in Linear Regression Model *M. Wissmann, Shalabh and H. Toutenburg*	**103**
Chapter 11	Predictive Inference from Two-parameter Weibull Model Given a Doubly Censored Sample of Hepatocellular Carcinoma of Liver Patients *Hafiz M.R. Khan, Ahmed Albatineh, Salem Tesfa, Nadine R. Jenkins and Nasar U. Ahmed*	**119**
Chapter 12	Parameters Estimation of the Log-logistic Distribution Using Ranked Set Sampling *I. Elbatal*	**133**
Chapter 13	Characterization of Some Discrete Families Based on Doubly Truncated Mean and Covariance *M. A. H. Khorasadizadeh, Rezaei Roknabadi and G. R. Borzadaran*	**143**
Chapter 14	Different Methods of Estimation for Marshall Olkin Exponential Distribution *Omar M. Bdair*	**155**
Chapter 15	Characterization of Beta and Gamma Distribution Based on the Doubly Truncated Mean Function *Zohdy M. Nofal*	**173**
Chapter 16	Estimation of the Seasonal ACF of PAR(1) Model in the Presence of Additive Outliers *A. A. Samadi and A. M. Al-Quraam*	**183**
Chapter 17	Regression Models for Poisson Data Based on Accelerated Lifetime Testing *Francisco Louzada-Neto and Vera Lucia Damasceno Tomazella*	**199**
Chapter 18	On Methods of Construction of Partially Balanced Incomplete Block Designs with Three-Associate-Class Association Scheme *Swati Raj, Jagdish Prasad, D. K. Ghosh and Sarla Pareek*	**209**
Chapter 19	Lower Generalized Order Statistics of Generalized Inverse Weibull Distribution *R. U. Khan, Benazir Zia and Haseeb Athar*	**219**
Chapter 20	Characterization of Continuous Distributions Conditioned on a Pair of Non-adjacent Generalized Order Statistics Using Meijer's G-function *M. J. S. Khan, A. H. Khan and M. I. Khan*	**237**
Chapter 21	Empirical Bayes Method on Changepoints Estimation of Tumor Growth Profiles in Xenograft Experiments *Wei Ning*	**249**
Index		**261**

PREFACE

This book makes a significant contribution to the advancement of statistical science. It contains research in many statistical designs, compares many statistical models, and includes theories that are oriented to real life problems.

The location and scale parameters in the distributional assumption of one-way regression model are eliminated by means of suitable transformations on the original observations. The transformed random variables are then shown in Chapter 1 to follow Student's t distributions with known parameters under the null hypothesis. The goodness-of-fit tests can then be performed on the transformed variables for testing the null hypothesis. The powers of the tests are estimated by Monte Carlo method for several alternatives.

In Chapter 2 the authors consider the problem of sample size determination for two-level mixed-effects logistic regression models for the analysis of longitudinal data. In this case, level 1 represents measurement occasion and level 2 represents subject. The authors provide an algorithm for sample size requirements (i.e., power characteristics) for a test of the treatment-by-time interaction and overall time trend that strictly maintains the type I error rate close to the nominal level. The methodology is illustrated both via simulation and an example from a now classic psychiatric randomized clinical trial.

Life tables explain the mortality experience of a group called cohort. The survivorship function plays a vital role in life tables. In developing countries with limited and scarce data, survivorship function is taken as an index of development and is used by policy makers and planners. Inefficient registration of vital events and geographically remote areas in many developing countries hinder the regular availability of accurate data. Hence mathematical modeling of such data is of utmost importance because of their role as development indicators. In Chapter 3, survivorship function of a developing country particularly Nepal is regressed on age and time variables. A parsimonious regression model with a very few regression coefficients, especially suited to such countries, keeping the constraints of limited and defective data is developed. The developed form satisfies all the criteria of a good model not only for developing countries like Nepal and India, but also for developed countries like Germany. A comparison of results is made between the data of countries with abundant and good quality data on one hand and limited and defective data on the other hand. Further the uncertainty in Nepalese data is measured by analyzing the error in the model parameters using Bayesian analysis.

In Chapter 4, the exponential distributions are characterized through an assumption on order statistics of the original random variable and the exponential random variables.

Chapter 5 deals with the problem of estimation of the stress-strength function $R=P(Y<X)$, when X and Y are two independent but not identically distributed random variables belonging to the generalized logistic distribution (GLD). Different estimators of R, namely, maximum likelihood, uniformly minimum variance unbiased and Bayes are derived in closed form. The Bayes estimates are based on conjugate and non-informative priors. Interval estimation of R (classical and Bayesian) are also discussed. A Monte Carlo simulation study is given for a comparisons between the different methods of estimation.

A new class of binary, square s-optimal row-column designs is proposed in Chapter 6. These designs are saturated in terms of eliminating two-way heterogeneity with an additive model. S-optimality criterion is used to select optimal designs. S-optimal designs are constructed for square designs that are treatment-connected with a relatively large number of nearly equi-replicated treatments. In light of connectedness, all treatment contrasts are estimable regardless of the row and column effects.

The s^n factorial design, where s is a prime or prime power, generally confounds a set of different orthogonal factorial effects with s^r blocks ($n > r > 1$) under one replicate. When this s^n factorial design in s^r blocks partially confounds different orthogonal sets of orthogonal factorial effects under the complete set of different orthogonal replicates it constitutes a BIB design which is called 'Multiply and Partially Confounded Factorial BIB Design' having similar properties, same result and detailed information of all factorial effects. The procedure is illustrated with examples in Chapter 7.

Through Chapter 8 the authors propose an extended version of hyper-Poisson distribution and derive its probability mass function, mean, variance and recurrence relations for probabilities, raw moments and factorial moments. Further, the authors discussed the estimation of the parameters of this extended version by the method of moments, method of mixed moments and method of maximum likelihood, and it is demonstrated with the help of a real data that the extended version fit the situations better than the existing model.

Bioequivalence is the degree to which clinically important outcomes after receiving a new (test) formulation resembles a well-established (reference) formulation. Bioequivalence trials form an integral part of the development programme of every drug product. Under these trials, the main interest of the experimenter is in the estimation of contrasts pertaining to test vs. test formulations and test vs. reference formulations with as high a precision as possible. Repeated measurements designs (RMDs) with two sets of formulations, one set of test and other of reference formulations, assuming the presence of residual effects are suitable for such trials. In Chapter 9, two classes of RMDs for bioequivalence trials have been proposed assuming the presence of residual effects of formulations. Variances of the estimates of contrasts pertaining to direct as well as residual effects of test formulations, reference formulations and test vs. reference formulations have been worked out for these designs. Variance of test vs. reference formulations and test vs. test formulations comparisons were found to be less as compared to reference vs. reference comparisons indicating that these designs are suitable for bioequivalence trials.

Chapter 10 discusses the role of categorical variable in the problem of multicollinearity in linear regression model. It exposes the diagnostic tool condition number to linear regression models with categorical explanatory variables and analyzes analytically as well as numerically how the dummy variables and choice of reference category can affect the degree of multicollinearity.

The purpose of Chapter 11 is to derive the predictive densities for future responses from a two-parameter Weibull model given a doubly censored sample. The predictive density for a single future response, bivariate future response, and a set of future responses are derived when the shape parameter α is either known or unknown. A numerical example of hepatocellular carcinoma of 69 liver patients is utilized to illustrate the predictive results for the stages of cancer. The survival days of 69 male and female patients with primary hepatocellular carcinoma of liver are analyzed. Four patients out of 69 patients are lost to follow-up and they are the first two and the last two observations of 69 patients. Furthermore, a sensitivity study of the mean and standard deviation of the predictive densities of survival days with respect to several values of the hyperparameter τ is performed when the shape parameter α is unknown. It is found that the rate of the males patients future survival days is lower than the females patients at the initial stage and higher at the later stage.

The log-logistic distribution is very similar in shape to the lognormal distribution but is mathematically more tractable and hence very useful in a wide variety of applications, especially analysis of survival data. It is also noted that the log-logistic distribution is a good alternative to the Weibull distribution, whose hazard function, while it may be increasing or decreasing, must be monotonic, whatever the values of its parameters. In Chapter 12 the authors derive the parameters estimator using the method of the maximum likelihood based on simple random samples, and ranked set sampling. Finally the authors use Montocarlo simulations to compare the results for the SRS data and the RSS data.

In Chapter 13, the authors generalize the characterizations of discrete lifetime distributions based on conditional doubly truncated mean and covariance which are related to the general failure rates, doubly truncated mean and doubly truncated variance of a lifetime random variable. In addition, the authors obtained the results for some general families and well-known distributions such as, modified power series, Ord, Katz and Telescopic form of distributions.

Marshall Olkin exponential distribution (MOE) has been recently studied quite extensively by several authors. In Chapter 14, the authors examine different methods of estimation for different parameters of the Marshall Olkin exponential distribution function. An extensive numerical simulation is used to compare the performances of these procedures.

In Chapter 15 the authors introduce a new characterization to beta and gamma distribution using conditional expectation in terms of failure rate and reversed failure rate. A new result is presented in this paper.

In Chapter 16 the 2003 works of Berkoun et al. for the estimation of the first lag autocorrelation (of the autoregressive model of order one (AR(1)) as well as that of Smadi et al. from 2009 for the estimation of the seasonal autocorrelation for the periodic autoregressive model of order one (PAR(1)) are considered. In the later work, the authors have generalized the estimators considered by the former towards the periodic case for the first lag seasonal autocorrelation. Here, the authors will again consider the PAR(1) model and the same estimators considered by Smadi et al., namely the moment estimator and two robust estimators. The authors will, however, generalize those estimators for the PAR(1) model but for all time lags. Some useful results about the theoretical seasonal autocorrelation function (SACF) of the PAR(1) model are derived. Then, a Monte-Carlo study is carried out to study the behaviors of all estimators in the presence of some additive outliers. The comparison of various estimators was done through the bias and MSE criterion. It is observed that the moment estimator is highly affected by the outliers whereas the other two estimators were less affected. A real application to river-flow time series is also presented.

The accelerated failure time model (AFT) is an important alternative to the Cox proportional hazards model (PH) in survival analysis. In Chapter 17 the authors present a natural extension of the conventional accelerated failure time model to accommodate recurrent events. This extension parallels the 1987 Lawless extensions of the classical Cox proportional hazard model. Like Lawless we formulate the effects of covariate on the intensity function of Poisson process for recurrent events. The proposed model retains the direct physical interpretation of the original accelerated failure time model in that the role of the covariates is to accelerate or decelerate the time to each recurrence. These include parametric approaches to model fitting, the authors consider estimate the vector of regression parameters under this model and the baseline intensity are considered specified up to a parameter θ. This methodology is illustrated with a known real data set.

Rectangular association scheme is three-associate-class association scheme. In Chapter 18 the authors have constructed rectangular association scheme using difference set for a balanced incomplete block design and symmetrical balanced incomplete block design. Here they have also derived some parametric relations of rectangular association scheme. Further based on these results, a table is also provided in the range of $4 \leq r, k \leq 20$.

In Chapter 19, some recurrence relations satisfied by single and product moments of lower generalized order statistics from generalized inverse Weibull distribution are derived and characterization of this distribution based on the conditional expectation of function of lower generalized order statistics and a recurrence relation for single moments are given. Further, some important deductions and computational works are carried out.

In Chapter 20, a generalized family of continuous distributions have been characterized through conditional expectation of generalized order statistics, conditioned on a pair of non-adjacent generalized order statistics using Meijer's G-function and its various deductions are discussed.

In Chapter 21, a method based on the empirical Bayes method is proposed to estimate the change point of a changepoint model, which is used to model the regrowth points of the tumor volume in xenograft experiment. Comparing with the other Bayes changepoint models, the advantage of using the proposed method is to estimate the parameters of the prior distributions through the marginal distribution of the data, instead of choosing values for the parameters vaguely in the prior distributions. The results show that the proposed method can discover the change pattern in the data successfully.

In: Research in Applied Statistical Science
Editor: Mohammad Ahsanullah, pp. 1-10
ISBN: 978-1-63321-818-5

Chapter 1

EDF Goodness-of-Fit Tests for Testing the Distributional Assumptions in One-Way Regression Model

***Dhanuja Kasturiratna*[1], *Truc T. Nguyen*[2] *and Arjun K. Gupta*[2]**
[1] Department of Mathematics and Statistics
Northern Kentucky University, Highland Heights, Kentucky, US
[2] Department of Mathematics and Statistics,
Bowling Green State University, Bowling Green, Ohio, US

Abstract

The location and scale parameters in the distributional assumption of one-way regression model are eliminated by means of suitable transformations on the original observations. The transformed random variables are then shown to follow Student's t distributions with known parameters under the null hypothesis. The goodness-of-fit tests can then be performed on the transformed variables for testing the null hypothesis. The powers of the tests are estimated by Monte Carlo method for several alternatives.

AMS 2000 Subject Classification: Primary 62E10, Secondary 62F03

Keywords: Conditional density function, UMVUE of the density function, Moment, EDF goodness-of fit tests, Power

1. Introduction

In the one-way classification with k treatments, we assume that the data are observed according to the additive model, where the error random variables are independently and identically distributed normally with mean zero and constant variance. Then to test whether a set of observed data are coming from the above regression model, we need to construct a test for testing the hypothesis that the observations of the treatment groups are normally distributed with same variance. A common practice of testing this hypothesis is to first test the normality of the samples separately and then test the equality of variance. In this case researches don't concern about the type I error, which is no longer the expected level unless they adjust the level of the tests. The motivation of characterization given in this paper is to find a transformation in the procedure to construct an exact empirical distribution function (EDF) goodness-of-fit test for testing the above hypothesis.

Characterizations based on the conditional moments have been studied by several authors. Rao (1967) gave characterization of normal distribution based on conditional expected value of one linear statistic given two other linear statistics. Singh and Oliker (1979) studied minimum variance unbiased estimation and characterization of densities of several popular distributions, and proposed a conjecture on how to characterize a normal distribution based on the UMVU estimate of its density function. Gupta and Varga (1990) studied the relationship between joint density and conditional densities . Wang, Gupta and Nguyen (1996) gave characterization theorems for some discrete distributions based on conditional structure. Gupta, Nguyen and Wang (1997) also gave characterization of normal and gamma distributions via conditional structure. Nguyen and Dinh (1998) gave characterization of normal distribution based on the third and, fourth conditional moments. Characterizations supporting goodness-of fit tests for normal distribution are given by several authors. Kotlarsky (1966) characterized the normal distribution by Student's law. Csargo, Seshadri and Yalovsky (1973) proposed exact EDF goodness-of fit tests for normality in the presence of unknown parameters. Nguyen and Dinh (2003) gave characterization of normal distribution based on the UMVU estimator of the density function which can be used as a transformation in the procedure to construct an EDF goodness-of-fit test for testing normality of a distribution.

In this paper, in section 2, characterization of normal distribution related to two samples based on second conditional moments, given by Kasturiratna, Nguyen and Gupta (2007), has been changed to a characterization based on Student's law. The characterization result is extended to k samples in section 3. Application of the characterization result to goodness-of-fit test is discussed in section 4. In section 5, Monte Carlo method is used to compute the power of EDF goodness-of fit tests for testing the normality and equivariance of samples.

2. Characterization of Normal Distribution Related to Two Samples Based on Student's Law

Let $X_i, \quad i = 1, \ldots, n$ and $Y_j, \quad j = 1, \ldots, m, \quad m, n \geq 2$, be two independent random samples from two normal distributions with means ξ and η, respectively, and with common variance σ^2. Then $Z_1 = \frac{\sum_{i=1}^{n} X_i}{n}, \quad Z_2 = \frac{\sum_{j=1}^{m} Y_j}{m}, \quad Z_3 = \sum_{i=1}^{n} (X_i - Z_1)^2 + \sum_{j=1}^{m} (Y_j - Z_2)^2$, are jointly complete and sufficient statistics for ξ, η, and σ^2.

Let $\underline{X}_{n-2} = (X_1, \ldots, X_{n-2})$ and $\underline{Y}_{m-1} = (Y_1, \ldots, Y_{m-1})$. The conditional joint density function of $\underline{X}_{n-2}, \underline{Y}_{m-1}$ given Z_1, Z_2, Z_3 is

$$f_{\underline{X}_{n-2}, \underline{Y}_{m-1} | Z_1, Z_2, Z_3}(\underline{x}_{n-2}, \underline{y}_{m-1} | z_1, z_2, z_3) =$$

$$K_1(n). \frac{1}{z_3^{\left(\frac{m+n-2}{2}-1\right)} \sqrt{2\left[z_3 - \sum_{j=1}^{m-1} y_j^2 - \left(mz_2 - \sum_{j=1}^{m-1} y_j\right)^2 + mz_2^2 + nz_1^2 - \sum_{i=1}^{n-2} x_i^2\right] - \left(nz_1 - \sum_{i=1}^{n-2} x_i\right)^2}},$$

$$\text{for} \quad (\underline{x}_{n-2}, \underline{y}_{m-1}, z_1, z_2, z_3) \in \mathcal{D} \tag{2.1}$$

where $K_1(n) = \frac{\sqrt{mn} \quad \Gamma\left(\frac{m+n-2}{2}\right)}{\pi^{\left(\frac{m+n-2}{2}\right)}}$ and $\mathcal{D} = \Bigg\{ (\underline{x}_{n-2}, \underline{y}_{m-1}, z_1, z_2, z_3) :$

$$2\left[z_3 - \sum_{j=1}^{m-1} y_j^2 - \left(mz_2 - \sum_{j=1}^{m-1} y_j\right)^2 + mz_2^2 + nz_1^2 - \sum_{i=1}^{n-2} x_i^2\right] > \left(nz_1 - \sum_{i=1}^{n-2} x_i\right)^2 \Bigg\}.$$

The UMVUE of the density function of X_1 at a given point x_1 is

$$f_{X_1|Z_1,Z_2,Z_3}(x_1|z_1, z_2, z_3) = K_2(n). \frac{\left[1 - \frac{n}{n-1}(x_1 - z_1)^2/z_3\right]^{\frac{m+n-5}{2}}}{\sqrt{z_3}}$$

$$\text{for } (x_1 - z_1)^2 < \frac{n-1}{n} z_3, \tag{2.2}$$

where $K_2(n) = \frac{\sqrt{n} \quad \Gamma\left(\frac{m+n-2}{2}\right)}{\sqrt{(n-1)\pi} \quad \Gamma\left(\frac{m+(n-1)-2}{2}\right)}$.

Similarly, the UMVUE of the density function of Y_1 at a given point y_1 is

$$f_{Y_1|Z_1,Z_2,Z_3}(y_1|z_1, z_2, z_3) = K_3(n). \quad \frac{\left[1 - \frac{m}{m-1}(y_1 - z_2)^2/z_3\right]^{\frac{m+n-5}{2}}}{\sqrt{z_3}}$$

$$\text{for } (y_1 - z_2)^2 < \frac{m-1}{m} z_3, \tag{2.3}$$

where $K_3(n) = \frac{\sqrt{m} \quad \Gamma\left(\frac{m+n-2}{2}\right)}{\sqrt{(m-1)\pi} \quad \Gamma\left(\frac{(m-1)+n-2}{2}\right)}$.

Kasturiratna, Nguyen and Gupta (2007) gave the following characterization of normal distribution related to two samples based on second conditional moments.

Let F and G be two nondegenerate distributions with finite second moments and let X_i, $i = 1, \ldots, n$, $n \geq 2$, Y_j, $j = 1, \ldots, m$, $m \geq 2$ be two independent random samples from F and G, respectively. Then F is a $N(\xi, \sigma^2)$ distribution and G is a $N(\eta, \sigma^2)$ distribution if and only if the joint conditional density function of $X_1, \ldots, X_{n-2}, Y_1, \ldots, Y_{m-1}$ given Z_1, Z_2, Z_3 is given by (2.1).

This characterization can be changed to a characterization using Student's law. Given Z_1, Z_2 and Z_3, the following consecutive one-to-one transformations are considered.

Let

$$u_i = \frac{x_i - z_1}{\sqrt{z_3}} \qquad i = 1, \ldots, n-2,$$

$$u_{n-2+j} = \frac{y_j - z_2}{\sqrt{z_3}} \qquad j = 1, \ldots, m-1. \tag{2.4}$$

Then the conditional distributions of $U_1, \ldots, U_k$ given Z_1, Z_2, Z_3 is obtained as

$$f_{(u_1,\ldots,u_k|z_1,z_2,z_3)} = K_4(n) \left[1 - \frac{n-k+1}{n-k} \sum_{i=1}^{k} u_i^2 - \frac{2}{n-k} \sum_{1 \leq i < i' \leq k} u_i u_{i'}\right]^{\frac{m+n-k-4}{2}}$$

for $\frac{n-k+1}{n-k}\sum_{i=1}^{k} u_i^2 + \frac{2}{n-k}\sum_{1\le i<i'\le k} u_i u_{i'} < 1,$

$$k = 1, \ldots, n-2, \tag{2.5}$$

where $K_4(n) = \dfrac{\sqrt{n}\quad \Gamma\left(\frac{m+n-2}{2}\right)}{\sqrt{(n-k)}\quad \pi^{\frac{k}{2}}\quad \Gamma\left(\frac{m+(n-k)-2}{2}\right)}$,

and the conditional distribution of $U_1, \ldots, U_{n-2}, U_{n-1}, \ldots, U_{n-2+l}$ given Z_1, Z_2, Z_3 is obtained as

$$f_{(u_1,\ldots,u_{n-2},u_{n-1},\ldots,u_{n-2+l}|z_1,z_2,z_3)} = K_5(n)\Bigg[1 - \frac{3}{2}\sum_{i=1}^{n-2} u_i^2 - \sum_{1\le i<i'\le n-2} u_i u_{i'} - \frac{m-l+1}{m-l}\sum_{j=1}^{l} u_{n-2+j}^2 - \sum_{1\le j<j'\le l} u_{n-2+j}u_{n-2+j'}\Bigg]^{\frac{m-l-2}{2}}$$

for $\frac{3}{2}\sum_{i=1}^{n-2} u_i^2 + \sum_{1\le i<i'\le n-2} u_i u_{i'} + \frac{m-l+1}{m-l}\sum_{j=1}^{l} u_{n-2+j}^2 + \sum_{1\le j<j'\le l} u_{n-2+j}u_{n-2+j'} \le 1$

$$l = 1, \ldots, m-1, \tag{2.6}$$

where $K_5(n) = \dfrac{\sqrt{mn}\quad \Gamma\left(\frac{m+n-2}{2}\right)}{\sqrt{(m-l)2}\quad \pi^{\frac{n+l-2}{2}}\quad \Gamma\left(\frac{m-l}{2}\right)}$.

We can see both (2.5) and (2.6) do not depend on Z_1, Z_2 and Z_3. Further, from (2.5) and (2.6) the conditional distribution of U_k given $U_1, \ldots, U_{k-1}, Z_1, Z_2, Z_3$ is obtained as

$$f_{(u_k|u_1,\ldots,u_{k-1},z_1,z_2,z_3)} = K_6(n)\frac{\left[1-\frac{n-k+1}{n-k}\sum_{i=1}^{k} u_i^2 - \frac{2}{n-k}\sum_{1\le i<i'\le k} u_i u_{i'}\right]^{\frac{m+n-k-4}{2}}}{\left[1-\frac{n-k+2}{n-k+1}\sum_{i=1}^{k-1} u_i^2 - \frac{2}{n-k+1}\sum_{1\le i<i'\le k-1} u_i u_{i'}\right]^{\frac{m+n-k-3}{2}}}$$

for $\frac{n-k+1}{n-k}\sum_{i=1}^{k} u_i^2 + \frac{2}{n-k}\sum_{1\le i<i'\le k} u_i u_{i'} < 1$, and

$\frac{n-k+2}{n-k+1}\sum_{i=1}^{k-1} u_i^2 + \frac{2}{n-k+1}\sum_{1\le i<i'\le k-1} u_i u_{i'} < 1,$

$$k = 1, \ldots, n-2, \tag{2.7}$$

where $K_6(n) = \dfrac{\sqrt{(n-k+1)}\quad \Gamma\left(\frac{m+n-k-1}{2}\right)}{\sqrt{(n-k)\pi}\quad \Gamma\left(\frac{m+n-k-2}{2}\right)}$,

and the conditional distribution of U_{n-2+l} given $U_1, \ldots, U_{n-2}, U_{n-1}, \ldots, U_{n-2+(l-1)}, Z_1, Z_2, Z_3$ is obtained as

$$f_{(u_{n-2+l}|u_1,\ldots,u_{n-2},u_{n-1},\ldots,u_{n-2+(l-1)},z_1,z_2,z_3)} =$$

$$\frac{K_7(n)\left[1-\frac{3}{2}\sum_{i=1}^{n-2} u_i^2 - \sum_{1\le i<i'\le n-2} u_i u_{i'} - \frac{m-l+1}{m-l}\sum_{j=1}^{l} u_{n-2+j}^2 - \frac{2}{m-l}\sum_{1\le j<j'\le l} u_{n-2+j}u_{n-2+j'}\right]^{\frac{m-l-2}{2}}}{\left[1-\frac{3}{2}\sum_{i=1}^{n-2} u_i^2 - \sum_{1\le i<i'\le n-2} u_i u_{i'} - \frac{m-l+2}{m-l+1}\sum_{j=1}^{l-1} u_{n-2+j}^2 - \frac{2}{m-l+1}\sum_{1\le j<j'\le l-1} u_{n-2+j}u_{n-2+j'}\right]^{\frac{m-l-1}{2}}}$$

$$\text{for } A + \tfrac{m-l+1}{m-l} \sum_{j=1}^{l} u_{n-2+j}^2 + \tfrac{2}{m-l} \sum_{1 \le j < j' \le l} u_{n-2+j} u_{n-2+j'} < 1, \text{ and}$$

$$A + \tfrac{m-l+2}{m-l+1} \sum_{j=1}^{l-1} u_{n-2+j}^2 + \tfrac{2}{m-l+1} \sum_{1 \le j < j' \le l-1} u_{n-2+j} u_{n-2+j'} < 1,$$

$$l = 1, \ldots, m-1, \tag{2.8}$$

where $K_7(n) = \frac{\sqrt{(m-l+1)}}{\sqrt{(m-l)\pi}} \frac{\Gamma\left(\frac{m-l+1}{2}\right)}{\Gamma\left(\frac{m-l}{2}\right)}$ and $A = \frac{3}{2}\sum_{i=1}^{n-2} u_i^2 + \sum_{1 \le i < i' \le n-2} u_i u_{i'}$.

Now consider the transformation

$$v_i = \frac{\sqrt{\frac{n-i+1}{n-i}} u_i + \frac{\sum_{j=1}^{i-1} u_j}{\sqrt{(n-i+1)(n-i)}}}{\sqrt{1 - \sum_{j=1}^{i-1} u_j^2 - \frac{\left(\sum_{j=1}^{i-1} u_j\right)^2}{n-i+1}}} \qquad \text{for } i = 1, \ldots, n-2,$$

$$v_{n-2+l} = \frac{\sqrt{\frac{m-l+1}{m-l}} u_{n-2+l} + \frac{\sum_{j=1}^{l-1} u_{n-2+j}}{\sqrt{(m-l+1)(m-l)}}}{\sqrt{1 - \frac{3}{2}\sum_{i=1}^{n-2} u_i^2 - \sum_{1 \le i < i' \le n-2} u_i u_{i'} - \sum_{j=1}^{l-1} u_{n-2+j}^2 - \frac{\left(\sum_{j=1}^{l-1} u_{n-2+j}\right)^2}{m-l+1}}}$$

$$\text{for } l = 1, \ldots, m-1. \tag{2.9}$$

Using (2.7) and (2.8), the conditional density function of V_k given $U_1, \ldots, U_{k-1}$, Z_1, Z_2, Z_3 is obtained as

$$f_{(v_k | u_1, \ldots, u_{k-1}, z_1, z_2, z_3)} = \frac{\Gamma\left(\frac{m+n-k-1}{2}\right)}{\sqrt{\pi}\ \Gamma\left(\frac{m+n-k-2}{2}\right)} \left[1 - v_k^2\right]^{\frac{m+n-k-4}{2}}$$

$$\text{for } v_k^2 < 1, \qquad k = 1, \ldots, m+n-3 \tag{2.10}$$

Notice that the density function (2.10) does not depend on $U_1, \ldots, U_{k-1}$. Now we can obtain the conditional joint density of $V_1, \ldots, V_{m+n-3}$ given Z_1, Z_2, Z_3, which is

$$f_{(v_1, \ldots, v_{m+n-3} | z_1, z_2, z_3)} = \prod_{j=1}^{m+n-3} \frac{\Gamma\left(\frac{m+n-j-1}{2}\right)}{\sqrt{\pi}\ \Gamma\left(\frac{m+n-j-2}{2}\right)} \left[1 - v_j^2\right]^{\frac{m+n-j-4}{2}}$$

$$\text{for } v_i^2 < 1. \tag{2.11}$$

Transforming V_i's to W_i's using

$$w_i = \sqrt{m+n-i-2} \frac{v_i}{\sqrt{1 - v_i^2}} \qquad i = 1, \ldots, m+n-3, \tag{2.12}$$

from (2.11), we can obtain the conditional density of $W_1, \ldots, W_{m+n-3}$ given Z_1, Z_2, Z_3 as

$$f_{(w_1,\ldots,w_{m+n-3}|z_1,z_2,z_3)}$$

$$= \prod_{j=1}^{m+n-3} \frac{\Gamma\left(\frac{m+n-j-1}{2}\right)}{\sqrt{\pi(m+n-j-2)}\ \Gamma\left(\frac{m+n-j-2}{2}\right)\left[1+\frac{w_j^2}{m+n-j-2}\right]^{\frac{m+n-j-1}{2}}}, \tag{2.13}$$

for all real values of $w_1, \ldots, w_{m+n-3}$.

From (2.13), given Z_1, Z_2 and Z_3, $W_i, i = 1, \ldots, m+n-3$, are conditionally independently distributed according to Student's t distribution with $m+n-i-2$ degrees of freedom, respectively.

We summarize these results in the following theorem.

Theorem 2.1. *Let F and G be two nondegenerate distributions with finite second moments and let X_i, $i = 1, \ldots, n$, $n \geq 2$, Y_j, $j = 1, \ldots, m$, $m \geq 2$ be two independent random samples from F and G, respectively. In the conditional model given Z_1, Z_2 and Z_3, let $W_i, i = 1, \ldots, m+n-3$ be obtained from $X_1, \ldots, X_{n-2}, Y_1, \ldots, Y_{m-1}$ by the sequence of one-to-one transformations defined by (2.4), (2.9), and (2.12). Then F is a $N(\xi, \sigma^2)$ distribution and G is a $N(\eta, \sigma^2)$ distribution if and only if $W_1, \ldots, W_{m+n-3}$ are conditionally jointly distributed according to (2.13), given Z_1, Z_2, Z_3.*

3. Characterization of Normal Distribution Related to k Samples Based on Student's Law

Kasturiratna, Nguyen and Gupta (2007) gave the following characterization of normal distribution related to k samples based on second conditional moments.

Let $X_{i1}, \ldots, X_{in_i}$, $i = 1, \ldots, k$, $n_i \geq 2$ be k random samples from nondegenerate distributions F_i, $i = 1, \ldots, k$ respectively, with a finite second moment. Let $Z_i = \frac{\sum_{j=1}^{n_i} x_{ij}}{n}$, $i = 1, \ldots, k$, $Z_{k+1} = \sum_{i=1}^{k}\sum_{j=1}^{n_i}(x_{ij} - z_i)^2$.Then F_i is a $N(\xi_i, \sigma^2)$, $i = 1, \ldots, k$ distributions if and only if the joint conditional density function of $X_{11}, \ldots, X_{1(n_1-2)}, X_{21}, \ldots, X_{2(n_2-1)}, \ldots, X_{k1}, \ldots, X_{k(n_k-1)}$ given $Z_1, \ldots, Z_k, Z_{k+1}$ is

$$f_{(x_{11},\ldots,x_{1(n_1-2)},x_{21},\ldots,x_{2(n_2-1)},\ldots,x_{k1},\ldots,x_{k(n_k-1)}|z_1,\ldots,z_k,z_{k+1})}$$

$$= \frac{K_9(n)}{z_{k+1}^{\left(\frac{n_1+\ldots+n_k-k}{2}-1\right)}} \cdot$$

$$\frac{1}{\sqrt{2\left[z_{k+1} - \sum_{i=2}^{k}\sum_{j=1}^{n_i-1} x_{ij}^2 - \sum_{i=2}^{k}\left(n_i z_i - \sum_{j=1}^{n_i-1} x_{ij}\right)^2 + \sum_{i=1}^{k} n_i z_i^2 - \sum_{j=1}^{n_1-2} x_{1j}^2\right] - \left(n z_1 - \sum_{j=1}^{n_1-2} x_{1j}\right)^2}},$$

$$\text{for} \quad 2\left[z_{k+1} - \sum_{i=2}^{k}\sum_{j=1}^{n_i-1} x_{ij}^2 - \sum_{i=2}^{k}\left(n_i z_i - \sum_{j=1}^{n_i-1} x_{ij}\right)^2 + \sum_{i=1}^{k} n_i z_i^2 - \sum_{j=1}^{n_1-2} x_{1j}^2\right]$$

$$> \left(n z_1 - \sum_{j=1}^{n_1-2} x_{1j}\right)^2, \tag{3.14}$$

where $K_9(n) = \dfrac{\sqrt{n_1.\dots.n_k}\quad \Gamma\left(\frac{n_1+\ldots+n_k-k}{2}\right)}{\pi^{\left(\frac{n_1+\ldots+n_k-k}{2}\right)}}$.

Corollary 3.1. *Let $X_{i1},\ldots,X_{in_i}, n_i \geq 2$ for $i = 1,\ldots,k$, be random samples from non-degenerate distributions F_i, $i = 1,\ldots,k$ respectively, with a finite second moment. In the conditional model given $Z_1,\ldots,Z_k$ and Z_{k+1}, let W_i, $i = 1,\ldots,(n_1+\ldots+n_k-k-1)$ be obtained by the sequence of one-to-one transformations defined by*

$$u_i = \frac{x_i - z_1}{\sqrt{z_{k+1}}}, \qquad i = 1,\ldots,n_1-2,$$

$$u_{n_1+\ldots+n_{s-1}-s+j} = \frac{x_{sj} - z_s}{\sqrt{z_{k+1}}}, \qquad s = 2,\ldots,k, \quad j = 1,\ldots,n_s-1. \tag{3.15}$$

$$v_i = \frac{\sqrt{\frac{n_1-i+1}{n_1-i}}u_i + \frac{\sum_{j=1}^{i-1} u_j}{\sqrt{(n_1-i+1)(n_1-i)}}}{\sqrt{1-\sum_{j=1}^{i-1} u_j^2 - \frac{\left(\sum_{j=1}^{i-1} u_j\right)^2}{n_1-i+1}}} \qquad i = 1,\ldots,n_1-2,$$

$$v_{n_1-2+l} = \frac{\sqrt{\frac{n_2-l+1}{n_2-l}}u_{n_1-2+l} + \frac{\sum_{j=1}^{l-1} u_{n_1-2+j}}{\sqrt{(n_2-l+1)(n_2-l)}}}{\sqrt{1-\frac{3}{2}\sum_{i=1}^{n_1-2} u_i^2 - \sum_{1\leq i<i'\leq n_1-2} u_i u_{i'} - \sum_{j=1}^{l-1} u_{n_1-2+j}^2 - \frac{\left(\sum_{j=1}^{l-1} u_{n_1-2+j}\right)^2}{n_2-l+1}}}$$

$$l = 1,\ldots,n_2-1,$$

$$v_{n_1+\ldots+n_{s-1}-s+l} =$$

$$\frac{\sqrt{\frac{n_s-l+1}{n_s-l}}u_{n_1+\ldots+n_{s-1}-s+l} + \frac{\sum_{j=1}^{l-1} u_{n_1+\cdots+n_{s-1}-s+j}}{\sqrt{(n_s-l+1)(n_s-l)}}}{\sqrt{1-\frac{3}{2}\sum_{i=1}^{n_1-2} u_i^2 - \sum_{1\leq i<i'\leq n_1-2} u_i u_{i'} - 2\sum_{j=3}^{s}\sum_{i=1}^{n_{j-1}-1} u_{n_1+\ldots+n_{j-2}-(j-1)+i}^2 - t_1 - t_2}}$$

$$s = 3,\ldots,k, \quad l = 1,\ldots,n_s-1, \tag{3.16}$$

where

$$t_1 = 2\sum_{j=3}^{s} \sum_{1\le i<i'\le n_{j-1}-1} u_{n_1+\ldots+n_{j-2}-(j-1)+i} u_{n_1+\ldots+n_{j-2}-(j-1)+i'},$$

$$t_2 = \sum_{j=1}^{l-1} u^2_{n_1+\ldots+n_{s-1}-s+j} - \frac{\left(\sum_{j=1}^{l-1} u_{n_1+\ldots+n_{s-1}-s+j}\right)^2}{n_s-l+1}.$$

$$w_i = \sqrt{n_1+\ldots+n_k-i-k}\frac{v_i}{\sqrt{1-v_i^2}} \qquad i = 1,\ldots,n_1+\ldots+n_k-k-1. \quad (3.17)$$

Then F_i is a $N(\xi_i, \sigma^2)$, $i = 1, \ldots, k$ distributions if and only if, given $Z_1, \ldots, Z_k$ and Z_{k+1}, W_i, $i = 1, \ldots, n_1+\ldots+n_k-k-1$ are conditionally independently distributed according to the Student's t distribution with $n_1+\ldots+n_k-i-k$ degrees of freedom, respectively.

4. Application to Goodness-of-Fit Tests

Let X_i, $i = 1, \ldots, n$, $n \ge 2$, Y_j, $j = 1, \ldots, m$, $m \ge 2$ be two independent random samples from nondegenerate distributions F and G with finite second moments , respectively. Consider testing the composite hypothesesH_0 : F is a $N(\xi, \sigma^2)$ and G is a $N(\eta, \sigma^2)$ versus H_a : not H_0, where ξ, η and σ^2 are unknown. Using the result of Theorem 2.1, it is equivalent to testing, H_0^* : Given Z_1, Z_2, Z_3, W_i, $i = 1, \ldots, m+n-3$ are independently distributed according to Student's t distribution with $m+n-i-2$ degrees of freedom, respectively. This simple hypothesis can be tested using any EDF statistic, such as Kolmogorov-Smirnov statistic, Kuiper statistic, Anderson-Darling statistic (See D'Agostino and Stephens (1986)). Kasturiratna, Nguyen and Gupta (2007) proposed another equivalent test for testing H_0 using $T = \frac{\sum_{i=1}^{n}(X_i-Z_1)^2/(n-1)}{\sum_{j=1}^{m}(Y_j-Z_2)^2/(m-1)}$, as the test statistic and reject H_0 at level α if and only if, $T \le c_1$ or $T \ge c_2$, where c_1 and c_2 are the $\left(\frac{\alpha}{2}\right)^{th}$ and $\left(1-\frac{\alpha}{2}\right)^{th}$ percentiles of the F distribution with $(n-1)$, $(m-1)$ degrees of freedom, respectively.

5. Power Study

In this section we estimate the power of the tests proposed for testing the normality and the equivariance of two samples. Power computation has been carried out for the following alternatives (a) exponential distribution, (b) lognormal distribution, (c) Student's t distribution, (d) Laplace distribution, (e) Weibull distribution, and (f) uniform distribution all with the same means and variance as the null distribution. However we present here tables for (a), (b) and (c) only. Two independent samples from normal distribution with unequal variances are also used as an alternative. The type I error of the tests are also calculated. We study how do the tests perform for change in means, change in variance and unequal sample sizes.Results for change in means are not included. The statistics to be used in the goodness-of-fit tests are the Kolmogorov-Smirnov statistic (D), the Kuiper statistic (V), the Anderson-Darling statistic (A^2) and the proposed T-statistic (T). Ten thousand Monte-Carlo samples of size n_1 and n_2 were drawn from each alternative and calculate the number of times of which were declared significant by the statistics quoted at level $\alpha = 0.1$ and $\alpha = 0.05$. The values corresponding to $\alpha = 0.05$ are always tabulated in the second line of the tables. Finally we extend our power study to k independent samples. We use the notation, $\mu_1, \mu_2, \ldots$ and $\sigma_1^2, \sigma_2^2, \ldots$ to represent means and variances of the distributions from which the samples are obtained respectively and σ^2 as the common variance. The results are presented in Tables 1 to 8, from which the following observations can be made.

(i) The EDF goodness-of-fit tests corresponding to the statistics D, V, A^2, and T, maintain very well the nominal level $\alpha = 0.1$ and $\alpha = 0.05$.

(ii) The power of the tests increases considerably as the sample size increases except for the test based on T statistic.

(iii) Out of the alternatives considered, the power of the tests does not get affected by change in mean or variance in exponential, Weibull and uniform distributions. But the power changes as the variance of Student's t distribution changes. Power of the tests are sensitive to both mean and variance when the alternative is lognormal distribution, specially in small sample sizes, but it does not change significantly for large sample sizes.

(iv) For unequal sample sizes, the power of the tests reduce substantially. But for some alternatives (exponential) it is not that significant.

(v) The Kuiper statistic (V) is recommended in testing this hypothesis especially for large sample sizes.

(vi) The T statistic is recommended for small sample sizes.

References

Csorgo, M., Seshadri V., Yalovsky, M. (1973) Some exact tests for normality in the presence of unknown parameters. *J. Roy. Statist. Soc., B* 35, 507-522.

D'Agostino, R.B., Stephens, M.A. (1986) *Goodness-of-fit Techniques*. Marcel Dekker, Inc., New york.

Gupta, A.K., Nguyen, T.T., Wang, Y. (1997) Characterizations of some continuous distributions. *J. Ital. Statist. Soc.*, 1, 59-65.

Gupta, A.K., Varga, T. (1990) Characterization of joint density by conditional densities. *Commun. Statist.-Theory Meth.*, 19, 4643-4652.

Kasturiratna, D., Nguyen, T.T., Gupta, A.K.(2007) Characterization of normal distribution related to two samples based on regression. *Metrika*, 65, 43-52.

Kotlarski, I. (1966) On characterizing the normal distribution by Student's law. *Biometrika*, 53, 603-606.

Nguyen, T.T., Dinh, K.T. (1998) Characterizations of normal distributions supporting goodness-of-fit tests based on sample skewness and sample kurtosis. *Metrika*, 48, 21-30.

Nguyen, T.T., Dinh, K.T. (2003) Characterizations of normal distributions and EDF goodness-of-fit tests. *Metrika*, 58, 149-157.

Rao, C.R. (1967) On some characterizations of the normal law. *Sankhya A*, 29, 1-14.

Singh, J., Oliker, V.I. (1979) On minimum variance unbiased estimation and characterization of densities. *Proc. Conf. Opt. Methods in Statist.*, Academic Press, pp.435-442.

Wang, Y., Gupta, A.K., Nguyen, T.T. (1996) Characterization theorems for some discrete distributions based on conditional structure. *The Canadian Journal of Statistics.*, 24, 257-262.

In: Research in Applied Statistical Science
Editor: Mohammad Ahsanullah, pp. 11-21
ISBN: 978-1-63321-818-5

Chapter 2

SAMPLE SIZE DETERMINATION FOR BETWEEN GROUP COMPARISONS IN MIXED-EFFECTS LOGISTIC REGRESSION MODELS FOR ANALYSIS OF LONGITUDINAL DATA

Dulal K. Bhaumik[1]***, Subhash Aryal***[1,2,*]***, Anup Amatya***[1] ***and Kush Kapur***[1]
[1]Center for Health Statistics
Department of Psychiatry, University of Illinois at Chicago, US
[2]University of North Texas Health Science Center
Fort Worth, Texas, US

Abstract

In this article we consider the problem of sample size determination for two-level mixed-effects logistic regression models for the analysis of longitudinal data. In this case, level 1 represents measurement occasion and level 2 represents subject. We provide an algorithm for sample size requirements (i.e., power characteristics) for a test of the treatment-by-time interaction and overall time trend that strictly maintains the type I error rate close to the nominal level. The methodology is illustrated both via simulation and an example from a now classic psychiatric randomized clinical trial.

Keywords: hierarchical models, type I error rate, power analysis, marginal maximum likelihood

1. Introduction

Randomized clinical trials (RCTs) are now standard for determining the efficacy of a new drug compared to existing drugs and/or placebo. In these studies subjects are randomized to different arms of the study (*e.g.*, placebo or drug; placebo or drug A or drug B etc) and followed over time. The main goal is to study the differential effects of these drugs over time, *i.e.*, group by time interaction. The data obtained from these longitudinal clinical studies

*E-mail address: subhash.aryal@unthsc.edu. Tel: (817) 735-2076 (Corresponding Author)

are correlated. While designing such studies several key questions need to be answered: how many subjects need to be recruited and how many time points are required to maintain a pre-specified level of significance (*e.g.*, 5%) and achieve the desired level of statistical power (*e.g.*, 80%, 90% etc), in order to test the primary hypothesis of interest.

To answer the questions posed above, Roy et al. (2007) derived sample size formulas and cost analysis for two-level and three-level nested designs in presence of attrition for testing the group by time interaction parameter in a linear mixed-effects model. Reference to other works in sample size determination for linear mixed-effects models can be found in Roy et al. (2007). For nonlinear models, such as logistic regression, there is a relatively extensive literature on power and sample size determination methods for generalized linear models; however the proposed methods are limited to fixed-effects models only. Generally, the parameters are estimated by the Generalized Estimating Equations (GEE) approach for both uncorrelated and correlated observations. Self and Mauritsen (1988) and Self, Mauritsen and Ohara (1992) described methods for approximating power/sample size of fixed-effects generalized linear models using non-central χ^2 approximation to the score test and likelihood ratio test respectively for finite covariate configuration. Shieh (2000) extended the work of Self, Mauritsen and Ohara (1992) to incorporate both finite and infinite covariate configurations and simplified the approximation of the noncentrality parameter of the noncentral χ^2 distribution for the likelihood ratio statistic for a fixed-effects generalized linear model. Hsieh, Bloch and Larsen (1998) proposed using sample size formulae for comparing means or proportion to calculate sample size requirements for fixed-effects logistic regression model. For multiple logistic regression models, they advocate adjusting the sample size requirement for simple logistic regression model with a variance inflation factor. Liu and Liang (1997) proposed a multivariate extension of the work of Self and Mauritsen (1988) for correlated observations and presented a methodology for determining power/sample size using the GEE approach.

Rochan (2005) also used the GEE approach for deriving sample size requirements for continuous and discrete longitudinal data. The author considered unequal allocation between treatment groups, staggered entry and attrition. Pan (2001) used the GEE approach for calculating sample size for correlated binary data. The main drawback of Pan's method is that it cannot incorporate additional covariates beside the treatment indicator.

Jung, Kang and Ahn (2001) developed a sample size determination method for testing binomial proportions when data are clustered. Jung and Ahn (2005) developed sample size formula for testing the change rates of binary repeated measurements for two group comparisons using GEE in the presence of missing data patterns such as independent missing and monotone missing and AR (1) correlated residual errors.

Note that all the aforementioned articles considered the problem of sample size determination for fixed-effects generalized linear models for independent/correlated observations. In contrast, we consider a hierarchical longitudinal model for comparing the rate of change in two groups where observations nested within subjects are measured at potentially different time points and the measurement at each time point is dichotomous. The goals of this article are (i) to test the group by time interaction parameter, (ii) to test the time trend parameter and (iii) to determine the necessary sample size required for testing (i) and (ii) for pre-specified type I error rate and statistical power for a *mixed-effects logistic regression model*.

As previously noted, most of the existing literature is based on the GEE approach. The performance of GEE in terms of controlling the type I error rate is satisfactory for large samples but is inadequate for small samples (Firth 1993). Kauermann and Carroll (2001) and Mancl and DeRouen (2001) have proposed adjusted expressions for the calculation of standard errors for the GEE approach that have somewhat better performance in small samples. Power calculation based upon the GEE approach without any adjustment for the standard errors will be inflated as will be the nominal type I error rate.

Several authors have investigated the problem of sample size determination for mixed-effects logistic regression models numerically (Moineddin, Matheson and Glazier 2007; Dang, Mazumdar and Houck 2008). The fundamental contribution of this article is to determine the necessary sample size in order to achieve a certain pre-specified level of power while maintaining the type I error rates at the nominal level for testing the time trend and the group by time interaction parameters in a two level mixed-effects logistic regression model with two random components (random intercept and random time trend) using the marginal maximum likelihood (MML) estimation approach. Note that the MML method provides consistent estimates with corresponding standard errors.

In Section 2 we introduce a mixed-effects logistic regression model and provide a brief description of the MML method of parameter estimation and propose a Wald type test statistic for testing fixed-effects parameters. In Section 3 we provide an algorithm for determining the sample size necessary to maintain the type I error rate while achieving the desired level of statistical power. In Section 4 we illustrate our method using a real dataset and in Section 5 we present a simulation study to evaluate our proposed methodology. We conclude our work with a general discussion.

2. Modeling and Estimation

Let y_{ij} be the outcome at the jth time point for the ith subject. We consider a sample with N subjects (*i.e.*, $i = 1, \cdots, N$) and the ith subject contributes n_i repeated observations (*i.e.*, $j = 1, \cdots, n_i$). We consider the case of binary outcome variable so y_{ij} either takes the value 1 or 0. Denoting the probability of a positive response by p_{ij}, then a general mixed-effects logistic regression model can be written as

$$log\left[\frac{p_{ij}}{1-p_{ij}}\right] = x'_{ij}\beta + z'_{ij}\upsilon_i \tag{1}$$

where x_{ij} is the vector of covariates, β is the vector of unknown regression coefficients, z_{ij} is the vector of covariates associated with the random-effects and υ_i is the vector of random-effects for the ith subject. We assume υ_i follows a multivariate normal distribution with mean $\mathbf{0}$ and covariance matrix Σ_υ. Expressing the random-effects vector $\upsilon_i = T\delta_i$ from (1) in standardized form we obtain,

$$log\left[\frac{p_{ij}}{1-p_{ij}}\right] = x'_{ij}\beta + z'_{ij}T\delta_i \tag{2}$$

$$\text{or} \quad p_{ij} = \Psi(x'_{ij}\beta + z'_{ij}T\delta_i) \tag{3}$$

where, TT' is the Cholesky factorization of the covariance matrix Σ_υ and $\Psi(\cdot)$ is the cumulative density function of the standard logistic distribution.

The likelihood function for an individual subject i is the product of likelihoods across the n_i repeated observations.

$$l(y_i|\delta) = \prod_{j=1}^{n_i} \left[\Psi(x'_{ij}\beta + z'_{ij}T\delta_i)\right]^{y_{ij}} \left[1 - \Psi(x'_{ij}\beta + z'_{ij}T\delta_i)\right]^{1-y_{ij}} \tag{4}$$

The marginal likelihood for subject i is given by

$$h(y_i) = \int_\delta l(y_i|\delta)\, g(\delta)\, d\delta, \tag{5}$$

where $g(\delta)$ represents the multivariate standard normal density. The full log-likelihood for N subjects is

$$L = \prod_{i=1}^{N} h(y_i) \tag{6}$$

$$\text{or} \qquad \log L = \sum_{i=1}^{N} \log h(y_i), \tag{7}$$

Let $\eta = (\beta, \delta)$. The first derivatives of the log-likelihood with respect to η are

$$\frac{\partial \log L}{\partial \eta} = \sum_{i=1}^{N} \frac{1}{h(y_i)} \frac{\partial h(y_i)}{\partial \eta}, \tag{8}$$

where

$$\frac{\partial h(y_i)}{\partial \eta} = \int_\delta \sum_{j=1}^{n_i} \frac{y_{ij} - \Psi(x'_{ij}\beta + z'_{ij}T\delta_i)}{\Psi(x'_{ij}\beta + z'_{ij}T\delta_i)(1 - \Psi(x'_{ij}\beta + z'_{ij}T\delta_i))}$$

$$\partial\Psi(x'_{ij}\beta + z'_{ij}T\delta_i) \frac{\partial\Psi(x'_{ij}\beta + z'_{ij}T\delta_i)}{\partial \eta}\, l(y_i|\delta)\, g(\delta)\, d\delta.$$

The integration on the random effects space (*i.e.*, δ) in the marginal likelihood equation in (5) can be approximated numerically using Gauss-Hermite quadrature to any practical degree of accuracy.

Therefore,

$$\frac{\partial \log L}{\partial \eta} = \sum_{i=1}^{N} \frac{1}{h(y_i)} \int_\delta \sum_{j=1}^{n_i} \frac{y_{ij} - \Psi(x'_{ij}\beta + z'_{ij}T\delta_i)}{\Psi(x'_{ij}\beta + z'_{ij}T\delta_i)(1 - \Psi(x'_{ij}\beta + z'_{ij}T\delta_i))} \partial\Psi(x'_{ij}\beta + z'_{ij}T\delta_i) \tag{9}$$

$$\frac{\partial(x'_{ij}\beta + z'_{ij}T\delta_i)}{\partial \eta}\, l(y_i|\delta)\, g(\delta)\, d\delta.$$

Here,

$$\frac{\partial(x'_{ij}\beta + z'_{ij}T\delta_i)}{\partial \beta} = x_{ij},$$

$$\frac{\partial(x'_{ij}\beta + z'_{ij}T\delta_i)}{\partial \mathrm{v}(T)} = [\mathbf{J_r}(\boldsymbol{\delta_i} \otimes \mathbf{z_{ij}})]'$$

As defined by Magnus (1988), vector v($\mathbf{T}$) contains $\left(\frac{p\times(p-1)}{2}\right)$ elements below and on the diagonal of a $p \times p$ matrix $\mathbf{T}$, and $\mathbf{J_p}$ is the $\left(\frac{p\times(p-1)}{2}\right) \times p^2$ elimination matrix containing 1s and 0s that removes the elements above the main diagonal.

Solution of the above likelihood equations can be obtained iteratively using the Newton-Raphson algorithm. In many cases, obtaining the second derivatives required to implement the Newton-Raphson algorithm is complicated. Alternatively, the Fisher scoring method can be used which does not require computation of the matrix of second derivatives. On the ith iteration, estimates for the vector of parameters, Θ, are improved by

$$\Theta_{i+1} = \Theta_i + I(\Theta_i)^{-1}\frac{\partial \log L}{\partial \Theta_i} \tag{10}$$

where

$$I(\Theta_i) = E\left[-\frac{\partial^2 \log L}{\partial \Theta_i \partial \Theta_i'}\right] = \sum_{i=1}^{N} \frac{1}{h^2(y_i)} \frac{\partial h(y_i)}{\partial \Theta_i} \left(\frac{\partial h(y_i)}{\partial \Theta_i}\right)' \tag{11}$$

To test the significance of each element of β, a Wald statistic is constructed by dividing the square of the parameter estimates by the corresponding diagonal element of the inverse of the information matrix $I(\Theta)$ *i.e.*, ,

$$z^2 = \left[\frac{\hat{\beta}_i}{se(\hat{\beta}_i)}\right]^2, \tag{12}$$

where, z^2 follows asymptotically a χ^2 distribution with 1 degree of freedom.

3. Methodology for Determining Sample Size

In longitudinal RCTs, the main goal is to study the effect of the treatment on the response process over time. Let us assume the N subjects are randomized to receive either drug or placebo. Let π be the allocation proportion for the active treatment group, so $n\pi$ subjects receive an active treatment and the remaining $n(1-\pi)$ subjects receive the control. Suppose we conduct a longitudinal study at T different time points and assume a linear time trend is present. Let U and T_R be the treatment and time trend matrices respectively. The order of U and T_R is 2×2 and $T \times 2$. In our study the treatment matrix U is $\begin{pmatrix} 1 & 0 \\ 1 & 1 \end{pmatrix}$ and the first column of the T_R matrix is $(1,\ldots,1)'$ and the second column is $(1,\ldots,T)'$. The first column of T_R captures the baseline effect and the second column represents the time trend $g(t)$. The columns of T_R are assumed to be linearly independent. Let y_{ist} denote the response at the tth time point($t = 1,\ldots,T$) of the ith subject ($i = 1,\ldots,N$) nested in the sth treatment group ($s = 1,2$). Then a plausible model for the response probability p_{ist} is,

$$logit(p_{ist}) = \beta_0 + \beta_1 g(t) + \beta_2 x_{is} + \beta_3 x_{is} g(t) + \delta_{0i} + \delta_{1i} g(t). \tag{13}$$

Here x_{is} is an indicator variable taking the value 1 if the subject is randomized to the active treatment condition and 0 otherwise and δ's are the random subject and time trends. We assume multivariate normality for the variance components. In (13), β_0 is the logit for placebo group at time-point $t = 0$, β_1 is the logit change in outcome due to time in the placebo group, β_2 is the difference in logit for active treatment group at time-point $t = 0$ and β_3 is the difference in logit due to time for the active treatment group. The main interest is in testing for a significant group by time interaction which indicates that the placebo and the treatment group are changing at different rates over time. The random-effects δ_{0i} and δ_{1i} represent a subject's deviation from their group intercept and group trend respectively. The main interest is in testing

$$H_0 : \beta_3 = 0 \quad vs \quad H_1 : \beta_3 \neq 0. \tag{14}$$

In some situations, studying the rate of change over time in a single group of subjects is of interest. In such situations we ignore the grouping variable and model (13) reduces to

$$logit(p_{it}) = \beta_0 + \beta_1 g(t) + \delta_{0i} + \delta_{1i} g(t) \tag{15}$$

Here $g(t)$ is the trend at time point t and δ's are the random-effects. In model (15), β_1 is the logit change in outcome due to time and describes the rate of change over time in terms of the probability of response. δ_{0i} represents the intercept deviation and δ_{1i} represents the slope deviation for subject i. The main interest is in testing

$$H_0 : \beta_1 = 0 \quad vs \quad H_1 : \beta_1 \neq 0. \tag{16}$$

There is no closed form solution for obtaining the sample size required to test (14) and (16) so we rely on the following iterative method.

1. Obtain MML estimates of the fixed-effects parameters $(\beta_0, \beta_1, \beta_2, \beta_3)$ and the covariance matrix of the random-effects $\Sigma_\delta = \begin{pmatrix} \sigma^2_{\delta_0} & \sigma_{\delta_0\delta_1} \\ \sigma_{\delta_0\delta_1} & \sigma^2_{\delta_1} \end{pmatrix}$ from a pilot study or from the existing literature.

2. Set the sample size N to a reasonably large number (*e.g.*, 70, 80).

3. Simulate correlated binary data using the estimates of the fixed-effects parameters and the covariance matrix of the random-effects.

4. Obtain the MML estimates of the fixed-effects parameters and the corresponding standard errors from the simulated data.

5. Obtain the type I error rates for testing the parameter of interest(*i.e.*, β_3 or β_1) using (12).

6. Adjust the sample size if the type I error rate is not close to the desired nominal level.

7. Adjust (12) under the alternative hypothesis and calculate power.

8. Increase the sample size until the desired level of power is obtained and check the type I error rate.

Even though we have presented two specific models in (13) and (15), the sample size method outlined above can be used for models which include a number of covariates and multiple random-effects. We used SuperMix to carry out our simulation study. SuperMix is a FORTRAN based software package designed by researchers in our group to analyze both linear and non-linear mixed-effects regression models for 2 and 3-level hierarchical data. We have added a data simulator to generate correlated binary data for this study. We implemented the aforementioned algorithm in 1/5 of the time in SuperMix as compared to Proc NLMIXED in SAS.

4. Illustration

To illustrate the testing procedure with an example, we consider a Psychiatric Research dataset presented by Hedeker and Gibbons (2006). The data were collected as part of the National Institute of Mental Health schizophrenia collaborative study on treatment-related changes in overall severity using the Inpatient Multidimensional Psychiatric Scale (IMPS) by Lorr and Klett (1966). For the mixed-effects logistic model described above we focus on Item 79 of the Inpatient Multidimensional Psychiatric Scale (IMPS). In this study subjects were randomized to either a placebo or one of three drug conditions (chlorpromazine, fluphenazine, or thioridazine). Hedeker and Gibbons (2006) merged the data from the three drug groups (chlorpromazine, fluphenazine, or thioridazine) to obtain one drug condition and contrasted it with the placebo condition. Note that this classic study is the last study of schizophrenia that had a placebo condition. To linearize the relation between the outcome variable and time, $g(t) = \sqrt{t-1}$ is considered. Next we present in Table 1 the results obtained from fitting model (13) to the NIMH Schizophrenia Collaborative Study data.

Table 1. Results from NIMH Schizophrenia Collaborative Study - Differential Effect of Time

	Estimates	SE	Z	$p<$
Intercept	5.9281	0.9463	6.26	.001
Drug (0 = placebo; 1 = drug)	0.2891	0.7433	0.39	.70
Time (sqrt week)	-1.4026	0.4734	-2.96	.001
Drug by Time	-1.6114	0.4804	-3.35	.001
$-2\log L = 1227.43$				

The variance covariance matrix of the random-effects $\hat{\Sigma}_\delta$ is $\begin{pmatrix} 6.9695 & -2.0961 \\ -2.0961 & 3.0816 \end{pmatrix}$. From Table 1 we see that the p-value for testing the drug by time interaction parameter is < 0.001 which indicates that the drug and the placebo groups change at different rates over time.

Next we considered data from the placebo group only as our interest is studying the personal trend rather than comparing groups. We fit model (15) to these data and the results are presented in Table 2.

Table 2. Results from NIMH Schizophrenia Collaborative Study - Time Trend

	Estimates	SE	Z	$p <$
Intercept	6.2488	0.7486	8.3473	.001
Time (sqrt week)	-2.7137	0.3632	-7.4716	.001
$-2\log L = 1260.4$				

The variance covariance matrix of the random-effects $\hat{\Sigma}_\delta$ is $\begin{pmatrix} 7.3408 & -2.3297 \\ -2.3297 & 3.4965 \end{pmatrix}$. From Table 2, we can conclude that there is a significant time trend ($p < 0.001$).

5. Simulation Study for Power Characteristics

The sample size calculation methodology we have proposed is based on an iterative solution. In order to evaluate the performance of our methodology, we conducted an extensive simulation study based upon 100,000 replicates of simulated data. We treated the estimates presented in Table 1 as the parameter values for model (13) and determined the sample size required to attain a power of at least 80% with $\alpha = 0.05$ for testing the group by time interaction parameter β_3 in model (13).

In the existing literature, authors did not report the type I error rates of their test statistics. In our investigation, we observed that the type I error rate while testing both the group by time interaction and the time trend parameter when the sample size is ≤ 60 is close to 10%. Any power calculation based upon sample sizes ≤ 60 are also inflated. Our goal is to maintain the type I error rate in the neighborhood of pre-specified value of 5% so we are only considering sample sizes for which the type I error is close to 5%. Next we report the type I error rates for different sample sizes for testing β_3 in (13).

Table 3. Sample Size and the corresponding type I error rate for testing β_3 in (13)

Total Sample Size (N)	60	100	150	200
Simulated type I error rate	0.0945	0.0653	0.0577	0.0509

From Table 3, we see that the type I error rate is inflated for sample sizes ≤ 100. The type I error rates decrease monotonically with increasing sample sizes. For example, the simulated type I error rate is 0.0577 for a sample of size 150 whereas it decreases only to 0.05 when the sample size is increased to 200. A researcher may prefer to have a sample of 150 rather than 200 even though the type I error rate is slightly above the pre-specified level of 0.0500. Similarly, in Table 4 below, we present the type I error rate for testing the significance of the β_1 parameter in (15).

Table 4. Sample Size and the corresponding type I error rate for testing β_1 in (15)

Total Sample Size (N)	60	100	150	200
Simulated type I error rate	0.0881	0.0520	0.0515	0.0501

Similar to Table 3, the type I error rate is inflated for sample size of 60. From Table 4, we see that the type I error rate close to 5% is achieved for sample size of 100 while testing the significance of β_1, whereas a sample of size 200 was required while testing the significance of β_3. Note that, there are four fixed-effect parameters involved in (13). On the other hand, in (15) there are only two fixed-effect parameters. We also studied the power characteristics of the test statistic and the power curves are presented in Figures 1 and 2.

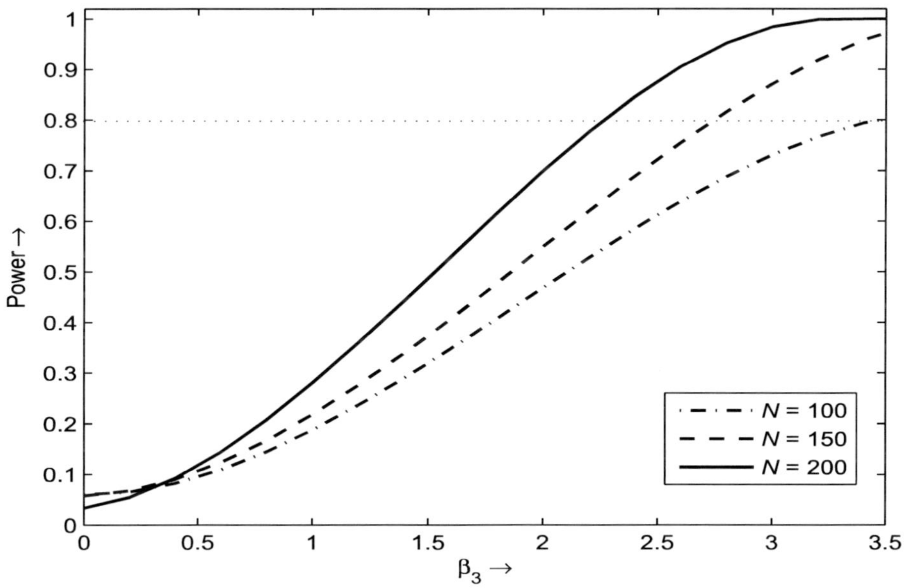

Figure 1. Power curves as function of β_3 for various values of N.

Figure 1 reveals that 80% power is achieved when β_3 is 2.17 for $N = 200$. In order to achieve the same level of power, β_3 must be 3.45 or more for $N = 100$. The general conclusion is that once the nominal type I error rate is achieved, the sample size requirement depends heavily and non-linearly on the effect size (a function of β_3).

Figure 2 also reveals similar power characteristics for testing β_1 in (15). 80% power is achieved when β_1 reaches .76 for $N = 200$. In order to achieve the same level of power β_1 must be at least 1.12 for $N = 100$. Note that for this model, the type I error rate is maintained even for $N = 100$. However the advantage for experimenters with $N = 200$ over $N = 100$ is in their ability to detect a significant time trend for a comparatively smaller effect size.

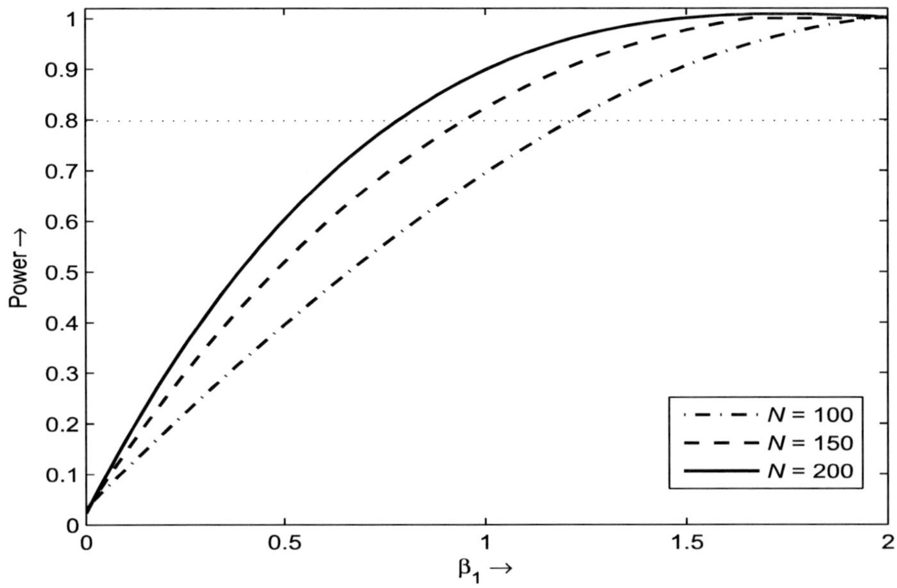

Figure 2. Power curves as function of β_1 for various values of N.

6. Conclusion

In this paper we present sample size/power calculation for non-linear mixed-effects logistic regression models. We propose an iterative methodology to study the sample size requirement to maintain the type I error rate at $\alpha = 0.05$ while testing the time trend and group by time interaction parameter. We observed that for very small sample sizes (< 60 subjects in each group when comparing two groups), the type I error rate is inflated. We increased sample size until the type I error rate stabilized around then nominal level. Next, we studied the power characteristics. Power increased monotonically and in a non-linear manner when sample size increased. In addition, when the number of fixed-effects parameters increased, the sample size needed to observe the nominal level of type I error rate also increased.

Acknowledgment

This work was supported in part by a grant from the National Institute of Mental Health (R01 MH069353-01).

References

Dang, Q., Mazumdar, S., and Houck, P. R. (2008). Sample size and power calculations based on generalized linear mixed models with correlated binary outcomes. *Computer Methods and Programs in Biomedicine* 91, 122–127.

Firth, D. (1993). Recent developments in quasi-likelihood methods. In *Proceedings ISI 49th session*, pages 341–358.

Hedeker, D. and Gibbons, R. D. (2006). *Longitudinal Data Analysis*. Hoboken: Wiley.

Hsieh, F. Y., Bloch, D. A., and Larsen, M. D. (1998). A simple method for sample size calculation for linear and logistic regression. *Statistics in Medicine* 17, 1623–1634.

Jung, S. H., Kang, S. H., and Ahn, C. (2001). Sample size calculations for clustered binary data. *Statistics in Medicine* 20, 1971–1982.

Jung, S. H. and Ahn, C. (2005). Sample size for a two-group comparison of repeated binary measurements using GEE. *Statistics in Medicine* 24, 2583–2596.

Kauermann, G. and Carroll, R. J. (2001). A note on the efficiency of sandwich covariance matrix estimation. *Journal of American Statistical Association* 96, 1387–1397.

Liu, G. and Liang, K. Y. (1997). Sample size calculations for studies with correlated observations. *Biometrics* 53, 937–947.

Lorr, M. and Klett, C. J. (1966). *Inpatient multidimensional psychiatric scale: Manual.*Palo Alto, CA: Consulting Psychologists Press.

Magnus, J. R. (1988). *Linear structures*. New York: Oxford University Press.

Mancl, L. A. and DeRouen, T. A. (2001). A covariance estimator for gee with improved small-sample properties. *Biometrics* 57, 126–134.

Moineddin, R., Matheson, F. I., and Glazier, R. H. (2007). A simulation study of sample size for multilevel logistic regression models. *BMC Medical Research* 7, 34.

Pan,W. (2001). Sample size and power calculations with correlated binary data. *Controlled Clinical Trials* 22, 211–227.

Rochon, J. (1998). Application of GEE procedures for sample size calculations in repeated measures experiments. *Statistics in Medicine* 17, 1643–1658.

Roy, A., Bhaumik, D. K., Aryal, S., Gibbons, R. D. (2007). Sample size determination for hierarchical longitudinal designs with differential attrition rates. *Biometrics* 63, 699–707. Doi: 10.1111/j.1541-0420.2007.00769.x.

Self, S. G. and Mauritsen, R. H. (1988). Power/sample size calculations for generalized linear models. *Biometrics* 44, 79–86.

Self, S. G., Mauritsen, R. H., and Ohara, J. (1992). Power calculations for liklihood ratio tests in generalized linear models. *Biometrics* 48, 31–39.

Shieh, G. (2000). On power and sample size calculations for likelihood ratio tests in generalized linear models. *Biometrics* 56, 1192–1196.

In: Research in Applied Statistical Science
Editor: Mohammad Ahsanullah, pp. 23-36
ISBN: 978-1-63321-818-5

Chapter 3

MATHEMATICAL MODELING OF MORTALITY FOR COUNTRIES WITH LIMITED AND DEFECTIVE DATA*

***Jyoti U. Devkota*[1,†] *and Radhey S. Singh*[2,‡]**
[1]Mathematics Group, Department of Natural Sciences
Kathmandu University, Nepal
[2]Department of Mathematics and Statistics, University of Guelph
Guelph, Ontario, Canada

Abstract

Life tables explain the mortality experience of a group called cohort. The survivorship function plays a vital role in life tables. In developing countries with limited and scarce data, survivorship function is taken as an index of development and is used by policy makers and planners. Inefficient registration of vital events and geographically remote areas in many developing countries hinder the regular availability of accurate data. Hence mathematical modeling of such data is of utmost importance because of their role as development indicators. In this paper, survivorship function of a developing country particularly Nepal is regressed on age and time variables. A parsimonious regression model with a very few regression coefficients, especially suited to such countries, keeping the constraints of limited and defective data is developed. The developed form satisfies all the criteria of a good model not only for developing countries like Nepal and India, but also for developed countries like Germany. A comparison of results is made between the data of countries with abundant and good quality data on one hand and limited and defective data on the other hand. Further the uncertainty in Nepalese data is measured by analyzing the error in the model parameters using Bayesian analysis.

Keywords: Probability of survival, defective data, confidence interval, mathematical model

---

* The research is supported in part by the Natural Sciences and Engineering Research Council of Canada, Grant No. 400045.
† E-mail address: devkotajb@gmail.com
‡ E-mail address: rssingh@uoguelph.ca (Corresponding author)

1. Introduction

Forecasting the vital events of developing countries is very challenging. In face of lack of proper awareness of the public towards proper and timely registration of vital events such as birth, death, migration, deterministic forecasts of these events do not give an appropriate indication of forecast uncertainty. In addition to these, apathy of the governmental organizations, administrative hurdles and lack of incentives to the public hinders timely registration of vital events. Census collects data every ten years. Information collected on vital event from the last ten years by the enumerators is faulty because of recall error, measurement error, rounding off errors etc. Mathematical Models are used very frequently to represent intensive data sets. In addition to that, in the case of countries with limited and defective data, it can be used to interpolate the missing data set, or to project for the year, for which the data is missing.

The objective was to formulate a parsimonious model, specially suited to the needs of such countries. Most of the countries of the developing world including Nepal suffer from the problem of limited and defective data. The model developed here is very simple in terms of parameters to be estimated. But despite of its simplicity, this model gave very good results for countries with limited data particularly Nepal and India. It can also explain a wide variety of mortality patterns. Further the parameters on which the model is based can be interpreted very well. Thus it fulfills all the criteria of a good model, which are simplicity, wide applicability and meaningful interpretation of the parameters. The data values have been obtained from the Population Monographs of the Central Bureau of Statistics of Nepal (1995, 2001) based on the data collected in Census 1991 and 2001. Central Bureau of Statistics is a governmental organisation working under the aegis Planning Commission of Nepal. It has been conducting Censuses at a regular interval of ten year since 1961. Whereas census in developing countries is intermittent counting of population, registration of vital events (e.g. births, deaths, marriage, and migration) normally keeps a continuous check on demographic changes. If registration of vital events is complete and accurate, it can serve as reliable source of demographic data. But in many countries in the developing world including Nepal the registration system is unreliable, with dearth in accuracy, timeliness, completeness and coverage. There is also lack of uniformity in collection, compilation and transmission of data which is different for rural and urban areas, and multiple registration agencies (e.g. health agency, panchayat agency, police agency and revenue agency).

This paper is based on the concept of separation of age and time effects by Lee and Carter (1992) and mathematical analysis of vital events by Devkota (2002, 2000b, 2000a) who modeled the rural urban differential in mortality and fertility and the parity progression respectively with applications to demographic data of Nepal.

In this paper a survivorship model is developed which specially suits the countries with very limited and defective data like Nepal and India. In addition to ease of application it has wider range of applicability and gives good fits not only for such developing countries like India and Nepal but also for countries with abundant and accurate data like Germany.

The layout of rest of this paper is as follows. In Section 2 we describe the methodology of our model. In Section 3 we interpret the parameters of the model. In Section 4 we discuss the advantages of the model. Section 5 gives the conclusion on the basis of graphs and tables, and in Section 6 we study the uncertainty of the data with the help of probabilistic analysis of the parameters of the developed model. Tables and graphs are presented in appendices 1 and 2 respectively at the end.

2. Methodology

Let $l_{x,t}$ be the probability of survival from birth to age x at year t. We use $l_{x,t}$ to construct other life table functions such as $\mu_{x,t}$, which is the intensity to survive at age x in year t. When plotted against x, $\ln(1/l_x)$ lies almost on a straight line (Figure I) for any particular year t. The developed model is also based on this linearity. The age and time effects are separated as in Lee and Carter (1992). In addition to this the weights $x/(105-x)$, the intensity to survive was considered. The intensity to survive increases with x and the change in the intensity to survive also increases with x as its first and second derivative is positive for $0 \le x < 105$.This implies that the intensity function has a minimum value at birth that is at x = 0 and increases with age. It is assumed that 105 is the upper limit of the survival age and human life.

3. The Model

When the age variable x and the time variable t are modeled against the probability of survival our proposed model is

$$l_{x,t} = e^{-(a_x+b_xk_t)\frac{x}{105-x}}, \tag{1}$$

where $l_{x,t}$ (probability of survival) lies between 0 and 1, a_x, b_x and k_t are three parameters.

When we substitute x = 0 and x = 105 in equation (1) and get, $1/l_0 = 1$ and $1/l_{105} = \infty$ respectively for any particular year t. These are the lower and upper bounds of $1/l_x$. Thus the factor $x/(105-x)$ keeps l_x in the model within the bounds 0 and 1. As an individual's intensity to survive also increases with age, $\frac{x}{105-x}$ is the suitable weight for this model.

Taking log on both sides of (1) we get $\ln(1/l_{x,t}) = a_x(x/(105-x)) + b_xk_t(x/(105-x)) + e_{x,t}$. The above equation when simplified further reduces to

$$\ln(\frac{1}{l_{x,t}}) = (a_x + b_xk_t)\frac{x}{105-x} + e_{x,t}. \tag{2}$$

Equation (2) is not a simple linear equation as all the terms are unknown. The model cannot be fit by ordinary regression methods, because we have only parameters to be estimated and the unknown index k_t. The singular value decomposition (SVD) can be used to find a least squares solution when applied to the matrix of logarithms of rates after the averages over time of the (log) age specific rates have been subtracted. The first right and left vectors and leading value of SVD after normalization provide unique solution. Besides, the parameters a_x, b_x, k_t do not have unique values. As it is not a model with two or more predictor variables problem multicollinearity doesn't exist. When, $a_x = a$, $b_x = b$ and $k_t = k$ are substituted in (2) we get $\ln(1/l_{x,t}) = (a+bk)(x/(105-x)) + e_{x,t}$. When $a_x = a - bc$, $b_x = b$ and $k_t = k + c$ are substituted in equation (2) we get

$$\ln(1/l_{x,t}) = (a - bc + bk + bc)(x/(105-x)) + e_{x,t}$$

or

$$\ln(1/l_{x,t}) = (a+bk)(x/(105-x)) + e_{x,t}.$$

If a, b and k satisfy (2) then $a - bc$, b and $k + c$ are also satisfy (2). Thus (2) does not have unique solution. To get distinct solution of (2), the following conditions are imposed. $\sum_x b_x = 1$ and $\sum_t k_t = 0$. It is assumed that the time effects cancel out each other in the long run so that $\sum_t k_t = 0$, whereas the age effect sums up to 1 over all ages so (like probability), $\sum_x b_x = 1$. Demoivre in 1725 used $x/(105-x)$ to hypothesize a linear relationship between age x and survivorship function of the life table l_x, where he fixed the limit of life at 85 years instead of 105 years , see Denny (1997). He perhaps made the earliest attempt to model a life table function. He applied the model $l(x) = l(0)\left(1 - \frac{x}{\alpha}\right)$, where α is an upper age limit.

The strategy pursued here is to generate a one parameter (k_t) family of age schedules, in the sense that the variation in one parameter generate the entire range of schedules in the family.

3.1. Model Construction

Now the parameters of our developed Model (1) have to be estimated. As the model expresses the reciprocal of the probability of survival as an exponential transformation of $(a_x + b_x k_t)(x/(105-x))$, to estimate the parameters of the model namely a_x, b_x and k_t, we follow the following steps:

Step 1. Taking the logarithmic transformation on both sides of (1) the model is reduced to a linear form.

Step 2. Now the parameter a_x is found by applying the condition $\sum_t k_t = 0$ in the following way: $l_{x,t} = e^{-(a_x + b_x k_t)(x/(105-x))}$

or $\ln(1/l_{x,t}) = a_x(x/(105-x)) + b_x k_t (x/(105-x))$

or $\sum_t \ln(1/l_{x,t}) - a_x \sum_t (x/(105-x)) = b_x (x/(105-x)) \sum_t k_t$.

Since $\sum_t k_t = 0$ we get

$$a_x = \left(\frac{105-x}{x}\right) \frac{\sum_t \ln(\frac{1}{l_{x,t}})}{\sum_t 1}.$$

Step 3. Then we calculate k_t under the condition $\sum_x b_x = 1$ in the following way. First $\sum_x \ln(1/l_{x,t}) - (x/(105-x)) \sum_x a_x = (x/(105-x)) \sum_x b_x k_t$ is obtained from (2), and then we get

$$k_t = \sum_x \left(\frac{105-x}{x}\right) \ln(\frac{1}{l_{x,t}}) - \sum_x a_x .$$

Step 4. Once the parameters a_x and k_t are estimated, b_x is calculated using singular value decomposition method. Now the model is reduced to

$$\ln(\frac{1}{l_{x,t}}) - a_x \left(\frac{x}{105-x}\right) = b_x k_t \frac{x}{105-x} + e_{x,t} .$$

Now a_x and k_t are known. The next step is taken towards the estimation of the parameter b_x at each x. Since the left hand side of the preceding equation is a matrix and so is $k_t \dfrac{x}{105-x}$, we use singular value decomposition method to find b_x as explained below in more generality. For a general matrix A, the singular value decomposition of A is $A = USV^T$, where U and V are orthogonal matrices and S is a diagonal matrix. Let A^+ be the pseudo inverse $A^+ = (USV^T)^+ = (V^T)^+ S^+ U^+$.

As the matrices V^T and U are orthogonal matrices, their pseudo-inverse are simply their ordinary inverses. In the case of any diagonal matrix D of order m X n, its pseudo – inverse denoted by D^+ of order n X m will have a reciprocal in the case of a non zero element of D and a zero in the case of a zero element. The computation procedure is given in Appendix A.

4. Interpretation of the Parameters

The visual summary of the parameters of Model (1) in Figures II and III show that a_x is the average over time t of the log of reciprocal of probability of survival. While comparing the three countries, it is known that for any particular year t $l_x(Nepal) < l_x(India) < l_x(Germany)$ and thus the values of a_x for Nepal are higher than a_x for India which is higher than that of Germany (Figure 3).

Further, b_x function is the relative sensitivity of the probability of survival to variation in k_t, the time effect. The younger the age the greater is the sensitivity to variation in k_t. The sensitivity at younger age is highest for Nepal as compared to India and Germany. The l_x function, however, for Germany is the least time sensitive, whereas for Indian female is most time sensitive. From (1) we get

$$\frac{d(\ln(\frac{1}{l_x}))}{dt} = b_x \frac{x}{105-x} \frac{dk_t}{dt}.$$

Note that b_x is positive which suggests that the probability of survival for that age decreases as $1/l_x$ increases. The smoothness in the movement of b_x is not exhibited by Nepal in comparison to Germany and India. This is an indicator of inaccuracy in the data collected.

The linear trend over time t is indicated by k_t. If k_t decreases linearly with time, then $\frac{dk_t}{dt}$ from above equation will be constant, which implies that $1/l_x$ will increase at exponential rate.

5. Advantages of the Model

The model satisfies all the characteristics of a good model namely

i) Simplicity: The model is very simple in the sense that the number of parameters to be estimated is only three, and at the same time, it is very effective in predicting the survivorship function.

ii) Ease: The fitting of this model does not require complicated computer program.

iii) Accuracy: This is very well suited to Nepalese needs. It has shown very good results not only for Nepal and India, which have limited and defective data, but also for Germany with highly accurate and abundant data.

6. Conclusion

The objective of this paper was to formulate a model specially suited to the developing countries' needs. Faults and errors in the population censuses and vital records are typical features of under developed country. In spite of such problems, the model showed quiet promising results. The low value of residual sum of square validates this fact, as is seen from Table 1. This is also depicted from Figures IV – VI. The chi square test of goodness of fit was conducted to test the fit of the developed model. The goodness of fit test showed a very good fit with very close agreement value between the observed and predicted values. The chi square values for males are .004492, .004492, .000341 and .001648 for the years 2001, 1991, 1986 and 1976 respectively. The chi square values for females are .012793, .007254, .001133 and .001679 for the years 2001, 1991, 1986 and 1976 respectively. The same trend was reflected in the chi square values for Germany and India. In the case of India the chi square values for males are as close as .00005081, .000048727, and .0000000055 for the years 1991, 1990 and 1988 respectively. The values for females are as close as .0000534, .0000523, .0000000201 for the years 1991, 1990 and 1988 respectively. The chi square test statistics of goodness of fit for German males were .008554, .003947, .002336, .006177, .003579, .001003, .003927, .000252, .009936 for the years 1994-96, 1992-94, 1982-84, 1979-81, 1977-79, 1972-74, 1967-69, 1964-66 and 1960-62 respectively. Similarly for German females the chi square values were .015977, .011275, .00111, .000456, .000374, .000943, .00842, .008335 for the years 1994-96, 1992-94, 1982-84, 1979-81, 1977-79, 1972-74, 1967-69, 1964-66 respectively. The model was successful in modeling not only the mortality scenes of developing countries like Nepal and India but also of developed countries like Germany. The model showed better results for India as compared to Nepal. Big age intervals and irregular data are some of the reasons behind relatively poor performance of this model in case of Nepal. In contrast to these countries, Germany has a record of accurate and abundant data. This model cannot be discussed in light of multicollinearity as it doesn't have two or more independent variables and hence a correlation between them doesn't arise. Further the objective behind development of this model is estimation and prediction and problem of multicollinearity doesn't play an important role here.

7. Probabilistic Analysis of the Parameters

The uncertainty about the quality of data on which this model is developed is more for a developing country like Nepal rather than for a developed country like Germany. Here 95 % confidence interval is constructed for each parameter and for each country. The length of the confidence interval is longer for countries with limited and defective data and shorter for countries with accurate and abundant data. A comparison is made between Germany with accurate and abundant data and Nepal with inaccurate and scarce data. The parameters are assumed to be normally distributed with mean equal to the obtained parameter values from the model and the standard deviation is taken as the max difference between the observed and predicted value. So, the standard deviation value is the value obtained in the worst possible case, see Daponte et al. (1997).

As illustrated in the previous sections the developed model showed very good results with very good values of R^2 and MSE. The fit was very good for Germany with R^2 as high as .999 and for Nepal .9768. The model has shown better results for Germany with accurate data and India with reasonable accurate data than Nepal. The Mean Absolute Deviation (MAD) between observed and expected l_x was used to guess the value of the standard deviation s.

For a given t and x, k_t, a_x and b_x are normally distributed, where $k_t \sim N(k_t^{'}, s^2)$. With $E(k_t) = k_t^{'}$ and $V(k_t) = s^2$, where $k_t^{'}$ is the value of k_t calculated from the data and s expresses the scattered ness in k_t given t and thus the uncertainty. Here s was estimated by MAD for the year t and the difference between observed and predicted l_x was used to guess the value of s. Higher value of s indicates higher uncertainty in the data. Also, $a_x \sim N(a_x^{'}, s_1^{2})$, and $b_x \sim N(b_x^{'}, s_2^{2})$ where $a_x^{'}$ and $b_x^{'}$ are the calculated values of a_x and b_x from the data. Since a_x and b_x are age dependent, the data on young ages and very old ages are not so reliable in such developing countries. Some individual age adjustments have been made in s_1 and s_2, where s_1 and s_2 are taken as maximum MAD, taking all the years into consideration. The standard deviation s expresses uncertainty, thus s for Germany for the same time t is much lower compared to s for Nepal.

For a given t,

$$P[E(k_t) - z_{\alpha/2}\sqrt{V(k_t)} < k_t < E(k_t) + z_{\alpha/2}\sqrt{V(k_t)}] = 1 - \alpha ,$$

and for a given x

$$P[E(a_x) - z_{\alpha/2}\sqrt{V(a_x)} < a_x < E(a_x) + z_{\alpha/2}\sqrt{V(a_x)}] = 1 - \alpha$$

and

$$P[E(b_x) - z_{\alpha/2}\sqrt{V(b_x)} < b_x < E(b_x) + z_{\alpha/2}\sqrt{V(b_x)})] = 1 - \alpha.$$

For constructing 95 % confidence interval for k_t, a_x and b_x, we have taken $z_{\alpha/2} = 1.96$.

7.1. Conclusion of the Probabilistic Analysis

The parameters of the developed model are assumed to be normally distributed with expected values equal to the calculated values and the variance proportional to the accuracy of the base population. As the data of Germany is very accurate in comparison to the data of Nepal and India, the length of the confidence interval was much smaller in the case of Germany than in the case of Nepal. As seen from Figure VII, the length of the confidence interval was much smaller in the case of Germany than in the case of Nepal. So, the parameter values obtained from the model based on the data will lie between the given intervals with 95% confidence. This result reflects the degree of trust on the obtained parameter values on the basis of the data from which it is calculated. In the case of countries with accurate and abundant data like Germany the MAD values are much smaller and hence the confidence interval is smaller. In the case of countries with inaccurate data the MAD is larger and thus the length of the confidence interval is longer and thus the degree of trust is lower.

Appendix A. Computational Procedure

We choose the normalization used above; the k's sum to 0 and b's to unity. The a_x must be equal to the weighted average of $\ln(\frac{1}{l_{x,t}})$. Furthermore, k_t must (very nearly) equal the weighted sum over age of $\ln(\frac{1}{l_{x,t}}) - a_x$, since the sum of b_x's has been chosen to be unity. All that now remains to be estimated are b_xs. Each b_x can be found by regressing $\ln(\frac{1}{lx,t}) - a_x$ on k_t separately for each age group x.

Appendix B. Tables

Table 1. Mean square error and R^2, Nepal, Germany, India

Country: Nepal	R^2		M.S.E.	
Years	Male	Female	Male	Female
2001	.9894	.9888	.000424	.00122
1991	.9709	.9948	.000735	.00017
1986	.9795	.9928	.001365	.00034
1976	.9961	.9996	.0002575	.000035
Country: Germany				
1994-96	.9983	.99618	.000154	.000213
1992-94	.999308	.99772	.0000648	.000132
1982-84	.99977	.999833	.0000242	.00001125
1979-81	.99951	.999861	.0000523	.0000099
1977-79	.99949	.99987	.000055	.0000096
1972-74	.99987	.999933	.0000143	.0000054
1967-69	.99947	.99904	.0000596	.000082
1964-66	.99996	.99979	.0000042	.0000172
1960-62	.99853	.99919	.00016	.0000701
Country: India				
1991	.999912	.99986	.0000021	.0000023
1990	.999917	.99987	.0000020	.0000022
1988	.999999	.99999	.0000021	.00000041

Appendix C. Figures

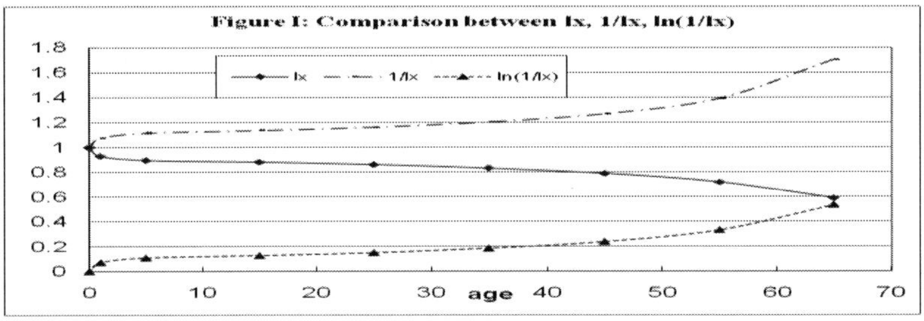

Figure I. l_x, $1/l_x$, $\ln(1/l_x)$ for the data of Nepal. It justifies the construction of the model.

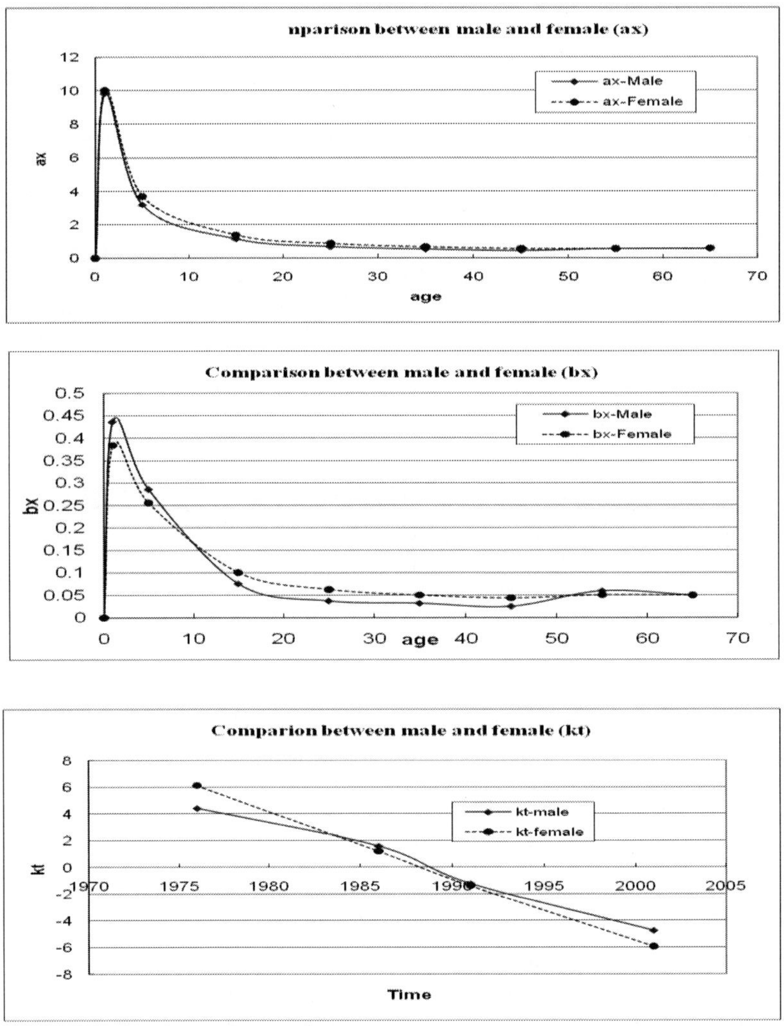

Figure II. Behavior of the parameters a_x, b_x and k_t male and female. Here the mathematical meaning of the parameters is represented. Where a_x and b_x represent the age effect and k_t represents the time effect.

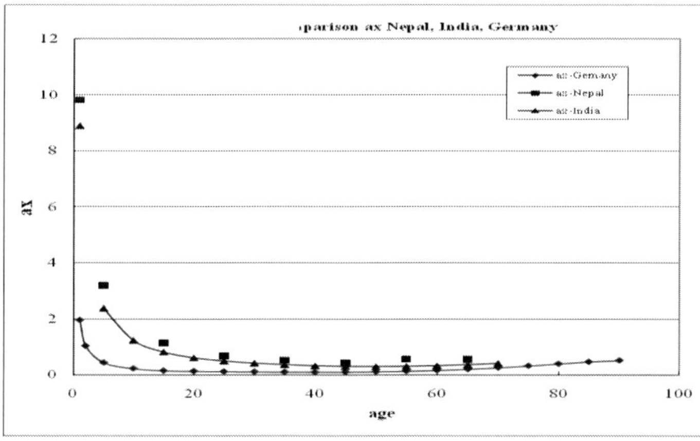

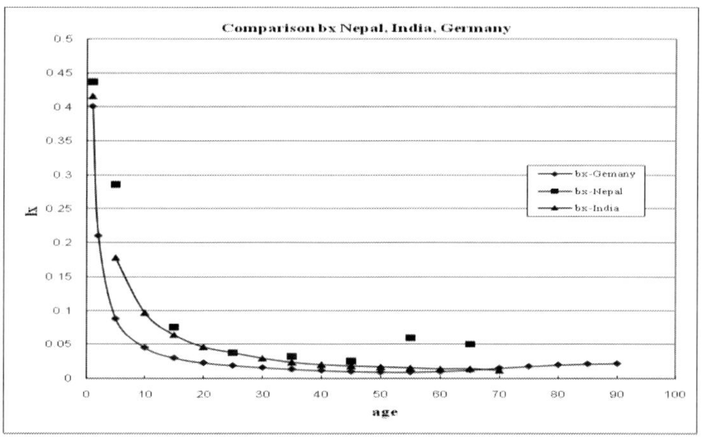

Figure III. Comparison of a_x, b_x and k_t between Germany, India and Nepal. It reflects the state of data in these three countries and the corresponding behavior of these two parameters which reflect the age effect. Here Nepal and India provide regional comparison where as, Nepal and Germany provides comparison on the basis of data accuracy and data availability.

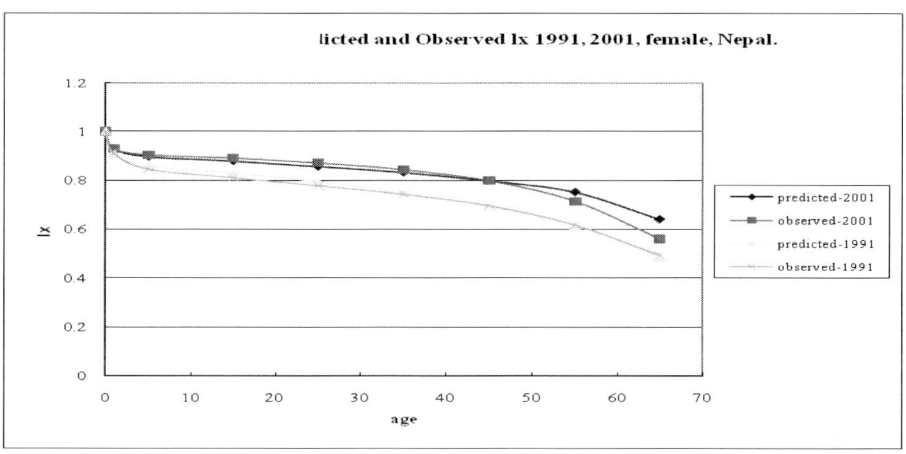

Figure IV. Continued on next page.

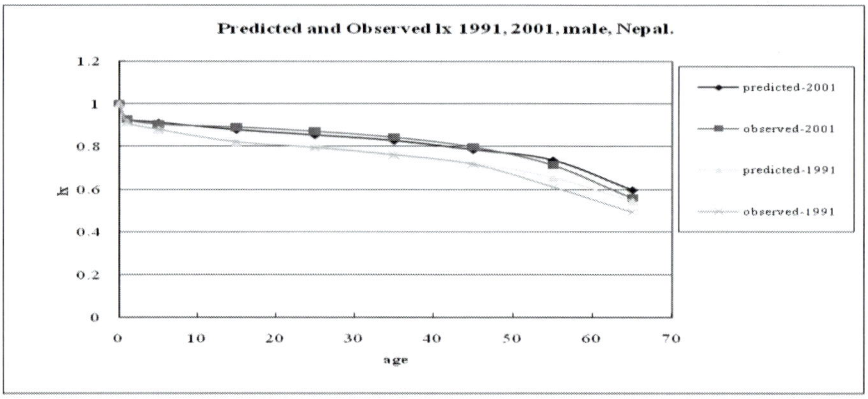

Figure IV. Comparison between observed and predicted l_x, Nepal, male and female. It reflects the level of fit of the model for Nepalese data.

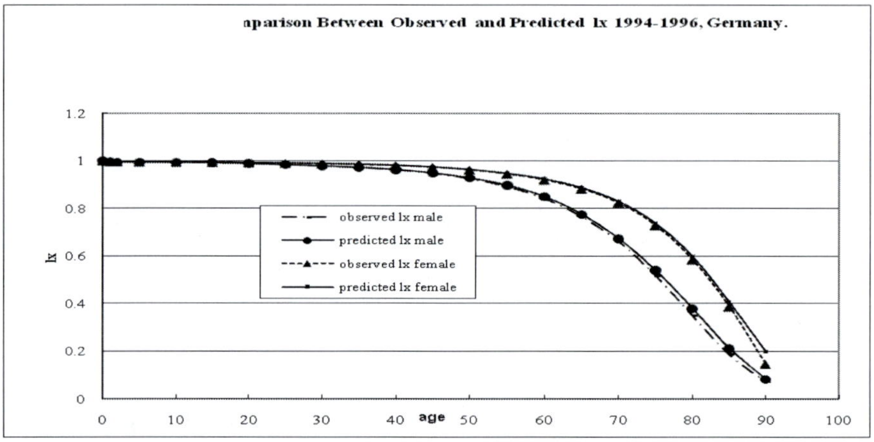

Figure V. Comparison between observed and predicted l_x, Germany, male and female. It reflects the level of fit of the model for Nepalese data.

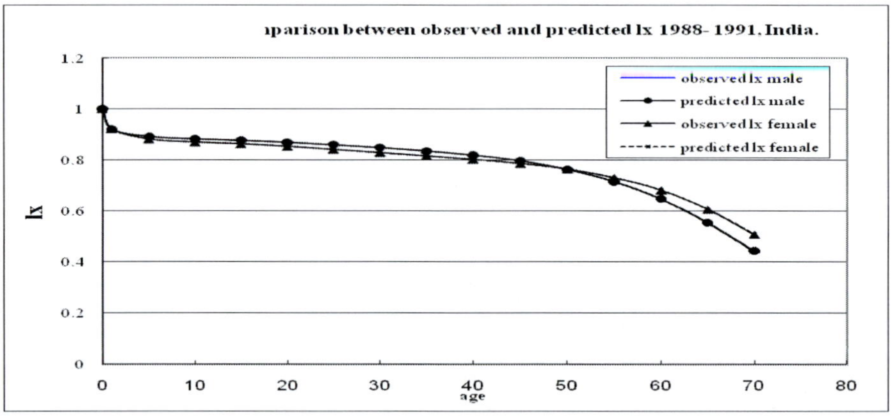

Figure VI. Comparison between observed and predicted l_x, India, male and female. It reflects the level of fit of the model for Nepalese data.

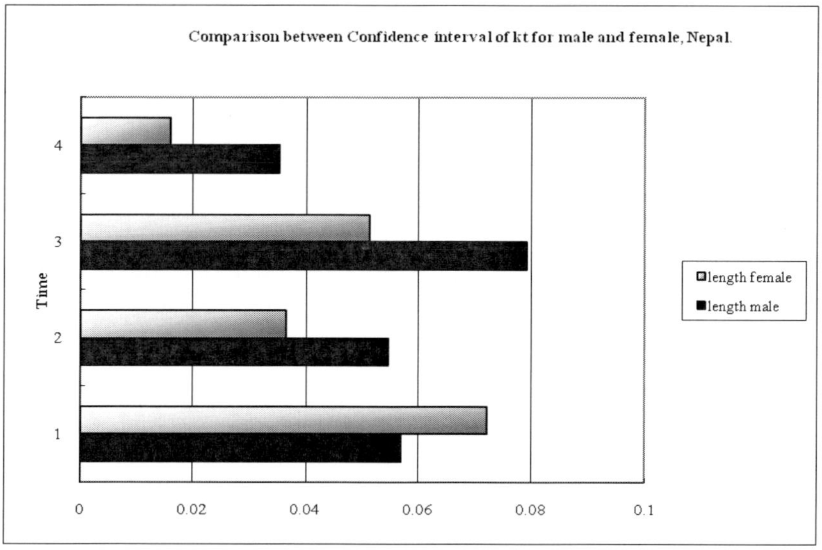

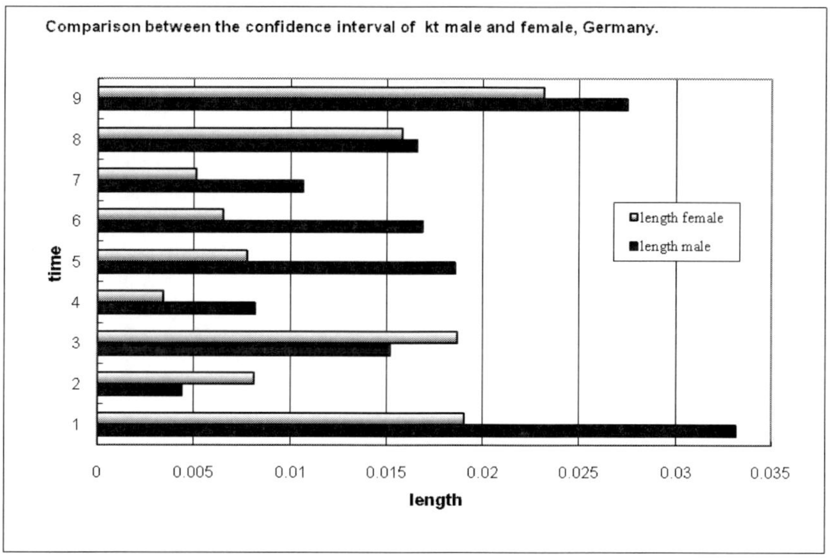

Figure VII. Comparison between the lengths of the confidence interval of the parameter k_t, which reflects the time effect. It compares the error in Nepal's data with that of Germany.

References

Central Bureau of Statistics, Nepal (1995). *Population Monograph of Nepal*, National Planning Commission Secretariat, Central Bureau of Statistics, Kathmandu, Nepal.

Central Bureau of Statistics, Nepal (2001). *Population Monograph of Nepal*, National Planning Commission Secretariat, Central Bureau of Statistics, Kathmandu, Nepal.

Daponte B. O., Kadane J.B. and Wolfson L. J. (1997). Baysian demography: projecting the Iraq Kurdish population 1977- 1990. *Journal of the American Statistical Association,* 92(440), 1256 – 1267.

Devkota, J.U. (2000a). Modeling and Projecting Nepal's Mortality and Fertility. Unpublished Ph.D. thesis, University of Osnabrueck, Department of Mathematics and Computer Science, Germany.

Devkota, J.U. (2000b). Calculation and Modeling of Parity Progression for Nepal. *The Nepali Mathematical Sciences Report*, 18 (1 & 2), 79-102.

Devkota, J.U. (2002). Rural and urban effects on mortality and fertility. *Archives of the Himalayan Environment,* 1, 16-24.

Denny C. (1997). A model of the probability of survival from birth. *Mathematical Computer Modeling*, 26(6), 69 – 78.

Lee R. D. and Carter L. R. (1992). Modeling and forecasting U. S. mortality. *Journal of the American Statistical Association*, 87(419), 659-672.

In: Research in Applied Statistical Science
Editor: Mohammad Ahsanullah, pp. 37-43
ISBN: 978-1-63321-818-5

Chapter 4

ON CHARACTERIZATIONS OF THE EXPONENTIAL DISTRIBUTIONS[*]

Mohammad Ahsanullah[1,†], Alexandre Berred[2,‡] and Valery Nevzorov[3,#]
[1]Rider University, Lawrenceville, New Jersey, US
[2]U.F.R. des Sciences et Techniques, Universite' du Havre, France
[3]Faculty of Mathematics and Mechanics, St-Petersburg State University, St-Petersburg, Russia

Abstract

The exponential distributions are characterized through an assumption on order statistics of the original random variable and the exponential random variables.

AMS 2000 subject classifications: Primary 62E10, secondary 62G30

Keywords: Characterizations, Exponential distribution, Logistic distribution, Representations, Order Statistics

1. Introduction

Consider a sequence of independent random variables $X, X_1, X_2, \ldots$ with a common continuous distribution function $F(x) = P(X \le x)$ and the corresponding order statistics $X_{1,n} \le \ldots \le X_{n,n}$, n $\ge$ 1. Denote by $\xi, \xi_1, \xi_2, \ldots$ a sequence of independent standard exponential random variables, i.e. $P(\xi \le x) = 1 - \exp(-x), x \ge 0,$ and by

[*] This work was partially supported by RFBR grants 10-01-00314 and 09-01-00808
[†] E-mail address: ahsan@rider.edu
[‡] E-mail address: alexandre.berred@univ-lehavre.fr
[#] E-mail address: vanev@mail.ru

$\xi_{1,n} \le \ldots \le \xi_{n,n}, n \ge 1,$ the associated order statistics. A random variable X is said to have the standard logistic distribution if its distribution function $H(x)$ is

$$H(x) = \frac{1}{1+\exp(-x)}, -\infty < x < \infty. \tag{1.1}$$

In several papers the logistic distribution was characterized using the following exponential order statistics representation (see David, Nagaraja (2003)).

$$\{\xi_{k,n}, k=1,\ldots,n\} \underline{\underline{d}} \{\sum_{j=1}^{k} \frac{\xi_j}{n-j+1}, k=1,\ldots,n\}, n \ge 1. \tag{1.2}$$

Lin, Hu (2008) have shown that under some additional assumptions on the distribution function F, the representation

$$X \underline{\underline{d}} X_{1,n} + \sum_{j=1}^{n-1} \frac{\xi_j}{j} \tag{1.3}$$

characterizes the logistic distribution $H(x-\mu). -\infty < \mu < \infty.$ Note that (1.3) can be written as

$$X \underline{\underline{d}} X_{1,n} + \xi_{n-1,n-1} \tag{1.4}$$

The symmetry with respect to zero of the standard logistic distribution allows to write (1.4) in the form

$$X \underline{\underline{d}} X_{n,n} - \xi_{n-1,n-1} \tag{1.5}$$

In Zhykov, Nevzorov (2010), it was shown that the relation

$$X + \xi \underline{\underline{d}} X_{n,n} \tag{1.6}$$

holds for some $n \ge 2$, if and only if

$$F(x) = \begin{cases} (1+\exp(-(n-1)(x-c)))^{-1/(n-1)}, x \ge c \\ 0, \text{otherwise}, \end{cases} \tag{1.7}$$

where an arbitrary constant can be taken as c.

Another characterization result was given in Berred, Nevzorov (2011). They showed that

$$X_{n,n} - \xi_{r,r} \underset{=}{d} X_{m,m} - \xi_{r-1.r-1} \tag{1.8}$$

holds for some $1 \le m < n$ and $r \ge 1$, if and only if

$$F(x) = \left(1 + \exp\left(-\frac{r(n-m)}{m}(x-\mu)\right)\right)^{-\frac{1}{n-m}}, -\infty < x < \infty, \tag{1.9}$$

where an arbitrary constant can be taken as μ.

Other generalizations of the above representations are given in Ahsanulah, Yanev, Onica (2010) and Wesolowski, Ahsanullah (2004).

In the rest of the paper, we consider other situations of the above representations.

2. Results

We now consider other version of the above representation.

Theorem 2.1. Suppose that the cumulative distribution function (cdf) *F*(*x*) is absolutely continuous (with respect to Lebesgue measure) with probability density function (pdf) f(x) and $-\infty < \lambda < \infty$, where λ= inf{x|F(x) >0}.Then the representation

$$X_{k,n} + \frac{\xi}{n+1} \underset{=}{d} X_{k+1,n+1} \tag{2.1}$$

holds for some $k, 1 \le k \le n$, if and only if

$$F(x) = \begin{cases} 1 - \exp(-(x-\lambda)), x \ge \lambda \\ 0, \text{ otherwise} \end{cases} \tag{2.2}$$

Proof. Using the characteristic function, we can obtain from (2.1)

$$\varphi_{k,n}(t) = \varphi_{k+1,n+1}(t)\ (1 - \frac{it}{n+1}) \tag{2.3}$$

For all t, $-\infty < t < \infty$, where $\phi_{k,n}(t)$ and $\phi_{k+1,n+1}(t)$ are the characteristic functions of $X_{k,n}$ and $X_{k+1,n+1}$ respectively.

Dividing both sides of (2.3) by $t (t \neq 0)$, and inverting, we obtain that

$$F_{k,n}(x)=F_{k+1,n+1}(x)+\frac{1}{n+1}F'_{k+1,n+1}(x)$$

where $F_{k,n}(x), F_{K+1,n+1}(x)$ and $\acute{F}_{K+1,n+1}$, are distribution functions of $X_{k,n}$, $X_{k+1,n+1}$ and probability density function of $X_{k+1,n+1}$ respectively. We have

$$F_{k,n}(x)-F_{k+1,n+1}(x)=\int_{\lambda}^{x}\frac{n!}{(k-1)!(n-k)!}(F(u))^{k-1}(1-F(u))^{n-k}f(u)du$$

$$-\int_{\lambda}^{x}\frac{n!}{k!(n-k)!}(F(u))^{k}(1-F(u))^{n-k}f(u)du=\frac{n!}{k!(n-k)!}(F(x))^{k}(1-F(x))^{n-k+1}$$

But

$$\frac{1}{n+1}F'_{k+1,n+1}(x)=\frac{n!}{k!(n-k)!}(F(x))^{k}(1-F(x))^{n-k}f(x)$$

Thus

$$\frac{f(x)}{1-F(x)}=1$$

Integrating both sides of the above equation and using the boundary condition $F(\lambda)=0$, we obtain (2.2).

Theorem 2.2. Suppose that the cumulative distribution function (cdf) $F(x)$ is absolutely continuous (with respect to Lebesgue measure) with probability density function (pdf) f(x) and $-\infty<\lambda<\infty$, where $\lambda=\inf\{x|F(x)>0\}$.Then the representation

$$X_{n-k,n}+\frac{\xi}{n+1}\stackrel{d}{=}X_{n-k+1,n+1} \tag{2.4}$$

holds for some $k, 1\le k<n$ if and only if the cumulative distribution function, F(x) is of the form (2.2).

Proof. By using the characteristic functions of the random variables in the representation (2.4), we obtain

$$\phi_{n-k,n}(t)=\phi_{n-k+1,n+1}(t)\left(1-\frac{it}{n+1}\right),\ -\infty<t<\infty, \tag{2.5}$$

where $\emptyset_{n-k,n}(t)$ and $\emptyset_{n-k+1,n+1}(t)$ are the characteristic functions of $X_{n-k,n}$ and $X_{n-k+1,n+1}$ respectively. Dividing both sides of (2.5) by $t\ (\neq 0)$ and inverting, we obtain

$$F_{n-k}(x) = F_{n-k+1,n+1}(x) + F'_{n-k+1,n+1}(x)$$

where $F_{n-k,n}(x) = F_{n-k+1,n+1}(x)$ and $F'_{n-k+1,n+1}(x)$ are the distribution functions of $X_{n-k,n}, X_{n-k+1,n+1}$ and probability density function of $X_{n-k+1,n+1}$ respectively.

We have

$$\begin{aligned} &F_{n-k,n}(x) - F_{n-k+1,n+1}(x) \\ &= \frac{n!}{(n-k-1)!\,k!} \int_\lambda^\mu (F(u))^{n-k-1}(1-F(u)^k f(u)du \\ &= \frac{(n+1)!}{(n-k)!k!} \int_\lambda^\mu (F(u))^{n-k}(1-F(u)^k f(u)du \\ &= \frac{n!}{(n-k)!k!} \int_\lambda^\mu (F(x))^{n-k}(1-F(x)^{k+1}. \end{aligned}$$

But $\frac{1}{n+1} \acute{F}_{n-k+1,n+1}(x) = \frac{n!}{k!(n-k)!} \; (F(x))^{n-k}(1-F(x)^k F(x)$

Hence

$$\frac{f(x)}{1-F(x)} = 1$$

Integrating both sides of the equation and using the boundary condition $F(x) = 0$, we obtain (2.2).

Wesolowski and Ahsanullah (2004) proved the following two results.

We assume that the cumulative distribution (cdf) function *F(x)* is absolutely continuous (with respect to Lebesgue measure) with $F(\lambda) = 0, -\infty < \lambda < \infty$,where $\lambda, -\infty < \lambda < \infty$ and $f(x) = \frac{d}{dx}F(x)$.

Result 1. The random variable X has the cdf $F(x)$ as given in (2.2) if and only if

$$X_{k+1,n} \overset{d}{=} X_{k,n} + \frac{\xi}{n-k} \tag{2.6}$$

Result 2. The random variable X has the cdf $F(x)$ as given in (2.2) if and only if

$$X_{k,n} + \frac{\xi_1}{n+1} \overset{d}{=} X_{k,n+1} + \frac{\xi_2}{n-k+1} \tag{2.7}$$

where ξ_1 and ξ_2 are independently and identically distributed with $F(x) = 1 - e^{-x}, x \geq 0$.

We will give for completeness a proof of result 2. The proof is different from that given in Wesolowski and Ahsanullah (2004).

Proof of Result 2. It is easy to see that if X has the cdf as given in (2.2) then $X_{k,n}+\frac{\xi_1}{n+1}\overset{d}{=}X_{k,n+1}+\frac{\xi_2}{n-k+1}$.

Suppose $X_{k,n}+\frac{\xi_1}{n+1}\overset{d}{=}X_{k,n+1}+\frac{\xi_2}{n-k+1}$, then using characteristic function, we obtain from (2.7)

$$\emptyset_{k,n}(t)\left(1-\frac{it}{n-k+1}\right)=\emptyset_{k,n+1}(t)\left(1-\frac{it}{n+1}\right) \qquad (2.8)$$

for all $t, -\infty < t < \infty$.

Dividing (2.8) by t $(\neq 0)$ and inverting, we obtain

$$F_{k,n}(x)+\frac{1}{n-k+1}F'_{n,k}(x)=F_{k,n+1}(x)+\frac{1}{n+1}F'_{k,n+1}(x),$$

where $F_{k,n}(x)$ and $F_{k,n+1}(x)$ are the cdfs of $X_{k,n}$ and $X_{k,n+1}$ respectively and

$F'_{k,n}(x)$ and $F'_{k,n+1}(x)$ are the pdfs of $X_{k,n}$ and $X_{k,n+1}$ respectively.

We know (see Ahsanullah and Wesolowski (2004)),

$$F_{k,n+1}(x)-F_{k,n}(x)=\binom{n}{k-1}(F(x))^k(1-F(x))^{n-k+1}. \qquad (2.9)$$

Further

$$\frac{1}{n-k+1}\acute{F}_{k,n}-\frac{1}{n+1}\acute{F}_{k,n+1}=\binom{n}{k-1}(F(x))^k(1-F(x))^{n-k}f(x) \qquad (2.10)$$

From (2.9) and (2.10), we obtain on simplification $\frac{f(x)}{1-F(x)}=\lambda$, for all x and λ, $-\infty < \lambda < x < \infty$.

Solving the above equation and using the boundary condition $F(\lambda)=0$, we get $F(x)$ as given in (2.2).

References

Ahsanullah, M. Yanev, G. P., Onica, C. (2010). *Characterizations of logistic distribution based on distributional properties of order statistics*. submitted.

Berred, A., Nevzorov, V. (2011). *On characterizations of the logistic family distributions*. Preprint.

David, H. A., Nagaraja, H. N. (2003). *Order Statistics*, 3rd Ed., Wiley, New Jersey.

Lin, G. D., Hu, C. Y. (2008). On characterizations of the logistic distribution. *J. Statist. Plann. Infer* 138, 1147-1157.

George, E. O., Mudholkar, G. S. (1981). A characterization of the logistic distribution by a sample median. *Ann. Inst. Statist. Math.* 33, part A, 125-129.

Wesolowski, J., Ahsanullah, M. (2004). Switching order statistics through random power contractions. *Australian & New Zealand Journal of Statistics*, 46, 297-303.

Zykov, V. O., Nevzorov, V. B. (2010). Some characterizations of families of distributions, including logistic and exponential ones, by properties of order statistics. *Zap. Nauchn. Sem POMI* 384, 176-181. (English translation in *Journal of Mathematical Sciences*, 176 (2), 2011, 203-206).

In: Research in Applied Statistical Science
Editor: Mohammad Ahsanullah, pp. 45-58
ISBN: 978-1-63321-818-5

Chapter 5

ESTIMATION OF $P[Y < X]$ FOR GENERALIZED LOGISTIC DISTRIBUTION

H. M. Okasha*
Al-Azhar University, Faculty of Science, Mathematics Department,
Nasr City, Cairo, Egypt

Abstract

This paper deals with the problem of estimation of the stress-strength function $R = P(Y < X)$, when X and Y are two independent but not identically distributed random variables belonging to the generalized logistic distribution (GLD). Different estimators of R, namely, maximum likelihood, uniformly minimum variance unbiased and Bayes are derived in closed form. The Bayes estimates are based on conjugate and non informative priors. Interval estimation of R (classical and Bayesian) are also discussed. A Monte Carlo simulation study is given for a comparisons between the different methods of estimation.

Keywords: Stress-Strength model, Generalized logistic distribution, uniformly minimum variance unbiased and Bayes, Asymptotic distributions, Monte Carlo simulation

1. Introduction

The standard logistic distribution has important uses in describing growth and as a substitute for the normal distribution. It has also attracted interesting applications in the modeling of the dependence of chronic obstructive respiratory disease prevalence on smoking and age, degrees of pneumoconiosis in coal miners, geological issues, hemolytic uremic syndrome data for children, physicochemical phenomenon, psychological issues, survival time of diagnosed leukemia patients, and weight gain data.

A generalized logistic distribution is proposed, based on the fact that the difference of two independent Gumbel-distributed random variables has the standard logistic distribution.

*E-mail address: hassanokasha@yahoo.com; Current address: King AbdulAziz University, Faculty of Science, Mathematics Department, P.O. Box 80200, Jeddah 21589, Saudi Arabia.

Balakrishnan and Leung (1988) defined the generalized logistic distribution (GLD) as one of three generalized forms of the standard logistic distribution. The GLD has received additional attention in estimating its parameters for practical usage (see Asgharzadeh (2006)). The two-parameter GLD (denoted by GL(λ, θ)) has the probability density function (pdf)

$$f(x;\lambda,\theta) = \lambda\theta(1+e^{-\theta x})^{-\lambda-1}e^{-\theta x}; \qquad -\infty < x < \infty, \quad \theta > 0, \quad \lambda > 0. \tag{1.1}$$

and has the cumulative distribution function (cdf)

$$F(x;\lambda,\theta) = (1+e^{-\theta x})^{-\lambda}; \qquad \theta > 0, \quad \lambda > 0, \tag{1.2}$$

Here λ and θ are the shape and scale parameters, respectively. The above GL distribution was originally proposed as a generalization of the logistic distribution by Johnson et al. (1995). For $\lambda = 1$, the GL distribution becomes the standard logistic and it is symmetric. The pdf of a GL distribution is an increasing function for $x < (\theta^{-1}\ln\lambda)$ and it is a decreasing function for $x > (\theta^{-1}\ln\lambda)$. Therefore it is a unimodal distribution with mode at $\ln\lambda)$. The density function of GL$(\lambda, \theta))$ is log-concave for all values of λ. One can also observe that for $\lambda > 1$, it is positively skewed and for $0 < \lambda < 1$, it is negatively skewed. For this, the proportionality constant λ can be represented as the skewness parameter. The hazard rate of GL distribution can be either bathtub type or an increasing function, depending on the shape parameter λ. For $\lambda < 1$, the hazard rate of GL$(\lambda, \theta))$ is bathtub type and for $\lambda \geq 1$, it has an increasing function. For more details, see Alkasasbeh and Raqab (2009).

The pdf in (1.1) has been obtained by compounding an extreme value distribution with a gamma distribution. Recently, different estimation procedures can be found in Balakrishnan(1990), Chen and Balakrishnan (1995) and Alkasasbeh (2007).

It is well known that the strength of a manufactured unit is a random variable. This fact is the basis of all reliability modeling. In stress strength model the second source of variability that needs to be considered is the stress conditions of the operating environment. If X represents the environmental stress on a unit and Y the strength of a unit, the stress strength reliability is then defned by $R = P(Y < X)$, R represents the probability that a manufactured unit works satisfactorily. The statistical formulation of the stress strength reliability appears to be given first by Birnbaum (1956). For an excellent review we refer to Kotz et al. (2003).

The main aim of this paper is to focus on the inference of the stress-strength parammeter $R = P(Y < X)$, under the assumption that X and Y are two independent but not identically random variables belonging to GLD with different shape parameters. Different estimators of R are derived, namely, maximum likelihood, uniformly minimum variance unbiased (UMVU) estimators are obtained in closed form. The Bayes estimate of R is computed based on conjugate and non informative priors. Interval estimation of R (classical and Bayesian) are also discussed. A Monte Carlo simulation study is given to compare the different methods of estimation.

2. Estimation of $R = P(Y < X)$

Let $X \sim GL(\alpha, \theta))$ be the strength of a component and $Y \sim GL(\beta, \theta)$ be the stress acting on it, where X and Y are independent random variables. The reliability of the component will be

$$\begin{aligned} R &= P(Y < X) = E\left(P(Y < X|X)\right), \\ &= E\left(\int_{-\infty}^{x} f(y)dy\right) = E\left([1 + \exp\{-\theta x\}]^{-\beta}\right), \\ &= \frac{\alpha}{\alpha+\beta}. \end{aligned} \tag{2.1}$$

If α and β are known then R is simply calculated using (2.1). We see that R does not depend on θ. However, if α and β are unknown and θ is known, the estimators of α and β depend on θ, and hence so does the estimator of R.

It is interesting to note that

(i) If T_1 and T_2 are two independent exponential random Variables with failure rates α and β respectively, then

$$P(T_2 < T_1) = \frac{\alpha}{\alpha+\beta} = P(Y < X) = R.$$

(ii) If U_1 and U_2 are two independent Pareto random Variables with parameters (α, θ) and (β, θ), respectively, with pdf densities are distributions

$$f(u_1) = \alpha\frac{\theta^{\alpha}}{u_1^{\alpha+1}}, \qquad u_1 > \theta, \quad \theta > 0,$$

and

$$f(u_2) = \beta\frac{\theta^{\beta}}{u_1^{\beta+1}}, \qquad u_1 > \theta, \quad \theta > 0,$$

we have

$$\begin{aligned} P(U_2 < U_1) &= P\left(\ln\left(\frac{U_2}{\theta}\right) < \ln\left(\frac{U_1}{\theta}\right)\right) \\ &= P(T_2 < T_1) = \frac{\alpha}{\alpha+\beta} = P(Y < X) = R, \end{aligned}$$

where $T_2 = \ln\left(\frac{U_2}{\theta}\right)$ is exponential random Variables with failure rates β. Therefore, the results obtained in this paper can be generalized to the exponential and Pareto distributions. The linear failure rate function has many applications. Another lifetime model that has a linear failure rate function is the Rayleigh distribution, which is a special case of the Weibull family. This type of situation would exist if failures result at random and from wear out or deterioration.

2.1. Maximum Likelihood Estimator of R

If α and β are unknown and θ is known the MLE, R_1, of R is given by

$$R_1 = \frac{\hat{\alpha}}{\hat{\alpha}+\hat{\beta}}, \tag{2.2}$$

where $\hat{\alpha}$ and $\hat{\beta}$ are the MLEs of α and β, respectively. Notice that R contains θ because $\hat{\alpha}$ and $\hat{\beta}$ are functions of θ.

For obtaining $\hat{\alpha}$ and $\hat{\beta}$ are argue as follows:

Suppose $\mathbf{x} \equiv (x_1, x_2, ..., x_n)$ and $\mathbf{y} \equiv (y_1, y_2, ..., y_m)$ be independent random samples from $GL(\alpha, \theta)$ and $GL(\beta, \theta)$, respectively. Therefore, when θ is known, the likelihood function of α and β for the observed sample is

$$l(\alpha, \beta; \mathbf{x}, \mathbf{y}) = \alpha^n \beta^m \theta^{n+m} e^{-(n\bar{x}+m\bar{y})\theta} e^{-(\alpha+1)W-(\beta+1)V}, \tag{2.3}$$

where

$$\bar{x} = \frac{\sum_{i=1}^{n} x_i}{n} \quad and \quad \bar{y} = \frac{\sum_{i=1}^{m} y_i}{nm},$$

and

$$\left.\begin{array}{l} W = \Sigma_{i=1}^{n} \ln(1 + e^{-\theta x_i}) \\ V = \Sigma_{j=1}^{m} \ln(1 + e^{-\theta y_j}) \end{array}\right\}. \tag{2.4}$$

The log-likelihood function is given by

$$\begin{aligned} L(\alpha, \beta; \mathbf{x}, \mathbf{y}) &= n \ln \alpha + (n+m) \ln \theta + m \ln \beta - (n\bar{x} + m\bar{y})\theta \\ &\quad -(\alpha+1)W - (\beta+1)V. \end{aligned} \tag{2.5}$$

Differentiating partially with respect to α, β and setting the results equal to zero we get

$$\frac{\partial L}{\partial \alpha} = \frac{n}{\alpha} - W, \tag{2.6}$$

and

$$\frac{\partial L}{\partial \beta} = \frac{m}{\beta} - V. \tag{2.7}$$

Solving (2.6) and (2.7), simultaneously, yields the MLEs of α and β, as

$$\hat{\alpha} = \frac{n}{W} \qquad \textbf{and} \qquad \hat{\beta} = \frac{m}{V}, \tag{2.8}$$

where W and V are given by (2.4).

2.1.1. Some Properties of R_1

Since θ are known, then R_1 can be put in the form

$$R_1 = \left[1 + \frac{m}{n}\frac{W}{V}\right]^{-1}. \tag{2.9}$$

Notice that if X is generalized logistic random variable with parameter (α, θ), then it is easy to show that $2\alpha W$ is distributed as chi-square with degree of freedom $(2n)$. Similarly, $2\beta V$ is distributed as chi-square with degree of freedom $(2m)$. Then

$$\frac{\alpha}{\beta}\left(\frac{1}{R_1} - 1\right) = \frac{\alpha m W}{\beta n V} \backsim F(2n, 2m), \tag{2.10}$$

we use this fact to study the properties of R_1. We can show that,

$$E(R_1) = \frac{\alpha}{\alpha + \frac{m\beta}{m-1}} \left[1 + \frac{m+n-1}{n(m-2)} \left(1 - \frac{\alpha^2}{\alpha + \frac{m\beta}{m-1}} \right)^2 \right], \tag{2.11}$$

and

$$\lim_{n,m \longrightarrow \infty} E(R_1) = R, \tag{2.12}$$

from (17), $R_1$is asymptotically unbiased. Also,

$$Var(R_1) = \frac{m+n-1}{n(m-2)} \left(1 - \frac{\alpha}{\alpha + \frac{m\beta}{m-1}} \right)^2 \left(\frac{\alpha}{\alpha + \frac{m\beta}{m-1}} \right)^2. \tag{2.13}$$

Then

$$\lim_{n,m \longrightarrow \infty} Var(R_1) = 0, \tag{2.14}$$

from (17) and (19) $R_1$is a consistant estimator for R. The MSE of R_1 is equal to $Var(R_1) + E(R_1)(E(R_1) - 2R) + R^2$, which tends to zero as $n, m \longrightarrow \infty$.

2.2. Uniformaly Minimum Variance Unbiased Estimator of R

In this subsection we obtain the Uniformaly minimum variance unbiased estimator, R_2, of R (UMVUE). Let X_i, $i = 1, \ldots, n$ and Y_j, $j = 1, \ldots, m$ be two independent random samples with size n and m drawn from GLD with parameters (α, θ) and (β, θ) respectively. Set $\xi_i = \ln\left(1 + e^{-\theta x_i}\right), i = 1, 2, \ldots n$ and $\eta_j = \ln\left(1 + e^{-\theta y_j}\right), j = 1, 2, \ldots, m$. We found

$$\xi = \Sigma_{i=1}^{n} \xi_i \qquad \text{and} \qquad \eta = \Sigma_{j=1}^{m} \eta_j,$$

are sufficient estimator for α and β respectively. Moreover (ξ, η) is jointly sufficient for (α, β). Let

$$T = \begin{cases} 1, & \textbf{if} \quad v_1 < w_1 \\ 0, & \textbf{if} \quad v_1 \geq w_1 \end{cases}, \tag{2.15}$$

then

$$\begin{aligned} E(T) &= P(v_1 < w_1) = P\left(\ln\left(1 + e^{-\theta x_1}\right) < \ln\left(1 + e^{-\theta y_1}\right)\right) \\ &= P(Y_1 < X_1) = R. \end{aligned} \tag{2.16}$$

Therefore T is an unbiased estimator for R. Then the UMVUE, R_3 for R is given by,

$$\begin{aligned} R_2 &= E(T|\xi, \eta) \\ &= P(Y_1 < X_1|\xi, \eta). \end{aligned} \tag{2.17}$$

Assuming that θ is known and making use of the results given by Tong (1974) we obtain

$$R_2 = \begin{cases} \sum_{i=0}^{n-1} a_i t^{-i}, & \textbf{if} \quad t \geq 1 \\ 1 - \sum_{j=0}^{m-1} b_j t^j, & \textbf{if} \quad t < 1 \end{cases}, \tag{2.18}$$

where

$$t = \frac{\eta}{\xi}, \tag{2.19}$$

and

$$a_i = \frac{(-1)^i \Gamma(m)\Gamma(n)}{\Gamma(m+i)\Gamma(n-i)}, \tag{2.20}$$

$$b_j = \frac{(-1)^j \Gamma(m)\Gamma(n)}{\Gamma(n+j)\Gamma(m-j)}. \tag{2.21}$$

Now, we find the MSE of R_2

$$\begin{aligned} \textbf{MSE}(R_2) &= E\left(R_2 - R\right)^2 \\ &= E\left(R_2\right)^2 - R^2. \end{aligned} \tag{2.22}$$

One can obtain

$$\begin{aligned} E\left(R_2\right)^2 &= \sum_{i=0}^{n-1}\sum_{k=0}^{n-1} a_i a_k E\left(T^{-i-k} | T \geq 1\right) P(T \geq 1) + P(T < 1) \\ &\quad -2\sum_{j=0}^{m-1} b_j E\left(T^j | T < 1\right) P\left(T < 1\right) + \sum_{i=0}^{m-1}\sum_{s=0}^{m-1} b_j b_s E\left(T^{i+s} | T < 1\right) P\left(T < 1\right). \end{aligned} \tag{2.23}$$

Equation (2.23) depends on the evaluation of $E\left(T^L | T < 1\right) P(T < 1)$ and $E\left(T^{-L} | T > 1\right) P(T > 1)$. To evaluate them, we obtain the p.d.f of T. Since, ξ, η are independent distributed as $\Gamma(n, \alpha)$ and $\Gamma(m, \beta)$ respectively. This implies that $\frac{n\beta}{m\alpha} T \backsim F(2m, 2n)$, so that the p.d.f of T is

$$H(t) = \frac{\left(\frac{\beta}{\alpha}\right)^m}{B\left(n, m\right)} t^{m-1} \left(1 + \left(\frac{\beta}{\alpha}\right) t\right)^{-n-m}, \qquad t > 0. \tag{2.24}$$

Where $B\left(n, m\right) = \frac{(n-1)!(m-1)!}{(n+m-1)!}$. Then for $L > 0$

$$\begin{aligned} E\left(T^L | T < 1\right) P(T < 1) &= \frac{\left(\frac{\beta}{\alpha}\right)^m}{B\left(n, m\right)} \int_0^1 t^{m+L-1} \left(1 + \left(\frac{\beta}{\alpha}\right) t\right)^{-n-m} dt \\ &= \frac{\left(\frac{\beta}{\alpha}\right)^m}{(m+L) B\left(n, m\right)} F_{2:1}\left(n+m; m+L; m+L+1, -\frac{\beta}{\alpha}\right), \end{aligned} \tag{2.25}$$

where $F_{2:1}\left(.,.;.;.\right)$ is the generlized hypergeometric function. Interchanging m with n and replacing $\frac{\beta}{\alpha}$ by $\frac{\alpha}{\beta}$ in (2.25), we have

$$E\left(T^{-L} | T > 1\right) P(T > 1) = \frac{\left(\frac{\beta}{\alpha}\right)^{-n}}{(n+L) B\left(n, m\right)} F_{2:1}\left(n+m; n+L; n+L+1, -\frac{\alpha}{\beta}\right). \tag{2.26}$$

To find $P(T < 1)$, put $L = 0$ in (2.25), we have

$$P(T < 1) = \frac{\left(\frac{\beta}{\alpha}\right)^m}{mB(n,m)} F_{2:1}\left(n+m; m; m+1, -\frac{\beta}{\alpha}\right). \quad (2.27)$$

Substituting from (2.25),(2.26), and (2.27) in (2.23) and then in (2.22) we obtain the MSE of R_2 respresenting by the hypergeometric function [see Gradshteyn and Ryzhik (2007)(formula 3.197(3))].

2.3. Bayes Estimation of R

We obtain Bayes estimation of R with respect to the mean square error loss function with respect to Conjugate prior distribution and non informative prior distributions. We show that the Bayes estimator of R with respect to non informative prior distribution is superior to that with respect to conjugate distribution.

2.3.1. Conjugate Prior Distribution

Let $X_i, i = 1, \ldots, n$ and $Y_j, j = 1, \ldots, m$ be two independent random samples with size n and m drawn from $GLD(\alpha, \theta)$ and $GLD(\beta, \theta)$ respectively. Suppose that θ is known and assumed that α and β have independent gamma priors with the pdf's

$$\pi(\alpha) = \frac{u_1^{v_1}}{\Gamma(v_1)} \alpha^{v_1-1} e^{-u_1\alpha}, \alpha > 0, \quad (v_1, u_1 > 0), \quad (2.28)$$

$$\pi(\beta) = \frac{u_2^{v_2}}{\Gamma(v_2)} \beta^{v_2-1} e^{-u_2\beta}, \beta > 0, \quad (v_2, u_2 > 0), \quad (2.29)$$

respectively. The posterior pdf's of α and β are as follows:

$$\begin{aligned} \Pi(\alpha, \beta|\mathbf{x}, \mathbf{y}) &= \frac{\alpha^{n+v_1-1}\beta^{m+v_2-1}e^{-(u_1+W)\alpha-(u_2+V)\beta}}{\int_0^\infty\int_0^\infty \alpha^{n+v_1-1}\beta^{m+v_2-1}e^{-(u_1+W)\alpha-(u_2+V)\beta}d\alpha d\beta}, \\ &= B_1\alpha^{n+v_1-1}\beta^{m+v_2-1}e^{-(u_1+W)\alpha-(u_2+V)\beta}, \end{aligned} \quad (2.30)$$

where W and V are given by (2.4) and

$$B_1 = \frac{(u_1+W)^{n+v_1}(u_2+V)^{m+v_2}}{\Gamma(n+v_1)\Gamma(m+v_2)}. \quad (2.31)$$

It follows, by using the method of transformation, that the distribution of $R = \frac{\alpha}{\alpha+\beta}$ and $S = \alpha$ is

$$\begin{aligned} \Pi_{S,R}(s, r|\mathbf{x}, \mathbf{y}) &= B_1 \frac{s}{r^2} s^{n+v_1-1} \left(\frac{s(1-r)}{r}\right)^{m+v_2-1} e^{-(u_1+W)s-\frac{1-r}{r}(u_2+V)s}, \\ &= B_1 r^{-m-v_2-1}(1-r)^{m+v_2-1} s^{(n+m)+(v_1+v_2)-1} e^{-\left[(u_1+W)+\frac{1-r}{r}(u_2+V)\right]s}. \\ & \qquad 0 < s < \infty, \quad 0 < r < 1 \end{aligned}$$

(2.32)

Therefore,

$$\begin{aligned}\Pi_R(r) &= \int_s \Pi_{S,R}(s,r|\mathbf{x},\mathbf{y})ds\\ &= B_1\frac{\Gamma(n+m+v_1+v_2)\,r^{-m-v_2-1}\,(1-r)^{m+v_2-1}}{\left[(u_1+W)+\frac{1-r}{r}(u_2+V)\right]^{(n+m)+(v_1+v_2)}},\\ &= B_1\frac{\Gamma[(n+m)+(v_1+v_2)]r^{n+v_1-1}\,(1-r)^{m+v_2-1}}{[((u_1-u_2)+(W-V))\,r+(u_2+V)]^{(n+m)+(v_1+v_2)}},\quad 0<r<1.\end{aligned} \tag{2.33}$$

Hence

$$\begin{aligned}R_3 &= E\,(R|\mathbf{x},\mathbf{y},\theta)=B_1\frac{\Gamma[(n+m)+(v_1+v_2)]}{(u_2+V)^{(n+m)+(v_1+v_2)}}\int_0^1 r^{n+v_1}\,(1-r)^{m+v_2-1}\\ &\quad\times[1-\frac{(u_2-u_1)+(V-W)}{u_2+V}r]^{-(n+m)-(v_1+v_2)}dr,\\ &= B_2F_{2:1}\left(n+m+v_1+v_2,n+v_1+1;n+m+v_1+v_2+1;\frac{(u_2-u_1)+(V-W)}{u_2+V}\right),\\ &\qquad 0<u_1+W<2(u_2+V),\end{aligned} \tag{2.34}$$

where

$$B_2=\left(\frac{n+v_1}{n+m+v_1+v_2}\right)\left(\frac{u_1+W}{u_2+V}\right)^{n+v_1}.$$

2.3.2. Non Informative Prior Distributions

Let $X_i, i=1,\ldots,n$ be a random samples with size n from $GL(\alpha,\theta)$. The prior distribution $\pi_1(\alpha)$, of α is proportional to $\sqrt{I(\alpha)}$, where $I(\alpha)$ is Fisher information matrix of the sample about α, and is given by

$$I(\alpha)=\frac{n}{\alpha^2}$$

. Thus

$$\pi_1(\alpha) \propto \frac{1}{\alpha}. \tag{2.35}$$

Similarly, if $Y_j, j=1,\ldots,m$ be a random samples with size m drawn from $GL(\beta,\theta)$,The prior distribution $\pi_2(\beta)$, of β will be

$$\pi_1(\alpha) \propto \frac{1}{\beta}. \tag{2.36}$$

Since α and β are independent, then the posterior pdf's of α and β are as follows:

$$\Pi(\alpha,\beta|\mathbf{x},\mathbf{y},\theta) \propto l(\alpha,\beta,\theta)\pi_1(\alpha)\pi_2(\beta). \tag{2.37}$$

From (2.3),(2.35), and (2.36) we see that

$$\Pi(\alpha,\beta|\mathbf{x},\mathbf{y}) = B_3\alpha^{n-1}\beta^{m-1}e^{-(\alpha W+\beta V)};\quad \alpha,\beta>0,$$

where W and V are given by (2.4) and

$$B_3 = \frac{W^n V^m}{\Gamma(n)\Gamma(m)}.$$

It follows, by using the method of transformation, that the distribution of $R = \frac{\alpha}{\alpha+\beta}$ and $T = \alpha$ is

$$\begin{aligned}\Pi_{T,R}(t, r|\mathbf{x}, \mathbf{y}) &= B_3 \frac{t}{r^2} t^{n-1} \left(\frac{t(1-r)}{r}\right)^{m-1} e^{-(W+\frac{1-r}{r}V)t}, \\ &= B_3 r^{-(m+1)} (1-r)^{m-1} t^{n+m-1} e^{-(W+\frac{1-r}{r}V)t}. \\ & \qquad 0 < t < \infty, \quad 0 < r < 1\end{aligned} \tag{2.38}$$

Therefore,

$$\begin{aligned}\Pi_R(r) &= \int_t \Pi_{T,R}(t, r|\mathbf{x}, \mathbf{y}) ds \\ &= B_3 r^{-(m+1)} (1-r)^{m-1} \frac{\Gamma(n+m)}{\left[W + \frac{1-r}{r}V\right]^{n+m}} \\ &= B_3 \frac{\Gamma(n+m) r^{n-1} (1-r)^{m-1}}{[(W-V)r + V]^{n+m}}, \qquad 0 < r < 1.\end{aligned} \tag{2.39}$$

Hence the Bayes estimator with respect to mean square error loss function will be

$$\begin{aligned}R_4 &= = B_3 \frac{\Gamma(n+m)}{V^{n+m}} \int_0^1 r^n (1-r)^{m-1} \left[1 - \frac{V-W}{V} r\right]^{-(n+m)} dr, \\ &= \vartheta_4 F_{2:1}\left(n+m, n+1; n+m+1; \frac{V-W}{V}\right), \\ & \qquad 0 < W < 2V,\end{aligned} \tag{2.40}$$

where

$$\vartheta_4 = \frac{n}{n+m}\left(\frac{W}{V}\right)^n. \tag{2.41}$$

3. Interval Estimation of R

3.1. Approximate Confidence Interval

Let X_i, $i = 1, \ldots, n$ and Y_j, $j = 1, \ldots, m$ be two independent random samples with size n and m drawn from $GLD(\alpha, \theta)$ and $GLD(\beta, \theta)$ respectively. The MLE R_1, in (2.2), is asymptotically normal with mean R and variance

$$\sigma_{R_1}^2 = \sum_{j=1}^{2}\sum_{i=1}^{2} \frac{\partial R}{\partial \theta_i}\frac{\partial R}{\partial \theta_j} I_{ij}^{-1},$$

where $(\theta_1, \theta_2) = (\alpha, \beta)$ and $(I_{ij}^{-1})^{th}$ is the $(i,j)^{th}$ element of the Fisher's information matrix, $I(\alpha, \beta)$, about the parameters (α, β), and

$$I(\alpha, \beta) = -\begin{bmatrix} \frac{n}{\alpha^2} & 0 \\ 0 & \frac{m}{\beta^2} \end{bmatrix}, \tag{3.1}$$

(see Rao (1965)). It can be seen that, $\sigma_{R_1}^2 = \frac{m+n}{mn} R^2(1-R)^2$.

Therefore $(1-\gamma)100\%$ asymptotic confidence interval of R would be (L_1, U_1), where

$$\left.\begin{aligned} L_1 &= R_1 - z_{(1-\gamma/2)}\sqrt{\tfrac{m+n}{mn}} R_1(1-R_1), \\ U_1 &= R_1 + z_{\gamma/2}\sqrt{\tfrac{m+n}{mn}} R_1(1-R_1), \end{aligned}\right\} \tag{3.2}$$

and $z_{(1-\gamma/2)}$ is the $(1-\gamma/2)^{th}$ quantile of the standard normal distribution and R_1 is given by (2.2).

3.2. Exact Confidence Interval

Let $X_1, ..., X_n$ and $Y_1, ..., Y_m$ be two independent random samples with size n, m from generalized logistic distribution with parameters α, β respectively. Since $w_i = -\ln(1+e^{-\theta x_i})$ and $v_j = -\ln(1+e^{-\theta y_j})$ are independent with GLD and, $2\alpha W = -2\alpha\sum_{i=1}^{n}\ln(1+e^{-\theta x_i})$ and $2\beta V = -2\beta\sum_{j=1}^{m}\ln(1+e^{-\theta y_j})$ are independent with chi-square distributions with degree of freedom $2n$ and $2m$. Thus R_1 in (2.2) could be rewritten as $R_1 = \left[1+\frac{\hat\beta}{\hat\alpha}\right]^{-1}$. Using (2.10) we can written

$$R_1 = \left[1+\frac{\beta}{\alpha}F_1\right]^{-1}, \tag{3.3}$$

where $F_1 = \frac{m\alpha W}{n\beta V}$ has F-distribution with $(2n, 2m)$ degrees of freedom. From equations (2.1) and (2.2) we can written

$$F_1 = \frac{mW}{nV}\frac{R}{1-R}.$$

Using F_1 as a pivotal quantity, we obtain a $(1-\gamma)100\%$ confidence interval for R as

$$a < F_1 = \frac{mW}{nV}\frac{R}{1-R} < b.$$

$$\frac{a}{\frac{mW}{nV}} < \frac{R}{1-R} < \frac{b}{\frac{mW}{nV}}. \tag{3.4}$$

a, b have F-distribution with degree of freedom $(2n, 2m)$ then the confidence interval will be

$$\frac{F_{(1-\frac{\delta}{2})}(2n, 2m)}{\frac{mW}{nV}} < \frac{R}{1-R} < \frac{F_{\frac{\delta}{2}}(2n, 2m)}{\frac{mW}{nV}}.$$

Therefore $(1-\gamma)100\%$ asymptotic confidence interval of R can be obtained and written as

$$L_2 < R < U_2,$$

where

$$\left.\begin{aligned} L_2 &= F_{(1-\frac{\delta}{2})}(2n, 2m)\left(\tfrac{mW}{nV} + F_{(1-\frac{\delta}{2})}(2n, 2m)\right)^{-1}, \\ U_2 &= F_{\frac{\delta}{2}}(2n, 2m)\left(\tfrac{mW}{nV} + F_{\frac{\delta}{2}}(2n, 2m)\right)^{-1}. \end{aligned}\right\} \quad (3.5)$$

3.3. Bayesian Credible Intervals

3.3.1. Conjugate Prior Distribution

Assuming that α and β are independent, we have see in subsec. 2.3 that the posterior distributions of α and β corresponding to gamma priors are gamma with parameters $(n + v_1, u_1 - W)$ and $(m + v_2, u_2 - V)$, respectively. Thus $2(n + v_1)$ and $2(m + v_2)$ are independent chi-squared radom variables each with $2(n+m+v_1+v_2)$ degree of freedom. From (2.1) we see that $\frac{\alpha}{\beta} = \frac{R}{1-R}$. Thus

$$F_2 = \frac{W}{V}\frac{R}{(1-R)} \quad (3.6)$$

is an F distributed random variable with $(2(n+v_1), 2(m+v_2))$ degrees of freedom, where W and V are give by (2.4).

Using F_2 in (3.6) as a pivotal quantity we obtain $(1-\gamma)100\%$ Bayes credible interval for R as (L_3, U_3), where

$$\left.\begin{aligned} L_3 &= F_{(1-\gamma/2)}(2(n+v_1), 2(m+v_2))\left(F_{(1-\gamma/2)}(2(n+v_1), 2(m+v_2)) + \tfrac{W}{V}\right)^{-1}, \\ U_3 &= F_{\gamma/2}(2(n+v_1), 2(m+v_2))\left(F_{\gamma/2}(2(n+v_1), 2(m+v_2)) + \tfrac{W}{V}\right)^{-1}. \end{aligned}\right\} \quad (3.7)$$

3.3.2. Non-informative Prior Distributions

we have see in subsec. 2.3 that assuming independent and non informative prior distribution for α and β, the posterior distribution of α and β are gamma with parameters $(n, \sum_{i=1}^{n} \ln(1 + e^{-\theta x_i})$ and $(m, \sum_{j=1}^{m} \ln(1 + e^{-\theta y_j})$, respectively. Thus $2\alpha W = 2\alpha \sum_{i=1}^{n} \ln(1+e^{-\theta x_i})$ and $2\beta V = 2\beta \sum_{j=1}^{m} \ln(1+e^{-\theta y_j})$ are independent with chi-square distributions with degree of freedom $2n$ and $2m$. Thus

$$F_3 = \frac{V}{W}\left(\frac{1}{R} - 1\right), \quad (3.8)$$

is F distributed random variable with $(2n, 2m)$ degrees of freedom. Using F_3 in (3.8) we obtain $(1-\gamma)100\%$ Bayes credible interval for R with lower and upper bouds exactly the same as those in (3.5).

4. Monte-Carlo Simulation and Comparisons

In this section, we conduct some numerical computations to compare the performances of the different estimators proposed in the previous sections. We conduct an extensive simulation study to compare the performances of the different methods in the risk under squared-error loss (MSE) for different sample sizes and for different parameter values. We firstly generate samples from GLD. Applying the the algorithms by steps are:

- For given values of the prior parameters (u_1, v_1), we generate α from the gamma prior density (2.28). For known $\theta = 1$ and the generated value α, we then generate random sample of size n from GLD($\alpha, \theta = 1$).
- Similarly, random sample of size m is generated from GLD($\beta, \theta = 1$), where β is generated from (2.29) with given values of (u_2, v_2).

  We notice that, random samples from GLD(α, θ) is generated by the simple method of transformation $X = (-1/\theta)(\ln(U^{-1/\alpha} - 1))$, where U is the uniform random variables. The codes of Maple[12] are used to generate from the gamma and uniform distributions.
- The estimates of R are then computed from (2.2), (2.22), (2.34) and (2.40).
- The quantities $(\hat{R} - R)^2$ are computed where $\hat{R}$ stands for an estimate of R and R defined by (2.1).
- The above steps are repeated 1000 times and the estimated risks (ER) computed by averaging the squared deviations over 1000 repetitions. The expression are of (ER) is

$$ER(\hat{R}) = \frac{1}{1000} \sum (\hat{R} - R)^2.$$

  The computational results are displayed in Tables (1) and (2).

Table 1. Estimated risks (ER) of the estimates of ML, $UMVU$, and Bayes (conjugate and non informative) when $(u_1 = 0.9, v_1 = 0.8, u_2 = 1.5, v_2 = 0.44)$

n	m	$ER(R_1)$	$ER(R_2)$	$ER(R_3)$	$ER(R_4)$
25	20	0.8924539592	0.8456049083	0.8896095284	0.8895898393
	25	0.8930097413	0.8960181504	0.8903893705	0.8900224403
	30	0.8934699157	0.9257962396	0.8910133596	0.8904160877
30	25	0.8933799308	0.8565503916	0.8910393301	0.8909773243
	30	0.8918445549	0.8943611085	0.8896531695	0.8893434299
	35	0.8937038578	0.9216108162	0.8916299453	0.8911694309
50	40	0.8937018588	0.8452999093	0.8922580058	0.8922668464
	50	0.8931360182	0.8946357736	0.8918105707	0.8916411298
	60	0.8930111197	0.9243194077	0.8917668570	0.8914786561

5. Concluding Remarks

1. From Table (1), (2) and (3) it may be observed that the estimated risks of the four estimates of R are decreasing when the samples size n and m are increasing.
2. The estimated risks of the Bayes estimates of R is less than the estimated risks of the corresponding maximum-likelihood estimate and UMVU estimates of R. Overall, the estimated risks of the Bayes estimates of R is the smallest estimated risks.

Table 2. Estimated risks (ER) of the estimates of ML, $UMVU$, and Bayes (conjugate and non informative) when $(u_1 = 1.9, v_1 = 1.9, u_2 = 2.9, v_2 = 3.9)$

n	m	$ER(R_1)$	$ER(R_2)$	$ER(R_3)$	$ER(R_4)$
25	20	0.5802546049	0.5075768061	0.5704947140	0.5780248882
	25	0.5802869946	0.5896559398	0.5662110133	0.5788258036
	30	0.5815944797	0.6705344047	0.5641840301	0.5794487677
30	25	0.5812947891	0.5171030037	0.5723553733	0.5799863194
	30	0.5770387954	0.5868390301	0.5657152401	0.5758595967
	35	0.5812078973	0.6569913321	0.5668158292	0.5794873964
50	40	0.5798684289	0.4912495526	0.5750962583	.5792430984
	50	0.5781832573	0.5816150996	0.5710294030	0.5774497351
	60	0.5778097914	0.6644370127	0.5689314968	0.5767407272

Table 3. Estimated risks (ER) of the estimates of ML, $UMVU$, and Bayes (conjugate and non informative) when $(u_1 = 4.0, v_1 = 3.5, u_2 = 3.7, v_2 = 5.9)$

n	m	$ER(R_1)$	$ER(R_2)$	$ER(R_3)$	$ER(R_4)$
25	20	0.7052010895	0.6090189330	0.5821968989	0.7026194154
	25	0.7056921228	0.7091660351	0.5681934733	0.7024535839
	30	0.7067597827	0.7794953499	0.5582674427	0.7031038685
30	25	0.7065288908	0.6287664601	0.5925299819	0.7042297073
	30	0.7030283633	0.7058501716	0.5800502971	0.7003398193
	35	0.7067648730	0.7690916848	0.5738706495	0.7037426742
50	40	0.7060749373	0.6076477241	0.6256509750	0.7047250070
	50	0.7047107253	0.7064119738	0.6150535989	0.7030585713
	60	0.7044067367	0.7758798091	0.6078779369	0.7025490093

References

Alkasasbeh Alkasasbeh, M.R. (2007). *Generalized Logistic Distribution: Estimation and Comparative Study.* Unpublished Thesis, University of Jordan.

Alkasasbeh Alkasasbeh, M.R. and Raqab, M.Z. (2009). Estimation of the generalized logistic distribution parameters: Comparative study. *Statistical Methodology.* 6, 262-279.

Asgharzadeh Asgharzadeh, A. (2006). Point and interval estimation for a generalized logistic distribution under progressive type II censoring. *Communications in Statistics-Theory and Methods.* 35, 1685. 1702.

Balakrishnan a Balakrishnan, N. (1991). *Handbook of the Logistic Distribution,* Marcel Dekker. New York.

Balakrishnan b Balakrishnan, N., (1990). Approximate maximum likelihood estimation for a generalized logistic distribution. *Journal of Statistical Planning and Inference.* 26, 221 236.

Balakrishnan Balakrishnan, N. and Leung, M.Y. (1988). Order statistics from the type I generalized logistic distribution. *Communications in Statistics-Simulation & Computation.* 17 (1), 25- 50.

Birnbaum Birnbaum Z.W. (1956). On a use of the MannWhitney statistic. *Proceedings of the Third Berkeley Symposium on Mathematics Statistics and Probability.* 1, 1317.

Chen Chen,G. and Balakrishnan, N. (1995). The infeasibility of probability weighted moments estimation of some generalized distributions, in: N. Balakrishnan (Ed.), *Recent Advances in Life-Testing and Reliability*, CRC Press, Boca Raton, FL, 565 573.

Gradshteyn Gradshteyn, I. S. and Ryzhik, I. M. (2007). *Tables of Integrals, Series and Products.* Corrected and Enlarged Edition. Prepared by Jeffrey, A., London: Academic Press, Inc.

Kotz Kotz, S., Lumelskii, Y., and Pensky, M. (2003). *The Stress-Strength Model and its Generalizations: Theory and Applications,* World Scientific Publishing Co. Pte. Ltd..

Rao Rao, C. R. (1965). *Liear Statistical Inference and its Applications.* New York: John Wiley and Sons.

Tong Tong, H. (1974). A note on the estimation of $P(Y < X)$ in the exponential case. *Technometrics.* 16, 625.

In: Research in Applied Statistical Science
Editor: Mohammad Ahsanullah, pp. 59-73
ISBN: 978-1-63321-818-5

Chapter 6

A CLASS OF SATURATED BINARY S-OPTIMAL ROW-COLUMN DESIGNS

Theophilus Ogunyemi
Department of Mathematics and Statistics
Oakland University, Rochester, Michigan, US

Abstract

A new class of binary, square s-optimal row-column designs is proposed. These designs are saturated in terms of eliminating two-way heterogeneity with an additive model. S-optimality criterion is used to select optimal designs. S-optimal designs are constructed for square designs that are treatment-connected with a relatively large number of nearly equi-replicated treatments. In light of connectedness, all treatment contrasts are estimable regardless of the row and column effects.

Keywords: Binary designs, s-optimality, row-column designs, saturated designs, treatment swaps

1. Introduction

Row-column designs are commonly used in agricultural and industrial experimentation to eliminate two-way heterogeneity. To compare v treatments $T_1, T_2, \ldots, T_v$ over b^2 experimental units, the units are laid out in the form of b rows and b columns. Any allocation of treatments to the units is called a row-column design. In this paper we consider only the case of square designs.

Let Y_{ij} be the observation from the (i, j)th cell, $i = 1, 2, \ldots, b$, and $j = 1, 2, \ldots, b$. The following additive model is considered in this paper.

$$Y_{ij} = \mu + \alpha_i + \beta_j + \tau_{d(i,j)} + \epsilon_{ij}, \tag{1}$$

where μ is the general mean, α_i is the effect of the ith row, β_j is the effect of the jth column, $d(i, j)$ denotes the treatment in the (i, j)th cell, τ_ℓ is the effect of the ℓth treatment T_ℓ, for $\ell = 1, 2, \cdots, v$, and ϵ_{ij}'s are independently and identically distributed random errors.

Following Shah and Sinha (1989), the information matrix C_d of a row-column design d has the following form under model (1)

$$C_d = R - \frac{1}{b}N_1N_1' - \frac{1}{b}N_2N_2' + \frac{1}{b^2}\mathbf{r}\mathbf{r}', \tag{2}$$

or, equivalently,

$$b^2C_d = b^2R - bN_1N_1' - bN_2N_2' + \mathbf{r}\mathbf{r}', \tag{3}$$

where $\mathbf{r}' = (r_1, \ldots, r_v)$ is the treatment replicate vector, $R = \text{diag}(r_1, r_2, \ldots, r_v)$ is a diagonal matrix with entries $r_1, \ldots, r_v$, $N_1 = (n_{ij}^{(r)})$ denotes the row incidence matrix of order $v \times b$, i.e., $n_{ij}^{(r)}$ is the number of times treatment T_i appears in the jth row, $N_2 = (n_{ij}^{(c)})$ denotes the column incidence matrix of order $v \times b$, and $n_{ij}^{(c)}$ is the number of times treatment T_i occurs in the jth column. In terms of matrix entries, equation (2) can be written as

$$c_{ij} = r_i\delta_{ij} - \frac{\lambda_{ij}}{b} - \frac{\mu_{ij}}{b} + \frac{r_ir_j}{b^2},$$

or, equivalently,

$$b^2c_{ij} = b^2r_i\delta_{ij} - b\lambda_{ij} - b\mu_{ij} + r_ir_j,$$

where, for $i, j = 1, 2, \ldots, v$, $\delta_{ij} = 1$ if $i = j$, 0 otherwise; $N_1N_1' = (\lambda_{ij})_{v\times v}$; $N_2N_2' = (\mu_{ij})_{v\times v}$. Note that, if all entries in N_1 and N_2 are either 0 or 1, that is, if design d is binary, then λ_{ij} and μ_{ij} are the numbers of times that treatments T_i and T_j appear in the same row and column, respectively.

One of the important goals in constructing row-column designs is to have every treatment contrast estimable. That is, the constructed design should be treatment-connected. John and Williams (1995, Chap. 5) has a detailed discussion on row-column designs. The purpose of this paper is to find s-optimal, treatment-connected designs in a specific class of row-column designs with the maximum number of treatments. A design is said to be s-optimal within a class of designs if it minimizes the trace of square of information matrix over the class of designs under consideration. The paper is arranged as follows. Section 2 introduces the new class of row-column designs. The main results are presented in Section 3 and concluding remarks are given in Section 4.

2. Saturated Row-Column Designs

The binary and square design class $\mathcal{D}$ considered in this paper are constructed as follows. Let $\nu = b^2 - 2b + 2$. Consider the following row-column layout to allocate ν treatments to the b^2 experimental units in $b(\geq 4)$ rows and $b(\geq 4)$ columns.

	1	2	3	$\cdots$	$b-1$	b
1	T_1	T_2	T_3	$\cdots$	T_{b-1}	η_1
2	T_b	T_{b+1}	T_{b+2}	$\cdots$	$T_{2(b-1)}$	η_2
3	T_{2b-1}	T_{2b}	T_{2b+1}	$\cdots$	$T_{3(b-1)}$	η_3
$\vdots$	$\vdots$	$\vdots$	$\vdots$	$\vdots$	$\vdots$	$\vdots$
$b-1$	$T_{(b-2)b-(b-3)}$	$T_{(b-2)b-(b-4)}$	$T_{(b-2)b-(b-5)}$	$\cdots$	$T_{\nu-1}$	η_{b-1}
b	θ_1	θ_2	θ_3	$\cdots$	θ_{b-1}	T_ν

a. The first $(b-1)$ rows and $(b-1)$ columns of the grid are filled with treatments T_1, T_2, $\ldots$, $T_{\nu-1}$ with one treatment per cell in a random manner.

b. For $1 \leq i \leq b-1$, η_i is any treatment in $\{T_1, T_2, \cdots, T_\nu\}$ that has not appeared in the ith row.

c. For $1 \leq j \leq b-1$, θ_j is any treatment in $\{T_1, T_2, \cdots, T_\nu\}$ other than those that are already in the jth column.

d. Due to the combinatorial complexity involved in searching for s-optimal binary designs in $\mathcal{D}$, we assume that no pair of treatments in the bth column swap rows and no pair of treatments in the bth row swap columns. For instance, treatments η_1 and η_2 swap rows if η_1 is any of the treatments in the second row and η_2 is any of the treatments in the first row. We call this swap a treatment row swap. A treatment column swap is similarly defined. Thus we are focusing on a subclass $\mathcal{D}^+$ of $\mathcal{D}$ that does not contain designs with treatment row and column swaps.

It is pertinent to mention at this stage that our inability to develop a mathematical tool to handle treatment swaps at the moment means we can only look at a subclass of $\mathcal{D}$ of designs without swaps. We denote the subclass by $\mathcal{D}^+$. However, from our extensive computer search, it seems that a design that is s-optimal in $\mathcal{D}^+$ is also s-optimal in $\mathcal{D}$. Because of the combinatorial complexity presented by swaps, extending the s-optimality from $\mathcal{D}^+$ to $\mathcal{D}$, and if combinatorially possible at all, will be a nontrivial challenge.

Since there are b^2 observations in the design, the total number of degrees of freedom is b^2-1. According to model (1), $b-1$ degrees of freedom are used to estimate row and column effects each. There are $b^2-1-2(b-1) = \nu - 1$ degrees of freedom left for treatment effects. Thus, the proposed design is saturated because $v = b^2 - 2b + 2$ is the maximum number of treatments that can be arranged in a row-column design to eliminate two-way heterogeneity. There are no degrees of freedom left to estimate the variance of process error.

Before we state the main theorem, we will first write the trace of the square of information matrix, $tr(C_d^2)$, in a simplified form. From (3) it follows that

$$\begin{aligned} C_d^2 &= (\text{Diag } \mathbf{r})(\text{Diag } \mathbf{r}) + \frac{1}{b^2}(N_1N_1')(N_1N_1') + \frac{1}{b^2}(N_2N_2')(N_2N_2') \\ &+ \frac{1}{b^4}(\mathbf{rr}')(\mathbf{rr}') - \frac{2}{b}(\text{Diag } \mathbf{r})(N_1N_1') - \frac{2}{b}(\text{Diag } \mathbf{r})(N_2N_2') \\ &+ \frac{2}{b^2}(\text{Diag } \mathbf{r})(\mathbf{rr}') + \frac{2}{b^2}(N_1N_1')(N_2N_2') - \frac{2}{b^3}(N_1N_1')(\mathbf{rr}') \\ &- \frac{2}{b^3}(N_2N_2')(\mathbf{rr}') \end{aligned} \tag{4}$$

Therefore, it follows that

$$\begin{aligned} b^4 tr(C_d^2) &= b^4\sum_{i=1}^{\nu} r_i^2 + b^2\sum_{i=1}^{\nu}\sum_{j=1}^{\nu}\lambda_{ij}^2 + b^2\sum_{i=1}^{\nu}\sum_{j=1}^{\nu}\mu_{ij}^2 + \sum_{i=1}^{\nu}\sum_{j=1}^{\nu} r_i^2 r_j^2 \\ &- 2b^3\sum_{i=1}^{\nu} r_i\lambda_{ii} - 2b^3\sum_{i=1}^{\nu} r_i\mu_{ii} + 2b^2\sum_{i=1}^{\nu} r_i^3 + 2b^2\sum_{i=1}^{\nu}\sum_{j=1}^{\nu}\lambda_{ij}\mu_{ij} \end{aligned}$$

$$
\begin{aligned}
&- 2b\sum_{i=1}^{\nu}\sum_{j=1}^{\nu} r_i r_j \lambda_{ij} - 2b\sum_{j=1}^{\nu}\sum_{i=1}^{\nu} r_i r_j \mu_{ij} \\
&= b^4\sum_{i=1}^{\nu} r_i^2 + \left(b^2\sum_{i=1}^{\nu} r_i^2 + b^2\sum_{i\neq j}^{\nu}\lambda_{ij}^2\right) + \left(b^2\sum_{i=1}^{\nu} r_i^2 + b^2\sum_{i\neq j}^{\nu}\mu_{ij}^2\right) \\
&+ \sum_{j=1}^{\nu}\sum_{i=1}^{\nu} r_i^2 r_j^2 + 2b^2\sum_{i=1}^{\nu} r_i^3 + \left(2b^2\sum_{i=1}^{\nu} r_i^2 + 2b^2\sum_{i\neq j}^{\nu}\lambda_{ij}\mu_{ij}\right) \\
&- 2b^3\sum_{i=1}^{\nu} r_i^2 - 2b^3\sum_{i=1}^{\nu} r_i^2 + \left(-2b\sum_{i=1}^{\nu} r_i^3 - 2b\sum_{i\neq j}^{\nu} r_j r_j \lambda_{ij}\right) \\
&+ \left(-2b\sum_{i=1}^{\nu} r_i^3 - 2b\sum_{i\neq j}^{\nu} r_j r_j \mu_{ij}\right) \\
&= \sum_{i=1}^{\nu}\sum_{j=1}^{\nu} r_i^2 r_j^2 + 2b(b-2)\sum_{i=1}^{\nu} r_i^3 + b^2(b^2-4b+4)\sum_{i=1}^{\nu} r_i^2 + b^2\sum_{i\neq j}^{\nu}\lambda_{ij}^2 + b^2\sum_{i\neq j}^{\nu}\mu_{ij}^2 \\
&+ 2b^2\sum_{i\neq j}^{\nu}\lambda_{ij}\mu_{ij} - 2b\sum_{i\neq j}^{\nu} r_i r_j \lambda_{ij} - 2b\sum_{i\neq j}^{\nu} r_i r_j \mu_{ij} \qquad (5)
\end{aligned}
$$

Next we introduce the layout of a design d in $\mathcal{D}^+$ that minimizes $tr(C_d^2)$. Assign treatments to the bth row such that $\theta_1 = T_{b+1},\ \theta_2 = T_{2b+1}, \ldots, \theta_{b-2} = T_{\nu-1},\ \theta_{b-1} = T_1$; and for the bth column, assign T_b to η_1, T_{2b} to $\eta_2, \cdots, T_{\nu-2}$ to η_{b-2}, T_{b-1}, T_{b-1} to η_{b-1}. We denote this s-optimal design in $\mathcal{D}^+$ by d^*. We must mention that the case $b = 5$ is the only exemption to the description of design d^*, as observed from our extensive computer search. See Tables 3 and 4. Though design d^* is not unique, but the diagonal treatments $T_1, T_{b+1}, \ldots, T_{\nu-1}$ are all common to the s-optimal designs. We next give the proof of the theorem on the s-optimality of design d^*.

3. Main Results

Theorem 3.1. *For any design d in $\mathcal{D}^+$ and, for $b > 6$, $b^4\left(tr\left(C_d^2\right) - tr\left(C_{d^*}^2\right)\right) \geq 0$, where d^* is an s-optimal design in $\mathcal{D}^+$.*

Proof. From (5) we have

$$
\begin{aligned}
b^4\left(tr\left(C_d^2\right) - tr\left(C_{d^*}^2\right)\right) &= \sum_{i=1}^{\nu}\sum_{j=1}^{\nu} r_i^2 r_j^2 + 2b(b-2)\sum_{i=1}^{\nu} r_i^3 + b^2\left(b^2-4b+4\right)\sum_{i=1}^{\nu} r_i^2 \\
&+ b^2\sum_{\substack{j=1\\ i\neq j}}^{\nu}\lambda_{ij}^2 + b^2\sum_{i\neq j}^{\nu}\mu_{ij}^2 + 2b^2\sum_{\substack{j=1\\ i\neq j}}^{\nu}\lambda_{ij}\mu_{ij} - 2b\sum_{\substack{j=1\\ i\neq j}}^{\nu} r_i r_j \lambda_{ij} \\
&- 2b\sum_{\substack{j=1\\ i\neq j}}^{\nu} r_i r_j \mu_{ij} - \sum_{i=1}^{\nu}\sum_{j=1}^{\nu}(r_i^*)^2(r_j^*)^2 - 2b(b-2)\sum_{i=1}^{\nu}(r_i^*)^3
\end{aligned}
$$

$$
\begin{aligned}
&- \quad b^2(b^2-4b+4)\sum_{i=1}^{\nu}(r_i^*)^2 - b^2\sum_{\substack{j=1\\ i\neq j}}^{\nu}(\lambda_{ij}^*)^2 - b^2\sum_{\substack{j=1\\ i\neq j}}^{\nu}(\mu_{ij}^*)^2 \\
&- \quad 2b^2\sum_{\substack{j=1\\ i\neq j}}^{\nu}\lambda_{ij}^*\mu_{ij}^* + 2b\sum_{\substack{j=1\\ i\neq j}}^{\nu}r_i^*r_j^*\lambda_{ij}^* + 2b\sum_{\substack{j=1\\ i\neq j}}^{\nu}r_i^*r_j^*\mu_{ij}^* \\
&= \quad 2b^2\sum_{i<j}^{\nu}\lambda_{ij}^2 + 2b^2\sum_{i<j}^{\nu}\mu_{ij}^2 + 4b^2\sum_{i<j}^{\nu}\lambda_{ij}\mu_{ij} - 4b\sum_{i<j}^{\nu}r_ir_j\lambda_{ij} \\
&- \quad 4b\sum_{i<j}^{\nu}r_ir_j\mu_{ij} - 2b^2\sum_{i<j}^{\nu}\left(\lambda_{ij}^*\right)^2 - 2b^2\sum_{i<j}^{\nu}\left(\mu_{ij}^*\right)^2 - 4b^2\sum_{i<j}^{\nu}\lambda_{ij}^*\mu_{ij}^* \\
&+ \quad 4b\sum_{i<j}^{\nu}r_i^*r_j^*\lambda_{ij}^* + 4b\sum_{i<j}^{\nu}r_i^*r_j^*\mu_{ij}^*, \qquad (6)
\end{aligned}
$$

using the fact that all the designs in $\mathcal{D}^+$ have the same number of treatments that are replicated twice. Our strategy of establishing s-optimality of design d^* over $\mathcal{D}^+$ is to first reduce the right hand side of (6) as much as possible to the point where we can apply partial order based on weak majorization.

Since treatment T_ν is assigned only to the (b,b)th cell for each design in $\mathcal{D}$, we have the following.

$$
\begin{aligned}
&\sum_{i<j}^{\nu}\lambda_{ij}^2 = \sum_{i<j}^{\nu-1}\lambda_{ij}^2 + b - 1; \; \sum_{i<j}^{\nu}\mu_{ij}^2 = \sum_{i<j}^{\nu-1}\mu_{ij}^2 + b - 1; \\
&\sum_{i<j}^{\nu}r_ir_j\lambda_{ij} = \sum_{i<j}^{\nu-1}r_ir_j\lambda_{ij} + 2(b-1); \; \sum_{i<j}^{\nu}r_ir_j\mu_{ij} = \sum_{i<j}^{\nu-1}r_ir_j\mu_{ij} + 2(b-1).
\end{aligned}
$$

Furthermore, for any s-optimal design d^* in $\mathcal{D}$, the following hold.

$$
\begin{aligned}
&\sum_{i<j}^{\nu-1}\left(\lambda_{ij}^*\right)^2 = \sum_{i<j}^{\nu-1}\left(\mu_{ij}^*\right)^2 = (b-1)(b^2-2)/2; \sum_{i<j}^{\nu-1}\lambda_{ij}^*\mu_{ij}^* = 4(b-1); \\
&\sum_{i<j}^{\nu-1}r_i^*r_j^*\lambda_{ij}^* = \sum_{i<j}^{\nu-1}r_i^*r_j^*\mu_{ij}^* = (b-1)(b^2+9b-8)/2.
\end{aligned}
$$

Substituting all of the preceding summations in (6), equation (6) becomes

$$
\begin{aligned}
b^4\left(t_r(C_d^2) - t_r(C_{d^*}^2)\right) &= 2b^2\sum_{i<j}^{\nu-1}\lambda_{ij}^2 + 2b^2\sum_{i<j}^{\nu-1}\mu_{ij}^2 + 4b^2\sum_{i<j}^{\nu-1}\lambda_{ij}\mu_{ij} \\
&-4b\sum_{i<j}^{\nu-1}r_ir_j\lambda_{ij} - 4b\sum_{i<j}^{\nu-1}r_ir_j\mu_{ij} - 2b^5 + 6b^4 + 20b^3 - 56b^2 + 32b. \qquad (7)
\end{aligned}
$$

Because of combinatorial complexity, the calculation of summations in (7) are not straight-forward. In order to calculate the summations in (7), we look into combinations of treatments within and between the following three sets of treatments.

Let Ω_{lr} represent the set of treatments in the bth row (last row) with the exemption of T_ν, Ω_{lc} represent the set of treatments in the bth column (last column) with the exemption of T_ν, and Ω_{un} represent the set of unreplicated treatments in a design d with the exemption of T_ν. Furthermore, let ω_{r_i} equal the number of treatments in Ω_{lr} that come from ith row, $i = 1, 2, \ldots, b-1$, and Ψ_{r_i} equal the number of replicated treatments in ith row, $i = 1, 2, \ldots, b-1$. ω_{c_j} and Ψ_{c_j}, $j = 1, 2, \ldots, b-1$, are similarly defined for the columns.

We illustrate the calculations of the summations in (7) by showing that of $\sum_{i<j}^{\nu-1} \lambda_{ij}^2$ in detail. Others are similarly calculated except for $\sum_{i<j}^{\nu-1} \lambda_{ij}\mu_{ij}$. For $\sum_{i<j}^{\nu-1} \lambda_{ij}^2$, we have the following decompositions. See Figs. 1, 2, and 3 in the appendix for the combinatorial graphs of $\sum_{\substack{i=1\\i<j}}^{\nu-1} (\lambda^* ij)^2$, $\sum_{\substack{i=1\\i<j}}^{\nu-1} r_i^* r_j^* \lambda_{ij}^*$, and $\sum_{\substack{i=1\\i<j}}^{\nu-1} \lambda_{ij}^* \mu_{ij}^*$ for d^* as examples. No row and column treatment swaps are allowed in these calculations. The summation $\sum_{i<j}^{\nu-1} \lambda_{ij}^2$ is decomposed as the sum of the following combinatorial summands:

$$\begin{aligned}
\text{Within } \Omega_{lr} : \quad & 2^2 \cdot \sum_{i=1}^{b-1} \binom{\omega_{r_i}}{2} + \left[\binom{b-1}{2} - \sum_{i=1}^{b-1} \binom{\omega_{r_i}}{2} \right] \\
& = \frac{b^2 - 3b + 2}{2} + 3 \sum_{i=1}^{b-1} \binom{\omega_{r_i}}{2} \\
\text{Within } \Omega_{lc} : \quad & \sum_{i=1}^{b-1} \binom{\Psi_{r_i} - \omega_{r_i}}{2} \\
\text{Within } \Omega_{un} : \quad & \sum_{i=1}^{b-1} \binom{b - \Psi_{r_i}}{2} \\
\text{Between } \Omega_{lr} \text{ and } \Omega_{lc} : \quad & \sum_{i=1}^{b-1} \omega_{r_i} \left(\Psi_{r_i} - \omega_{r_i} \right) \\
\text{Between } \Omega_{lr} \text{ and } \Omega_{un} : \quad & \sum_{i=1}^{b-1} \omega_{r_i} \left(b - \Psi_{r_i} \right) = b^2 - b - \sum_{i=1}^{b-1} \omega_{r_i} \Psi_{r_i} \\
\text{Between } \Omega_{lc} \text{ and } \Omega_{un} : \quad & \sum_{i=1}^{b-1} \left(\Psi_{r_i} - \omega_{r_i} \right) \left(b - \Psi_{r_i} \right). \qquad (8)
\end{aligned}$$

We have used the fact that $\sum_{i=1}^{b-1} \omega_{r_i} = b - 1$ in the second to the last summation. The calculation of $\sum_{i<j}^{\nu-1} r_i r_j \lambda_{ij}$ easily follows from that of $\sum_{i<j}^{\nu-1} \lambda_{ij}^2$. The decomposition of $\sum_{i<j}^{\nu-1} \mu_{ij}^2$ is similar to that of $\sum_{i<j}^{\nu-1} \lambda_{ij}^2$, though the diagram for $\sum_{i<j}^{\nu-1} \mu_{ij}^2$ is more entangled than that of $\sum_{i<j}^{\nu-1} \lambda_{ij}^2$. As for the summation $\sum_{i<j}^{\nu-1} r_i r_j \mu_{ij}$, its decomposition follows easily from that of $\sum_{i<j}^{\nu-1} \mu_{ij}^2$. The only combinatorially intractable summation is that of $\sum_{i<j}^{\nu-1} \lambda_{ij}\mu_{ij}$, but it happens that, for any design d in $\mathcal{D}^+$,

$$\sum_{i<j}^{\nu-1} \lambda_{ij}\mu_{ij} = \sum_{i<j}^{\nu-1} \lambda_{ij}^* \mu_{ij}^* + n = 4(b-1) + n, \qquad (9)$$

where $n = \sum_{i<j}^{\nu-1} \lambda_{ij}\mu_{ij} - 4(b-1)$, a non-negative integer. See Fig. 3.

Some simplifications are done in order to express $\sum_{i<j}^{\nu-1} \lambda_{ij}^2$ as a function of ω's and Ψ's.

$$\begin{aligned}
\sum_{i<j}^{\nu-1} \lambda_{ij}^2 &= \left(3\sum_{i=1}^{b-1}\binom{\omega_{r_i}}{2} + \frac{b^2-3b+2}{2}\right) + \left(b(b-1) - \sum_{i=1}^{b-1}\omega_{r_i}\Psi_{r_i}\right) \\
&+ \left(b(3(b-1)) - \sum_{i=1}^{b-1}\Psi_{r_i}^2 - b(b-1) + \sum_{i=1}^{b-1}\omega_{r_i}\Psi_{r_i}\right) + \sum_{i=1}^{b-1}\binom{b-\Psi_{r_i}}{2} \\
&+ \sum_{i=1}^{b-1}\binom{\Psi_{r_i}-\omega_{r_i}}{2} + \left(\sum_{i=1}^{b-1}\omega_{r_i}\Psi_{r_i} - \sum_{i=1}^{b-1}\omega_{r_i}^2\right) \\
&= \frac{7b^2-9b+2}{2} + 3\sum_{i=1}^{b-1}\binom{\omega_{r_i}}{2} + \sum_{i=1}^{b-1}\binom{b-\Psi_{r_i}}{2} + \sum_{i=1}^{b-1}\binom{\Psi_{r_i}-\omega_{r_i}}{2} \\
&+ \sum_{i=1}^{b-1}\omega_{r_i}\Psi_{r_i} - \sum_{i=1}^{b-1}\Psi_{r_i}^2 - \sum_{i=1}^{b-1}\omega_{r_i}^2 \\
&= \frac{7b^2-9b+2}{2} + \frac{3}{2}\sum_{\substack{i=1\\ \omega_{r_i}>1}}^{b-1}\omega_{r_i}^2 - \frac{3}{2}\sum_{\substack{i=1\\ \omega_{r_i}>1}}^{b-1}\omega_{r_i} + \left(\frac{b^2-b}{2}\right)\sum_{\substack{i=1\\ b-\Psi_{r_i}>1}}^{b-1} 1 \\
&- \left(\frac{b-1}{2}\right)\sum_{\substack{i=1\\ b-\Psi_{r_i}>1}}^{b-1}\Psi_{r_i} + \frac{1}{2}\sum_{\substack{i=1\\ b-\Psi_{r_i}>1}}^{b-1}\Psi_{r_i}^2 - \sum_{\substack{i=1\\ \Psi_{r_i}-\omega_{r_i}>1}}^{b-1}\omega_{r_i}\Psi_{r_i} - \frac{1}{2}\sum_{\substack{i=1\\ \Psi_{r_i}-\omega_{r_i}>1}}^{b-1}\Psi_{r_i} \\
&+ \sum_{\substack{i=1\\ \Psi_{r_i}-\omega_{r_i}>1}}^{b-1}\omega_{r_i}^2 + \frac{1}{2}\sum_{\substack{i=1\\ \Psi_{r_i}-\omega_{r_i}>1}}^{b-1}\omega_{r_i} + \sum_{i=1}^{b-1}\omega_{r_i}\Psi_{r_i} - \sum_{i=1}^{b-1}\Psi_{r_i}^2 - \sum_{i=1}^{b-1}\omega_{r_i}^2. \qquad (10)
\end{aligned}$$

$$\begin{aligned}
\sum_{i<j}^{\nu-1} r_i r_j \lambda_{ij} &= \left(4\sum_{i=1}^{b-1}\binom{\omega_{r_i}}{2} + 2(b^2-3b+2)\right) + 2\left(b(b-1) - \sum_{i=1}^{b-1}\omega_{r_i}\Psi_{r_i}\right) \\
&+ 2\left(b\,(3(b-1)) - \sum_{i=1}^{b-1}\Psi_{r_i}^2 - b(b-1) + \sum_{i=1}^{b-1}\omega_{r_i}\Psi_{r_i}\right) + \sum_{i=1}^{b-1}\binom{b-\Psi_{r_i}}{2} \\
&+ 4\sum_{i=1}^{b-1}\binom{\Psi_{r_i}-\omega_{r_i}}{2} + 4\left(\sum_{i=1}^{b-1}\omega_{r_i}\Psi_{r_i} - \sum_{i=1}^{b-1}\omega_{r_i}^2\right) \\
&= 7b^4 - 41b^3 + 50b^2 - 16b + (6b^2-16b)\cdot\frac{1}{2}\sum_{\substack{i=1\\ \omega_{r_i}>1}}^{b-1}\omega_{r_i}^2 \\
&- (6b^2-16b)\cdot\frac{1}{2}\sum_{\substack{i=1\\ \omega_{r_i}>1}}^{b-1}\omega_{r_i} + (2b^2-4b)\left(\frac{b^2-b}{2}\right)\sum_{\substack{i=1\\ \Psi_{r_i}<b-1}}^{b-1} 1
\end{aligned}$$

$$
\begin{aligned}
&- \quad (2b^2-4b)\left(\frac{2b-1}{2}\right)\sum_{\substack{i=1\\ \Psi_{r_i}<b-1}}^{b-1}\Psi_{r_i}+(2b^2-4b)\cdot\frac{1}{2}\sum_{\substack{i=1\\ \Psi_{r_i}<b-1}}^{b-1}\Psi_{r_i}^2\\
&+ \quad (2b^2-16b)\cdot\frac{1}{2}\sum_{\substack{i=1\\ \Psi_{r_i}>\omega_{r_i}+1}}^{b-1}\Psi_{r_i}^2-(2b^2-16b)\sum_{\substack{i=1\\ \Psi_{r_i}>\omega_{r_i}+1}}^{b-1}\omega_{r_i}\Psi_{r_i}\\
&- \quad (2b^2-16b)\cdot\frac{1}{2}\sum_{\substack{i=1\\ \Psi_{r_i}>\omega_{r_i}+1}}^{b-1}\Psi_{r_i}+(2b^2-16b)\cdot\frac{1}{2}\sum_{\substack{i=1\\ \Psi_{r_i}>\omega_{r_i}+1}}^{b-1}\omega_{r_i}^2\\
&+ \quad (2b^2-16b)\cdot\frac{1}{2}\sum_{\substack{i=1\\ \Psi_{r_i}>\omega_{r_i}+1}}^{b-1}\omega_{r_i}+(2b^2-16b)\sum_{i=1}^{b-1}\omega_{r_i}\Psi_{r_i}\\
&+ \quad (8b-2b^2)\sum_{i=1}^{b-1}\Psi_{r_i}^2+(16b-2b^2)\sum_{i=1}^{b-1}\omega_{r_i}^2. \qquad (11)
\end{aligned}
$$

The expressions for $\sum_{i<j}^{\nu-1}\mu_{ij}^2$ and $\sum_{i<j}^{\nu-1}r_ir_j\mu_{ij}$ are similar to those of $\sum_{i<j}^{\nu-1}\lambda_{ij}^2$ and $\sum_{i<j}^{\nu-1}r_ir_j\lambda_{ij}$, respectively. After making substitutions of the expressions into (7) and a long and tedious simplification, (7) becomes

$$
\begin{aligned}
b^4\left(tr(C_d^2)-tr(C_{d^*}^2)\right) &= 2b^2\sum_{i=1}^{b-1}\omega_{r_i}^2+2b^2\sum_{j=1}^{b-1}\omega_{c_j}^2-2b\sum_{i=1}^{b-1}\Psi_{r_i}^2-2b\sum_{j=1}^{b-1}\Psi_{c_j}^2\\
&\quad -4b^3+4b^2(n+10)-36b, \qquad (12)
\end{aligned}
$$

for $n=0,1,\ldots$. In the preceding simplification, we have used the facts that $\sum_{i=1}^{b-1}\omega_{r_i}=\sum_{j=1}^{b-1}\omega_{c_j}=b-1$ and $\sum_{i=1}^{b-1}\Psi_{r_i}=\sum_{j=1}^{b-1}\Psi_{c_j}=3(b-1)$.

In order to complete the proof, we need to show that the right hand side of (12) is non-negative for any design d in $\mathcal{D}^+$, and is zero for an s-optimal design d^*. That is, we want to show that

$$
\begin{aligned}
b^4\left(tr(C_d^2)-tr(C_{d^*}^2)\right) &= b\sum_{i=1}^{b-1}\omega_{r_i}^2+b\sum_{j=1}^{b-1}\omega_{c_j}^2-\sum_{i=1}^{b-1}\Psi_{r_i}^2-\sum_{j=1}^{b-1}\Psi_{c_j}^2\\
&\quad -2b^2+2b(n+10)-18\geq 0, \qquad (13)
\end{aligned}
$$

and is equal to zero for an s-optimal design d^*. For an s-optimal design d^*, the inequality in (13) is an equality, since for a design d^*, $\omega_{r_i}=1$ for $i=1,\ldots,b-1$; $\omega_{cj}=1$ for $j=1,2,\ldots,b-1$; $\Psi_{r_i}=3$ for $i=1,2,\ldots,b-1$; $\Psi_{c_j}=3$ for $j=1,2,\ldots,b-1$; and $n=0$.

Inequality (13) is proved by first splitting it into two inequalities, namely,

$$
-b^2+b(n+10)-9+b\sum_{i=1}^{b-1}\omega_{r_i}^2-\sum_{i=1}^{b-1}\Psi_{r_i}^2\geq 0
$$

and

$$-b^2 + b(n+10) - 9 + b\sum_{j=1}^{b-1} \omega_{c_j}^2 - \sum_{j=1}^{b-1} \Psi_{c_j}^2 \geq 0. \quad (14)$$

We will only prove the first inequality in (14). The proof of the second inequality is similar. In proving the first inequality, we will exploit some of the properties of d^*. In addition, we will consider two sets of values for b, namely, $6 \leq b \leq 9$ and $b > 9$. For $b = 4$ and 5, all the s-optimal designs d^* are tabulated in Tables 1-4. For an s-optimal design d^*,

$$-b^2 + b(n+10) - 9 + b\sum_{i=1}^{b-1} \left(\omega_{r_i}^*\right)^2 - \sum_{i=1}^{b-1} \left(\Psi_{r_i}^*\right)^2 = 0 \quad (15)$$

for all $b \geq 6$ and $n = 0$. This follows from the statement following inequality (13) above. That is, for a design d^* and $n = 0$, we have

$$b\sum_{i=1}^{b-1} \left(\omega_{r_i}^*\right)^2 - \sum_{i=1}^{b-1} \left(\Psi_{r_i}^*\right)^2 = b^2 - 10b + 9 = -(-b^2 + 10b - 9),\ 6 \leq b \leq 9.$$

Thus, for $6 \leq b \leq 9$ and a design d^*,

$$-b^2 + 10b - 9 = \alpha_b \geq 0 \text{ and } b\sum_{i=1}^{b-1} \left(\omega_{r_i}^*\right)^2 - \sum_{i=1}^{b-1} \left(\Psi_{r_i}^*\right)^2 = -\alpha_b \leq 0, \quad (16)$$

where α_b is a non-negative integer that depends on b. On the other hand, for $b > 9$, we have

$$-b^2 + 10b - 9 = -\beta_b \leq 0 \text{ and } b\sum_{i=1}^{b-1} \left(\omega_{r_i}^*\right)^2 - \sum_{i=1}^{b-1} \left(\Psi_{r_i}^*\right)^2 = \beta_b \geq 0, \quad (17)$$

where β_b is a nonnegative integer that depends on b.

Next we use the properties of d^* stated in (16) and (17) in the following argument. The term $b\sum_{i=1}^{b-1} \omega_{r_i}^2 - \sum_{i=1}^{b-1} \Psi_{r_i}^2$ can be negative or positive when $6 \leq b \leq 9$. When it is positive, then there is nothing to prove as the first inequality in (14) that we want to prove would hold. When it is negative, we will consider two cases: There are instances when (i) $b\sum_{i=1}^{b-1}(\omega_{r_i}^*)^2 - \sum_{i=1}^{b-1}(\Psi_{r_i}^*)^2 \leq b\sum_{i=1}^{b-1} \omega_{r_i}^2 - \sum_{i=1}^{b-1} \Psi_{r_i}^2 \leq 0$ and instances when (ii), $b\sum_{i=1}^{b-1} \omega_{r_i}^2 - \sum_{i=1}^{b-1} \Psi_{r_i}^2 \leq b\sum_{i=1}^{b-1}(\omega_{r_i}^*)^2 - \sum_{i=1}^{b-1}(\Psi_{r_i}^*)^2 \leq 0$. From case (i), it follows that, for $6 \leq b \leq 9$ and $n \geq 0$,

$$\begin{aligned} b^2 - 10b + 9 + \left(-b^2 + b(n+10) - 9\right) &\leq b\sum_{i=1}^{b-1} \omega_{r_i}^2 - \sum_{i=1}^{b-1} \Psi_{r_i}^2 + \left(-b^2 + b(n+10) - 9\right) \\ &< -b^2 + b(n+10) - 9, \end{aligned}$$

that is,

$$0 < bn \leq b\sum_{i=1}^{b-1} \omega_{r_i}^2 - \sum_{i=1}^{b-1} \Psi_{r_i}^2 + \left(-b^2 + b(n+10) - 9\right) \leq -b^2 + b(n+10) - 9.$$

Thus, with regard to case (i), the first inequality in (14) is proved. Next we simplify the inequality under case (ii) as follows. Under case (ii) we have instances when

$$b\sum_{i=1}^{b-1}\omega_{r_i}^2 - \sum_{i=1}^{b-1}\Psi_{r_i}^2 \leq b\sum_{i=1}^{b-1}(\omega_{r_i}^*)^2 - \sum_{i=1}^{b-1}(\Psi_{r_i}^*)^2 \leq 0.$$

That is,

$$b\sum_{i=1}^{b-1}\omega_{r_i}^2 - \sum_{i=1}^{b-1}\Psi_{r_i}^2 \leq b^2 - 10b + 9 \leq 0, \text{ for } 6 \leq b \leq 9. \tag{18}$$

We will return to (18) later.

For $b > 9, b\sum_{i=1}^{b-1}(\omega_{r_i}^*)^2 - \sum_{i=1}^{b-1}(\Psi_{r_i}^*)^2 = b(b-1) - 9(b-1) = b^2 - 10b + 9 \geq 0$, as discussed in (17). But the term, $-b^2 + b(n+10) - 9$, can be positive or negative when $b > 9$. When it is positive, then there is nothing to prove as the first inequality in (14) is satisfied. When it is negative, then we have $-b^2 + 10b - 9 \leq -b^2 + b(n+10) - 9$ for all $n \geq 0$. From this inequality it follows that

$$\begin{aligned}&\left(-b^2 + 10b - 9\right) + \left(b\sum_{i=1}^{b-1}\omega_{r_i}^2 - \sum_{i=1}^{b-1}\Psi_{r_i}^2\right)\\ &\leq \left(-b^2 + b(n+10) - 9\right) + \left(b\sum_{i=1}^{b-1}\omega_{r_i}^2 - \sum_{i=1}^{b-1}\Psi_{r_i}^2\right).\end{aligned} \tag{19}$$

Assuming that $b\sum_{i=1}^{b-1}(\omega_{r_i}^*)^2 - \sum_{i=1}^{b-1}(\Psi_{r_i}^*)^2 \leq b\sum_{i=1}^{b-1}\omega_{r_i}^2 - \sum_{i=1}^{b-1}\Psi_{r_i}^2$, then we can write (19) as

$$\begin{aligned}&\left(-b^2 + 10b - 9\right) + \left(b\sum_{i=1}^{b-1}(\omega_{r_i}^*)^2 - \sum_{i=1}^{b-1}(\Psi_{r_i}^*)^2\right)\\ &\leq \left(-b^2 + b(n+10) - 9\right) + \left(b\sum_{i=1}^{b-1}\omega_{r_i}^2 - \sum_{i=1}^{b-1}\Psi_{r_i}^2\right),\end{aligned}$$

that is,

$$0 \leq -b^2 + b(n+10) - 9 + b\sum_{i=1}^{b-1}\omega_{r_i}^2 - \sum_{i=1}^{b-1}\Psi_{r_i}^2,$$

which again proves the first inequality in (14). The proof for the second inequality in (14) is similar. In order to complete the proof, we need to show that, for $b > 9$ and $n \geq 0$, the assumption immediately after equation (19) holds, that is,

$$0 \leq b\sum_{i=1}^{b-1}(\omega_{r_i}^*)^2 - \sum_{i=1}^{b-1}(\Psi_{r_i}^*)^2 \leq b\sum_{i=1}^{b-1}\omega_{r_i}^2 - \sum_{i=1}^{b-1}\Psi_{r_i}^2$$

That is, for $b > 9$,

$$0 \leq b^2 - 10b + 9 \leq b\sum_{i=1}^{b-1}\omega_{r_i}^2 - \sum_{i=1}^{b-1}\Psi_{r_i}^2 \tag{20}$$

Thus, from (18) and (20), the first inequality in (14) is proved under the conditions that

$$b\sum_{i=1}^{b-1}\omega_{r_i}^2 - \sum_{i=1}^{b-1}\Psi_{r_i}^2 \leq b^2 - 10b + 9 \leq 0, \ \ 6 \leq b \leq 9$$

and

$$0 \leq b\sum_{i=1}^{b-1}\omega_{r_i}^2 - \sum_{i=1}^{b-1}\Psi_{r_i}^2 \leq b^2 - 10b + 9, \ \ b > 9. \tag{21}$$

In order to prove the inequalities in (21), we need weak majorization. See Olkin and Marshall (1979). Introduce a partial order on the vectors $\boldsymbol{\omega}$ and $\boldsymbol{\Psi}$ as follows. Arrange the components of $\boldsymbol{\omega}$ and $\boldsymbol{\Psi}$ in a non-decreasing order. For $\boldsymbol{\omega}' = (\omega_1, \omega_2, \ldots, \omega_{b-1})$ and $\boldsymbol{\Psi}' = (\Psi_1, \Psi_2, \ldots, \Psi_{b-1})$, we say $\boldsymbol{\Psi} \prec_w \boldsymbol{\omega}$, that is, $\boldsymbol{\Psi}$ is majorized weakly by $\boldsymbol{\omega}$, if $\Psi_1 \leq \omega_1, \ \Psi_1 + \Psi_2 \leq \omega_1 + \omega_2, \ldots, \Psi_1 + \Psi_2 + \cdots + \Psi_{b-1} \leq \omega_1 + \omega_2 + \cdots + \omega_{b-1}$. This is the so-called weak majorization order. In addition, functions that preserve the ordering of majorization are said to be Schur-convex, that is, Schur-increasing. If ϕ is convex on an interval $I \subset R$, then it is Schur-convex on I. If ϕ is an increasing, convex function on an interval I and $\boldsymbol{\Psi} \prec_w \boldsymbol{\omega}$, then $\sum_{i=1}^{b-1}\phi(\Psi_i) \leq \sum_{i=1}^{b-1}\phi(\omega_i)$. For example, take $\phi(x) = x^2$.

Either by majorization we have briefly discussed in the preceding paragraph or by the generalization of majorization, also discussed in Olkin and Marshall (1979), the inequalities in (21) will be proved. Rewrite the two inequalities in (21) as

$$\begin{aligned} &\sum_{i=1}^{b-1}\omega_{r_i}^2 - \sum_{i=1}^{b-1}\left(\left(\frac{\Psi_{r_i}}{\sqrt{b}}\right)^2 - 1 + \frac{9}{b}\right) \leq 0, \qquad 6 \leq b \leq 9 \\ &\sum_{i=1}^{b-1}\omega_{r_i}^2 - \sum_{i=1}^{b-1}\left(\left(\frac{\Psi_{r_i}}{\sqrt{b}}\right)^2 - 1 + \frac{9}{b}\right) \geq 0, \qquad b > 9. \end{aligned} \tag{22}$$

For $6 \leq b \leq 9$, $\boldsymbol{\omega}_r \prec_w \boldsymbol{\Psi}_r^+$, and for $b > 9$, $\boldsymbol{\Psi}_r^+ \prec_w \boldsymbol{\omega}_r$, where $\boldsymbol{\omega}_r' = (\omega_1, w_2, \ldots, w_{b-1})$ and $\Psi^{+\prime} = \left(\left(\left(\frac{\Psi_{r_i}}{\sqrt{b}}\right)^2 - 1 + \frac{9}{b}\right)^{1/2}, \ldots, \left(\left(\frac{\Psi_{r_{b-i}}}{\sqrt{b}}\right)^2 - 1 + \frac{9}{b}\right)^{1/2}\right)$. Since $\phi(\omega_{r_i}) = \omega_{r_i}^2$ and

$\phi(\omega_{r_i}) = \phi\left(\left(\frac{\Psi_{r_i}^2}{b} - 1 + \frac{9}{b}\right)^{1/2}\right) = \frac{\Psi_{r_i}^2}{b} - 1 + \frac{9}{b}$ are convex and increasing for a given b, the inequalities in (21) are proved. With the inequalities proved, we conclude the proof of the main theorem.

4. Concluding Remarks

Remark 1. We have only discussed of binary, square and saturated row-column designs without row or column treatment swaps. Designs with swaps are extremely difficult, if not impossible, to handle mathematically because of intractable combinatorial complexity the presence of swaps presents. We will investigate whether there is a mathematical trick to

Table 1. 4x4 s-optimal designs d^*

4th Row	Corresponding 4th Columns			4th Row	Corresponding 4th Columns		
T_5	T_4	T_8	T_7	T_9	T_4	T_6	T_7
T_9	T_8	T_3	T_2	T_1	T_8	T_7	T_2
T_1	T_3	T_4	T_6	T_5	T_3	T_2	T_6

Table 2. 4x4 s-optimal designs d^*

4th Column	Corresponding 4th Rows			4th Column	Corresponding 4th Rows		
T_5	T_2	T_6	T_3	T_9	T_4	T_8	T_7
T_9	T_6	T_7	T_4	T_1	T_8	T_3	T_2
T_1	T_7	T_2	T_8	T_5	T_3	T_4	T_6

Table 3. 5x5 s-optimal designs d^*

5th Row	Corresponding 5th Columns											
T_6	T_5	T_{12}	T_{12}	T_{15}	T_3	T_8	T_8	T_9	T_4	T_4	T_7	T_{13}
T_{11}	T_{12}	T_2	T_5	T_5	T_9	T_9	T_{14}	T_8	T_{10}	T_{13}	T_{13}	T_4
T_{16}	T_2	T_5	T_{15}	T_2	T_{14}	T_3	T_9	T_{14}	T_{13}	T_7	T_{10}	T_{10}
T_1	T_{15}	T_{15}	T_2	T_{12}	T_8	T_{14}	T_3	T_3	T_7	T_{10}	T_4	T_7

5th Row	Corresponding 5th Columns											
T_{16}	T_2	T_2	T_5	T_{15}	T_3	T_9	T_{14}	T_{14}	T_7	T_{10}	T_{10}	T_{13}
T_1	T_{12}	T_{15}	T_{15}	T_2	T_{14}	T_3	T_3	T_8	T_{10}	T_4	T_7	T_7
T_6	T_{15}	T_5	T_{12}	T_{12}	T_8	T_8	T_9	T_3	T_4	T_7	T_{13}	T_4
T_{11}	T_5	T_{12}	T_2	T_5	T_9	T_{14}	T_8	T_9	T_{13}	T_{13}	T_4	T_{10}

overcome the complexity due to swaps in future work so as to extend this work from $\mathcal{D}^+$ to $\mathcal{D}$. But, for now, it looks hopeless. With the presence of swaps, direct calculation of $tr(C_d^2)$ through the usual manipulation of matrices is meaningless. The alternative combinatorial route can be very messy, if not intractable.

Remark 2. The symmetry, induced by square row-column designs, makes the calculations less traumatic. The case $b \neq k$, that is, the more general rectangular row-column designs, will be investigated in future work.

Remark 3. Unlike most row-column designs in the literature, the proposed s-optical designs are not equi-replicated, but they are nearly so. Furthermore, most row-column designs with equireplication of treatments such as Latin squares and Youden squares have relatively small number of treatments, which may not be practical for a large-scale industrial

Table 4. 5x5 s-optimal designs d^*

5th Column	Corresponding 5th Rows											
T_6	T_2	T_{12}	T_{15}	T_{15}	T_3	T_9	T_{14}	T_{14}	T_4	T_{10}	T_{13}	T_{13}
T_{11}	T_{15}	T_2	T_2	T_5	T_{14}	T_3	T_3	T_8	T_{13}	T_4	T_4	T_7
T_{16}	T_5	T_5	T_{12}	T_2	T_8	T_8	T_9	T_3	T_7	T_7	T_{10}	T_4
T_1	T_{12}	T_{15}	T_5	T_{12}	T_9	T_{14}	T_8	T_9	T_{10}	T_{13}	T_7	T_{10}

5th Column	Corresponding 5th Rows											
T_{16}	T_2	T_5	T_5	T_{12}	T_3	T_8	T_8	T_9	T_4	T_7	T_7	T_{10}
T_1	T_{12}	T_{12}	T_{15}	T_5	T_9	T_9	T_{14}	T_8	T_{10}	T_{10}	T_{13}	T_7
T_6	T_{15}	T_2	T_{12}	T_{15}	T_{14}	T_3	T_9	T_{14}	T_{13}	T_4	T_{10}	T_{13}
T_{11}	T_5	T_{15}	T_2	T_2	T_8	T_{14}	T_3	T_3	T_7	T_{13}	T_4	T_4

process. Our saturated row-column designs could well be used in high throughput screening where more than $100,000$ compounds have to be tested on rectangular microplates in order to identify potentially leading candidates rapidly and accurately.

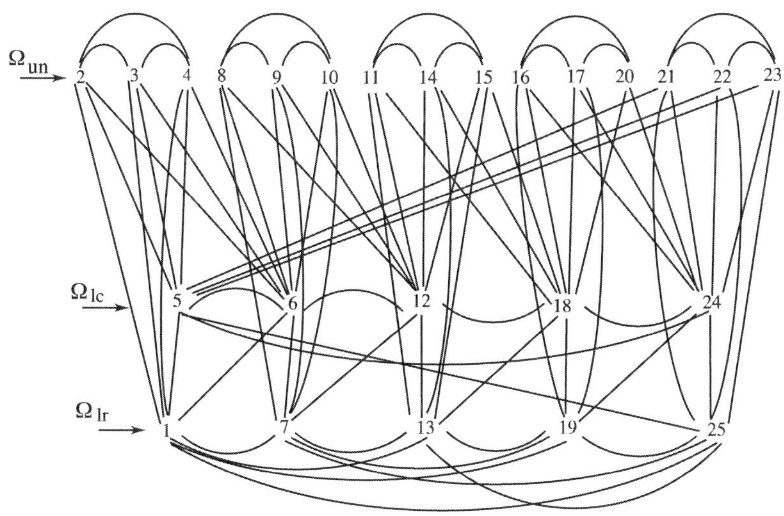

Figure 1. Graph of $\sum_{\substack{i=1 \\ i<j}}^{\nu-1} \left(\lambda_{ij}^*\right)^2$ for s-optimal design d^* when $b = 6$ as an illustration:

Ω_{lr} : the set of treatments in the last row, not including T_ν.

Ω_{lc} : the set of treatments in the last column, not including T_ν.

Ω_{un} : the set of non-replicated treatments in design d^*, not including T_ν.

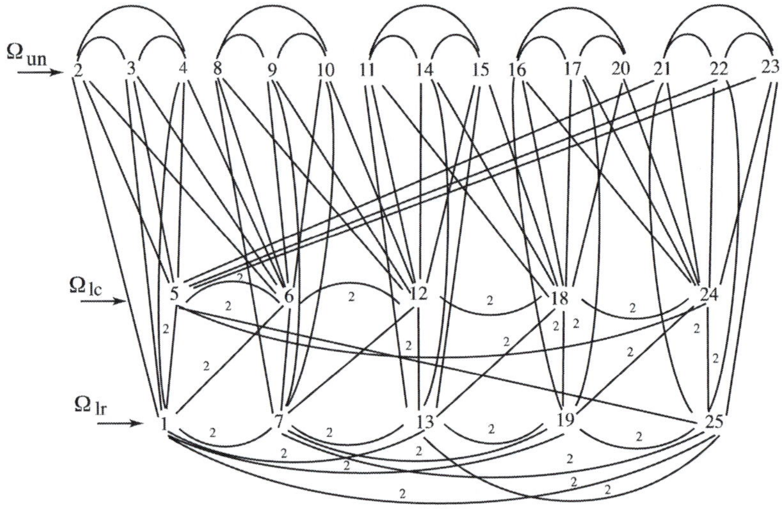

Figure 2. Graph for $\sum_{\substack{i=1\\i<j}}^{\nu-1} r_i^* r_j^* \lambda_{ij}^*$ for s-optimal d^* when $b = 6$ as an illustration:

Ω_{lr} : the set of treatments in the last row, not including T_ν.
Ω_{lc} : the set of treatments in the last column, not including T_ν.
Ω_{un} : the set of non-replicated treatments in design d^*, not including T_ν.
An edge with "2" means $r_i^* = r_j^* = 2$.

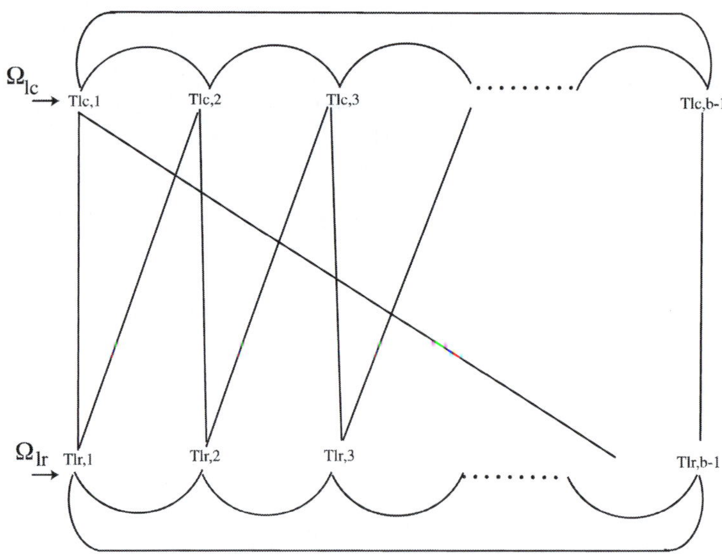

Figure 3. $\sum_{\substack{l=1\\i<j}}^{\nu-1} \lambda_{ij}^* \mu_{ij}^* = 2(b-1) + (b-1) + (b-1) - 4(b-1)$ Graph of $\sum_{\substack{l=1\\i<j}}^{\nu-1} \lambda_{ij}^* \mu_{ij}^*$ of an s-optimal design d^* Ω_{lr} : the set of treatments in the last row, not including T_ν.

References

Marshall, A. W. and Olkin, I. (1974). Inequalities: Theory of majorization and its applications. *Mathematics in Science and Engineering,* Vol. 143. Academic Press. New York.

Ogunyemi, T, O., Rao, C. R. and Rao, M. B. (2000). On improving the quality of minimal s-optimal row-column designs. *Journal of Applied Statistical Science*, 10(1), 37–46.

Shah, K. R. and Sinha, B. K. (1989). Theory of optimal designs. *Lecture Notes in Statistics* 54, Springer, Berlin.

Qu X. and Ogunyemi, T. O. (2008). A class of saturated row-column designs. Submitted to *Journal of Statistical Planning and Inference.*

Qu X. and Ogunyemi, T. O. (2008). On the connectedness of saturated square designs. Submitted to *Journal of Statistics and Probability Letters.*

In: Research in Applied Statistical Science
Editor: Mohammad Ahsanullah, pp. 75-82
ISBN: 978-1-63321-818-5

Chapter 7

MULTIPLY AND PARTIALLY CONFOUNDED FACTORIAL BIB DESIGNS

M. Shamsuddin and M. Albassam
Department of Statistics, King Abdulaziz University, Saudi Arabia

Abstract

The s^n factorial design, where s is a prime or prime power, generally confounds a set of different orthogonal factorial effects with s^r blocks $(n > r > 1)$ under one replicate. When this s^n factorial design in s^r blocks partially confounds different orthogonal sets of orthogonal factorial effects under the complete set of different orthogonal replicates it constitutes a BIB design which is called 'Multiply and Partially Confounded Factorial BIB Design' having similar properties, same result and detailed information of all factorial effects. The procedure is illustrated with examples.

Keywords: Blocking, Orthogonal Replicates, Parameters, Treatments

1. Introduction

Bose (1947), Bose and Kishan (1940), Bailey (1977), Kempthorne (1952), Hicks (1973), Anderson and McLean (1974), Das and Giri (1984) and many others constructed the different confounded s^n factorial designs by use of finite geometry, where s is a prime or prime power of levels. The balanced incomplete block design was first introduced by Yates (1937). Afterwards it is developed by Fisher (1942), Bose (1939, 1942), Shrikhande (1952), Raghavarao (1971) and many others. Both BIBD and factorial designs can be obtained by use of finite fields and ideas of parallel pencils. Yates (1937) and later on, Bose (1947) discussed the construction procedure of factorial designs. Yates (1936), Bose and Manvel (1984), Mukhopadhyay (1999) and others constructed the different BIB designs with the help of orthogonal series of Latin squares and different finite plane geometries like EG(2,s) and PG(2, s) , where s is a prime or prime power. John (1971) showed two examples of orthogonal replicates using 2^3 and 2^4 factorial effects in the case of BIB designs. But none has showed the direct method of construction of BIB designs using all possible partially confounded factorial designs as the complete set of orthogonal replicates. In order to show the alternative of BIB designs with the factorial designs Shamsuddin and Al-Marshadi (2009) constructed the s^n factorial BIB design in s blocks by developing the complete set of different partially confounded s^n factorial designs in s blocks as the complete set orthogonal replicates. But they made s^n factorial BIB designs in s blocks under the bigger set of complete orthogonal replicates.. In this paper, authors has generalized the factorial designs as the alternatives to the BIB designs by use of the different partially confounded s^n factorial design in s^r blocks as the orthogonal replicates of the smaller complete set and showed that though both designs provide same result, but the partially confounded factorial designs are more informative than any BIB design. The design as a whole is called the multiply and partially confounded factorial BIB design. Section 2 reveals construction and analysis of the concern design, examples are given in section 3 and the last section draws the conclusion.

2. Construction and Analysis

In constructing the partially confounded s^n factorial designs in s^r blocks, where s is a prime or prime power and $n > r > 1$, in the complete orthogonal replicates as the alternative shapes of different BIB designs we are to clarify first the chronological ideas: i) easy selection of complete orthogonal replicates, ii) partially confounded s^n factorial designs in s^r blocks under each replicate, iii) how these factorial designs are constructed as the alternatives to different BIB designs.

2.1. The Easy Selection of the Complete Set of Orthogonal Replicates

The s^n factorial design contains $(s^n-1)/(s-1)$ factorial effects in total and it confounds simultaneously a group of new $(s^r-1)/(s-1)$ factorial effects in s^r blocks in each replicate, so there are in total $(s^n-1)/(s^r-1)$ groups of factorial effects partially confounded respectively in $(s^n-1)/(s^r-1)$ orthogonal replicates, provided that $(s^n-1)/(s-1)$ must be multiple of $(s^r-1)/(s-1)$. Since all factorial effects are orthogonal to each other and each replicate in its s^r blocks confounds a new group of $(s^r-1)/(s-1)$ factorial effects, all replicates are orthogonal to each other and there are in total $(s^n-1)/(s^r-1)$ orthogonal replicates which constitute a complete set.

When the multiple or simultaneous (group) confounding are supposed to perform in each s^r blocks of s^n designs partially under each of the complete orthogonal replicates, it is difficult and also time consuming to choose the complete set of orthogonal replicates consisting of different groups of new and uncommon orthogonal factorial effects in different designs like 2^4 factorial design in 4 blocks, 2^6 design in 4 or 8 blocks, 3^4 designs in 3^2 blocks, etc. In order to choose the unique number of orthogonal replicates of different groups of uncommon factorial effects of the partially confounded design, we may follow a simple and easy rotational technique within the group in the following examples of s^n partially confounded design.

2.1.1. Selection of Different Orthogonal Replicates of s^n Factorial Design in s^r Blocks

The unique number of orthogonal replicates of the partially confounded s^n factorial design in s^r blocks can be possible, if and only if $[(s^n-1)/(s-1)]/ [(s^r-1)/(s-1)]$ is a natural and unique number. For r > 1 the round figure of this number is obtained for only some partially confounded designs like 2^4 factorial design in 4 blocks, 2^6 design in 4 or 8 blocks, 3^4 design in 9 blocks, etc. But if r =1, it is possible for all types of partially confounded s^n factorial designs in s blocks [Shamsuddin and Al–Marshadi 2009].

As for example, the 2^4 factorial design in 2^2 blocks confounds 2^2-1 factorial effects in one replicate, so there are in total [$(2^4-1)/(2^2-1) = 5$] number of orthogonal replicates of this partially confounded design. Let the 2^4 design having 15 factorial effects confounds three effects A, B, AB in 4 blocks in replicate I and C, D, CD in replicate II, and by use of the self-explanatory rotational technique we have the remaining other orthogonal replicates Rep III, Rep IV, and Rep V as follows:

Rep I	Rep II	Rep II(rotation I),	Rep II (rotation II)
A	C	D	CD
B	D	CD	C
AB	CD	C	D

Rep I	x Rep II	=Rep III;	Rep I	x Rep II (rota I)	=Rep IV;	Rep I	x Rep II (rota II)	=Rep V
A	C	AC	A	D	AD	A	CD	ACD
B	D	BD	B	CD	BCD	B	C	BC
AB	CD	ABCD	AB	C	ABC	AB	D	ABD

Hence, 5 orthogonal replicates with their respective factorial effects are Rep I: A, B, AB; Rep II: C, C, CD; Rep III: AC, BD, ABCD; Rep IV: AD, BCD, ABC; Rep V: ACD, BC, ABD.

Thus the 9 orthogonal replicates (each with confounded 7 factorial effects) of partially confounded 2^6 factorial design in 2^3 blocks are: R I: A,B,C,AB, BC, AC,ABC; RII:D, E, F, DE, EF, DF, DEF; RIII: AD, BE, CF, ABDE, BCEF, ACDF, ABCDEF; RIV: AE, BF, CDE,ABEF, BCDF, ACDEF, ABCD RV: AF, BDE, CEF, ABDF,BCDEF, ACD,ABCE; RVI: ADE, BEF, CDF, ABDEF, BCD, ACE, ABCF; RVII: AEF, BDF, CEF, ABD, BCE, ACF, ABCDE; RVIII: ADF, BDEF, CD, ABE, BCF, ACDE, ABCEF; RIX: ADEF, BD, CE,ABF, BCDE, ACEF, ABCDF.

The 21 orthogonal replicates (each with confounded 3 factorial effects) of the partially confounded 2^6 factorial design in 2^2 blocks: R-1:A,B,AB; R-2:C,D,CD; R-3:E,F,EF; R-4: AC,BD,ABCD; R-5: AD, BCD, ABC; R-6: ACD, BC, ABD; R-7: AE, BF, ABEF, R-8: AF, BEF, ABE; R-9: AEF, BE, ABF; R-10: CE, DF, CDEF; R-11: CF,DEF, CDE; R-12: CEF, DE, CDF; R-13: ACE, BDF, ABCDEF; R-14: ACF, BDEF, ABCDE;R-15:ACEF,BDE,ABCDF; R-16: ADE,BCDF,ABCEF; R-17: ACDE, BCF,ABDEF; R-18:ADF,BCDEF,ABCE; R-19: ACDEF,BCE,ABDF; R-20: ADEF,BCDE,ABCF; R-21 ACDF, BCEF, ABDE.

The 10 orthogonal replicates (each with confounded 4 factorial effects) of partially confounded 3^4 factorial design in 3^2 blocks: R-1: A, B, AB, AB^2; R-2: C, D, CD, CD^2; R-3: AC, BD, ABCD, AB^2CD^2; R-4: AC^2, BD^2, ABC^2D^2, AB^2C^2D; R-5: AD, BCD, $ABCD^2$, AB^2C; R-6: AD^2, BC^2D^2, ABC^2D, AB^2C^2; R-7: ACD, BCD^2, ABC, AB^2D; R-8: AC^2D^2, BC^2D, ABC^2, AB^2D^2; R-9: ACD^2, BC, ABD, AB^2CD; R-10:AC^2D, BC^2,ABD^2, $AB^2C^2D^2$.

2.2. The Partially Confounded s^n Factorial Designs in s^r Blocks in Different Replicates

By Bose (1947) the s^r blocks representing one orthogonal replicate of s^n factorial designs confounds the $(s^r-1)/(s-1)$ factorial effects in which any r factorial effects are independent are represented by use finite geometry and idea of parallel pencil and simply given by the following equation set

$$ax = u_j,\ j= 0,1, ..,s^r-1;$$

where, $a = [a_{i1}\ a_{i2} \ldots a_{in}]$, $x = [\ x_1\ x_2 \ldots\ x_n]'$ $u_o= [0\ 0 \ldots 0]'$, $u_1=[1\ 0 \ldots 0]'$, …, $u_{s^r-1}=[s\text{-}1\ s\text{-}1 \ldots s\text{-}1]'$, all elements u_j, a_{ik}, $x_k \in GF(s)$; k=1,2,..,n; i=1,2,..,r.

The coefficients in the matrix a $=[a_{i1}\ a_{i2} \ldots a_{in}\ ]$ are according to the coefficients of the equations of the independent factorial effects confounded in the blocks of the individual replicate. Thus the coefficients a_{ik} are different in different replicates in light of equations of the independent confounded effects. So the number of orthogonal replicates is the number of equation sets.

2.3. The s^n Factorial Designs Partially Confounding in s^r Blocks in a Complete Set of Replicates Can Be Shown as the Alternative to a BIB Design

In order to construct the partially confounded factorial BIB designs we are to relate first the different partially confounded s^n factorial designs with different BIB designs with the help of following theorems.

Theorem 2.3.1. *Any s^n factorial design in s^r blocks confounding partially different groups of equal numbered different orthogonal factorial effects respectively under the complete set of orthogonal replicates will be a balanced incomplete block design with parameters $v=s^n$, $b =s^r(s^n-1)/(s^r-1)$, $t=(s^n-1)/(s^r-1)$, $k=s^{n-r}$, $\lambda = (s^{n-r}-1)/(s^r-1)$.*

Proof 2.3.1. The BIB design possesses the parameters v treatments, b blocks, r replicates, k number of treatments in each block and same pair of treatments in λ blocks. The s^n factorial design in s^r blocks contains s^n (= v treatments in BIB design), s^{n-r} treatment (= k treatment) in each block confounding partially $(s^r-1)/(s-1)$ factorial effects under each of $(s^n-1)/(s^r-1)$ orthogonal replicates which is same as t orthogonal replicates in a BIB design. Since the factorial design under one orthogonal replicate contains s^r blocks, so under $(s^n-1)/(s^r-1)$ orthogonal replicates it contains $s^r(s^n-1)/(s^r-1)$ blocks = b blocks in BIB design. In the BIB design $\lambda = r(k-1)/(v-1) = [(s^n-1)/(s^r-1)](s^{n-r}-1)/(s^n-1) =(s^{n-r}-1)/(s^r-1)$ in the factorial design.

Theorem 2.3.2. *The ss (all factorial effects) in a factorial design is same as the ss (adjusted treatment) in BIB design.*

Proof 2.3.2. Shamsuddin and Al-Marshadi (2009) has proved that the s^n factorial design in s blocks under several orthogonal replicates provides BIB designs with same ss(treatment). Here in this paper the s^n factorial design in s^r under $(s^n-1)/(s^r-1)$ orthogonal replicates consisting of s^n (=v) treatments and b blocks [= $s^r(s^n-1)/(s^r-1)$] with s^{n-r} treatment (= k treatment) in each block is exactly same as the BIB design having similar properties with respect to treatments, blocks , block size, replicates and the same pair of treatments in λ block. Both designs conducted within the same structural layout plan with same data provide the same total ss's and block ss's and by [Shamsuddin and et al (2009)] also insert the factorial treatment (all factorial effects) ss's equal to the BIB adjusted treatment ss.

3. Example

3.1. The 2^4 Factorial Design Partially Confounded in Complete Orthogonal Replicates Is Shown as BIB Design

The 2^4 factorial design has 16 treatment combinations 0000, 1000, 0100, 1100, 0010, 1010, 0110, 1110, 0001, 1001, 0101, 1101, 0011, 1011, 0111, 1111, and 15 factorial effects A, B, AB, C, AC, BC, ABC, D, AD, BD, ABD, CD, ACD, BCD, ABCD. Five orthogonal replicates are shown as follows with 4 blocks in each replicate of 3 factorial effects confounded partially.

Rep I

$$\begin{matrix} x_1 = \\ x_2 = \end{matrix} \left.\begin{matrix}0\\0\end{matrix}\right\} \left.\begin{matrix}1\\0\end{matrix}\right\} \left.\begin{matrix}0\\1\end{matrix}\right\} \left.\begin{matrix}1\\1\end{matrix}\right\}$$

B1	B2	B3	B4
0000	1000	0100	1100
0010	1010	0110	1110
0001	1001	0101	1101
0011	1011	0111	1111

A,B, AB Confounding

Rep II

$$\begin{matrix} x_3 = \\ x_4 = \end{matrix} \left.\begin{matrix}0\\0\end{matrix}\right\} \left.\begin{matrix}1\\0\end{matrix}\right\} \left.\begin{matrix}0\\1\end{matrix}\right\} \left.\begin{matrix}1\\1\end{matrix}\right\}$$

B5	B6	B7	B8
0000	0010	0001	0011
1000	1010	1001	1011
0100	0110	0101	0111
1100	1110	1101	1111

C,D, CD Confounding

Rep III

$$\begin{matrix} x_1 + x_3 = \\ x_2 + x_4 = \end{matrix} \left.\begin{matrix}0\\0\end{matrix}\right\} \left.\begin{matrix}1\\0\end{matrix}\right\} \left.\begin{matrix}0\\1\end{matrix}\right\} \left.\begin{matrix}1\\1\end{matrix}\right\}$$

B9	B10	B11	B12
0000	1000	0001	1100
1010	0010	1011	0110
0101	1101	0100	1001
1111	0111	1110	0011

AC,BD, ABCD Confounding

Rep IV

$$\begin{matrix} x_1 + x_4 = \\ x_1 + x_2 + x_3 = \end{matrix} \left.\begin{matrix}0\\0\end{matrix}\right\} \left.\begin{matrix}1\\0\end{matrix}\right\} \left.\begin{matrix}0\\1\end{matrix}\right\} \left.\begin{matrix}1\\1\end{matrix}\right\}$$

B13	B14	B15	B16
0000	0001	0100	1000
0110	0111	0010	1110
1101	1100	1001	0101
1011	1010	1111	0011

AD,ABC, BCD Confounding

Rep V

$$\begin{matrix} x_2 + x_3 = \\ x_1 + x_3 + x_4 = \end{matrix} \left.\begin{matrix}0\\0\end{matrix}\right\} \left.\begin{matrix}1\\0\end{matrix}\right\} \left.\begin{matrix}0\\1\end{matrix}\right\} \left.\begin{matrix}1\\1\end{matrix}\right\}$$

B17	B18	B19	B20
0000	0100	1000	1100
1110	1010	0110	0010
0111	0011	1111	1011
1001	1101	0001	0101

BC,ACD, ABD Confounding

This is a BIB design with parameter [v=16, b=20, k=4, t=5, λ=1] as well as a 2^4 factorial design with 16 treatment combinations, 15 factorial effects and 20 blocks in 5 orthogonal replicates.

The jth adjusted treatment in BIB design = $Q_j = Y_{.j} - \frac{1}{k} \sum_{i=1}^{20} n_{ij} Y_{i.}$

where $[n_{ij}]$ = incident matrix, i=1,2,…,20; j=1,2,…,16.

Using the hypothetical data (Table 3.1.1), the adjusted treatment ss = $\frac{k}{\lambda v} \sum_j Q_j^2 = \frac{4}{1\times 16}(3991.625) = 997.906.$

From the factorial point of view, the design has following factorial effects ss's:

SS(A)= 76.562;	SS(B)=203.062;	SS(AB)=1.00;	SS(C)=260.015
SS(D)=415.140;	SS(CD)=4.515;	SS(AC)=3.062;	SS(BD)=6.25;
SS(ABCD)=10.562;	SS(BC)=4.515	SS(ABD)=0.0156;	SS(ACD)=3.515;
SS(AD)=1.5662;	SS(ABC)=7.562;	SS(BCD)=0.5062	

The total treatment ss in factorial design given by the total factorial effects:

SS's= SS(A)+ SS(B)+ SS(AB)+ SS(C)+ SS(AC)+ SS(BC)+ SS(ABC)+ SS(D)+ SS(AD) + SS(BD) + S(ABD)+ SS(CD)+ SS(ACD)+ SS(BCD)+ SS(ABCD)= 997.9

Hence, SS's due to all factorial effects in 2^4 factorial = SS(adjusted treatment) in BIB design. Since both designs have same structural layout plan and same blocks with block-contents, their total SS's are same and block SS's are same with same error SS's. The factorial 2^4 design possessing 15 separate and distinct factorial effects SS's equivalent to the adjusted treatment SS in BIB design is more informative than the BIB design.

3.2. Some Selected s^n Factorial Designs with BIB Character

Factorial design	Total factorial effects	Block per replicate	Confo-unded effect/ Rep	Total repli-cates	Total blocks	BIB [Parameter]	Remark
2^3 design	7	2	1	7	16	BIB [v=8,b=16,t=7, k=4,λ=3]	
2^4 design	15	2	1	15	30	BIB [v=16,b=30,t=15, k=8,λ=7]	
2^4 design	15	4	3	5	20	BIB [v=16,b=20,t=5, k=4,λ=1]	more usable
2^5 design	31	2	1	31	62	BIB [v=32,b=62,t=31, k=16,λ=15]	
2^6 design	63	2	1	63	126	BIB [v=64,b=126,t=63, k=32,λ=31]	
2^6 design	63	4	3	21	84	BIB [v=64,b=84,t=21, k=16,λ=5]	usable
2^6 design	63	8	7	9	72	BIB [v=64,b=72,t=9, k=8,λ=1]	more usable
3^3 design	13	3	1	13	39	BIB [v=27,b=39,t=13, k=9,λ=4]	
3^4 design	40	3	1	40	120	BIB [v=81,b=120,t=40, k=27,λ=13]	
3^4 design	40	9	4	10	90	BIB [v=81,b=90,t=10, k=9,λ=1]	more usable

Table 3.1.1. The 2^4 factorial BIB design with the data set

		1	2	3	4	5	6	7	8	9	10	11	12	13	14	15	16	
	Block	0000	1000	0100	1100	0010	1010	0110	1110	0001	1001	0101	1101	0011	1011	0111	1111	$Y_{i.}$
Rep 1	1	1				3				4				10				18
	2		2				7				10				11			30
	3			3				8				9				13		33
	4				7				12				14				18	51
Rep 2	5	1	2	3	5													11
	6					2	6	7	9									24
	7									3	8	9	10					30
	8													9	10	12	17	48
Rep 3	9	1					5					8					16	30
	10		2			4							9			11		26
	11			3					9	4					9			25
	12				5			6			7			9				27
Rep 4	13	1						5					9		10			25
	14				5		6			4						13		28
	15			3		3					6						16	28
	16		2						10			8		9				29
Rep 5	17	1							9		7					13		30
	18			3			5						7	7				22
	19		2					6		4							17	29
	20				5	4						9			10			28
	$Y_{.j}$	5	10	15	27	16	29	32	49	19	38	43	49	44	50	62	84	572

Conclusion

Every factorial design under the complete set of orthogonal replicates can be converted to a BIB design with the same properties and result. But this factorial design is more informative than BIB design.

References

Anderson, V.L. and McLean, R.A. (1974). *Design of Experiments: A Realistic Approach*, Marcel Dekker, Inc. New York.

Bailey, R.A. (1977). Pattern of Confounding in Factorial Design. *Biometrika*, 64, 597-603.

Bose R.C. (1939). On the Construction of Balanced Incomplete Block Designs. *Annals of Eugenics*, 9, 358-399.

Bose R.C. (1942). On some new Series of Balanced Incomplete Block Designs. Bull. *Calcutta Math. Soc*., 34, 17-31.

Bose R.C. (1947). Mathematical theory of the symmetrical factorial design. *Sankhya*, 8, 107-166.

Bose R.C. and Kishan, k. (1940). On the Problem of Confounding in General Symmetrical Factorial Designs. *Sankhya*, 5, 21-26.

Bose, R. C. and Manvel, B (1984). *Introduction to Combinatorial Theory*, John Wiley& Sons, New York.

Das, M.N. and Giri, N.C. (1984). *Analysis and Design of Experiments*. Dun. Publication, India.

Fisher, R.A. (1942). The theory of confounding in factorial experiments in relation to the theory of groups. *Annals of Eugenics* ,11, 341-353.

Hick, C.R. (1973). *Fundamental Concepts in Design of Experiments*. 2nd Ed. Rin and Win., New York.

John, P.W.M. (1971). *Statistical Design and Analysis of Experiments*. Mac. Company, New York.

Kempthorne, O. (1952). *The Design and Analysis of Experiments*. John Wiley& Sons, New York.

Mukhopadhyay, P. (1999). *Applied Statistics*. Books and Allied (P) Ltd, 8/1 C. D. Lane, Calcutta.

Raghavarao, D. (1971). *Construction and Combinatorial Problems in Design of Experiments*. John Wiley & Sons, New York.

Shrikhande, S.S. (1952). On the Dual of Some Balanced Incomplete Block Designs. *Biometrika*, 8, 66-72.

Shamsuddin, Md. and Al-Marshadi, A (2009). The Balanced Incomplete Block s^n Factorial Designs. *J. App. Prob. and Statistics*, 4, 99-104,USA.

Yates, F. (1936). Incomplete Randomized Blocks. *Annals of Eugenics*, 7, 121-140

Yates, F. (1937). Design and Analysis of Factorial Experiments. *Tech. Comm.,* 35, Imperial Bur. Soil Society, London.

In: Research in Applied Statistical Science
Editor: Mohammad Ahsanullah, pp. 83-91
ISBN: 978-1-63321-818-5

Chapter 8

AN EXTENDED VERSION OF HYPER POISSON DISTRIBUTION AND SOME OF ITS APPLICATIONS

***C. Satheesh Kumar*[1,*] *and B. Unnikrishnan Nair*[2]**
[1]Department of Statistics, University of Kerala, India
[2]Department of Statistics, M. S. M. College, Kayamkulam, India

Abstract

Through this paper we propose an extended version of hyper-Poisson distribution and derive its probability mass function, mean, variance and recurrence relations for probabilities, raw moments and factorial moments. Further we discussed the estimation of the parameters of this extended version by the method of moments, method of mixed moments and method of maximum likelihood, and it is demonstrated with the help of a real data that the extended version fit the situations better than the existing model.

Keywords: Confluent hypergeometric series, Displaced Poisson distribution, Factorial moments, Hermite distribution, Probability generating function

Mathematics Subject Classification Primary: 60E05, 60E10, Secondary 33C20

1. Introduction

Bardwell and Crow (1964) has obtained a two-parameter generalization of the Poisson distribution, namely the hyper-Poisson family, which has the following probability mass function (p. m. f).

$$f(x) = P(X = x) = \frac{\Gamma(\lambda)}{\phi(1;\lambda;\theta)} \frac{\theta^x}{\Gamma(\lambda + x)} \tag{1.1}$$

[*] E-mail address: drcsatheeshkumar@gmail.com

where $\lambda > 0$, $\theta > 0$; x = 0,1,2,... and $\phi\,(a;b;x)$ is the confluent hypergeometric series (cf. Mathai and Saxena, 1973 or Slater, (1960). When $\lambda = 1$, the hyper-Poisson distribution reduces to Poisson distribution and when λ is a positive integer, the distribution is known as the displaced Poisson distribution considered by Staff (1964). Bardwell and Crow (1964) termed the distribution as sub-Poisson when $\lambda < 1$ and super–Poisson when $\lambda > 1$. Various methods of estimation of the parameters of the distribution were discussed in Bardwell and Crow (1964) and Crow and Bardwell (1965). Some queuing theory with hyper-Poisson arrivals has been worked out by Nisida (1962). Roohi and Ahmad (2003a) attempted estimation of the parameters of the hyper-Poisson distribution using negative moments. Roohi and Ahmad (2003b) derived expressions for ascending factorial moments and further obtained certain recurrence relations for negative moments and ascending factorial moments of the hyper-Poisson distribution. Kemp (2002) developed q-analogue of the distribution and Ahmad (2007) introduced and studied Conway-Maxwell hyper-Poisson distribution.

In this article, we obtain an extended version of the hyper-Poisson distribution and call it as "the extended hyper-Poisson distribution" or in short "the EHP distribution". In section 2 we establish that the EHP distribution possess a stopped sum structure. In section 3, we present some of its properties by deriving expressions for its probability mass function, mean and variance. There we also obtain certain recurrence relations for probabilities, raw moments and factorial moments. Moreover the estimation of the parameters of the EHP distribution has been discussed in section 4 by the method of moments, the method of mixed moments using first observed frequency and the method of maximum likelihood. The results concerning further statistical inference in connection with the EHP distributions are given in the sequel.

2. Derivation of the EHP Distribution

Let $\{X_n,\ n \geq 1\}$ be a sequence of independent and identically distributed random variables, where X_n has the following probability generating function (p. g. f), in which *m* is a positive integer.

$$h(z) = \alpha z + (1-\alpha) z^m, \tag{2.1}$$

in which $\alpha = \theta_1(\theta_1 + \theta_2)^{-1}, \theta_1 > 0$ and $\theta_2 \geq 0$. Consider a non-negative integer valued random variable Y having EHP distribution with p. m. f. (1.1), which has the following p. g. f, in which $\theta = \theta_1 + \theta_2$.

$$\Lambda(z) = \delta\,\phi\,(1;\lambda;\theta\, z)\quad , \tag{2.2}$$

where $\delta = [\phi\,(1;\lambda;\theta)]^{-1}$. Assume that $Y, X_1, X_2, \ldots$ are statistically independent and let $T_0 = 0$. Define $T_Y = \sum_{n=0}^{Y} X_n$. Then the p.g.f. of T_Y is

$$\begin{aligned} H(z) &= E(z^{T_Y}) \\ &= E_Y\{E[z^{T_Y}/Y]\} \\ &= \Lambda(h(z)) \\ &= \delta\phi\ (1;\lambda;\theta_1 z+\theta_2 z^m). \end{aligned} \tag{2.3}$$

We define a distribution with p. g. f. (2.3), as 'the extended hyper-Poisson distribution' or in short 'the EHP distribution'. When λ is a positive integer, the EHP distribution reduces to an extended version of the displaced Poisson distribution of Staff(1964).The EHP distribution with $\lambda = 1$ is the well-known generalized Hermite distribution of Gupta and Jain (1974),which reduces to Hermite distribution of Kemp and Kemp (1965) when $m = 2$.

3. Properties

Let W be a random variable distributed as the EHP distribution with p. g. f.

$$\begin{aligned} H(z) &= \delta\phi\ (1;\lambda;\theta_1 z+\theta_2 z^m) \\ &= \sum_{r=0}^{\infty} f_r(1,\lambda)z^r, \end{aligned} \tag{3.1}$$

in which $f_r(1,\lambda) = P(W = r)$, $r = 0,1,2,...$ and $\delta = [\phi\ (1;\lambda;\theta_1+\theta_2)]^{-1}$. On expanding (3.1) and equating the coefficients of z^r, we have the following result.

Result 3.1. The probability mass function (p. m. f) $f_r(1,\lambda)$ of the EHP distribution with p. g. f (3.1) is the following, for r = 0, 1, 2, …

$$f_r(1,\lambda) = \delta \sum_{n=0}^{\left[\frac{r}{m}\right]} \frac{[r-(m-1)n]!\theta_1^{r-mn}\theta_2^{n}}{(\lambda)_{r-(m-1)n}(r-mn)!n!}, \tag{3.2}$$

where $(a)_0 = 1, (a)_n = a(a+1)...(a+n-1)$, for $n \geq 1$, and $[k]$ denote the integer part of k.

Further we have the following results.

Result 3.2. The mean and variance of the EHP distribution with p. g. f. (3.1) are as follows:

$$E(W) = \frac{\psi_1}{\lambda}(\theta_1 + m\theta_2), \tag{3.3}$$

$$Var(W)=\frac{1}{\lambda}\left(\frac{2\ \psi_2}{\lambda+1}-\frac{\psi_1{}^2}{\lambda}\right)(\theta_1+m\theta_2)^2+\frac{\psi_1}{\lambda}(\theta_1+m^2\theta_2), \tag{3.4}$$

where

$$\psi_j=\delta\phi\,(1+j;\lambda+j;\theta_1+\theta_2),\quad j=1,2. \tag{3.5}$$

The proof is simple and hence omitted.

Result 3.3. The following is a simple recurrence relation for the probabilities $f_r(1,\lambda)$ of the EHP distribution with p. g. f (3.1), for $r \geq (m-1)$.

$$f_{r+1}(1,\lambda)=\frac{\psi_1}{\lambda(r+1)}\{\theta_1 f_r(2,\lambda+1)+m\theta_2 f_{r-m+1}(2,\lambda+1)\}, \tag{3.6}$$

in which ψ_1 is as defined in (3.5).

Proof. On differentiating (3.1) with respect to z, we get the following.

$$\sum_{r=0}^{\infty}(r+1)f_{r+1}(1,\lambda)z^r=\frac{\delta}{\gamma}(\theta_1+m\theta_2 z^{m-1})\phi\,(2;\lambda+1;\theta_1 z+\theta_2 z^m)\ , \tag{3.7}$$

Replacing λ by $\lambda+1$ in (3.1) to obtain the following.

$$\delta^*\phi\,(2;\lambda+1;\theta_1 z+\theta_2 z^m)=\sum_{r=0}^{\infty}f_r(2,\lambda+1)z^r, \tag{3.8}$$

where $\delta^* = [\phi(2;\lambda+1;\theta_1+\theta_2)]^{-1}$. Relations (3.7) and (3.8) together lead to the following relationship:

$$\sum_{r=0}^{\infty}(r+1)f_{r+1}(1,\lambda)z^r=\frac{\psi_1}{\lambda}\left\{\theta_1\sum_{r=0}^{\infty}f_r(2,\lambda+1)z^r+m\theta_2\sum_{r=0}^{\infty}f_r(2,\lambda+1)z^{r+m-1}\right\} \tag{3.9}$$

Now, on equating coefficients of z^r on both sides of (3.9) we get (3.6).

Result 3.4. The following is a recurrence relation for factorial moments $\mu_{[n]}(1,\lambda)$ of the EHP distribution, for $n \geq 1$.

$$\mu_{[n+1]}(1,\lambda)=\frac{\Lambda_1}{\lambda}\left[\theta_1\mu_{[n]}(2,\lambda+1)+m\,\theta_2\sum_{r=0}^{m-1}\binom{m-1}{r}n^{(r)}\mu_{[n-r]}(2,\lambda+1)\right], \quad (3.10)$$

in which $\mu_{[0]}(1,\lambda)=1$, $n^{(r)}=n(n-1)(n-2)...(n-r+1)$, for any positive integer r and $n^{(0)}=1$.

Proof. The factorial moment generating function $F_W(t)$ of the EHP distribution with p. g. f (3.1) is the following

$$\begin{aligned}F_W(t)&=H(1+t)\\&=\delta\,\phi\,[1;\lambda;\theta_1(1+t)+\theta_2(1+t)^m]\\&=\sum_{n=0}^{\infty}\mu_{[n]}(1,\lambda)\frac{t^n}{n!}\end{aligned} \quad (3.11)$$

Differentiate (3.11) with respect to t to obtain

$$\frac{\delta}{\lambda}[\theta_1+m\theta_2(1+t)^{m-1}]\phi\,[2;\lambda+1;\theta_1(1+t)+\theta_2(1+t)^m]=\sum_{n=1}^{\infty}\mu_{[n]}(1,\lambda)\frac{t^{n-1}}{(n-1)!}. \quad (3.12)$$

By using (3.11) with λ replaced by $\lambda+1$ we get the following from (3.12).

$$\begin{aligned}\sum_{n=0}^{\infty}\mu_{[n+1]}(1,\lambda)\frac{t^n}{n!}&=\frac{\psi_1}{\lambda}[\theta_1+m\theta_2(1+t)^{m-1}]\sum_{n=0}^{\infty}\mu_{[n]}(2,\lambda+1)\frac{t^n}{n!}\\&=\frac{\psi_1}{\lambda}\theta_1\sum_{n=0}^{\infty}\mu_{[n]}(2,\lambda+1)\frac{t^n}{n!}+\frac{\psi_1}{\lambda}m\;\theta_2\sum_{n=0}^{\infty}\sum_{r=0}^{m-1}\binom{m-1}{r}\mu_{[n]}(2,\lambda+1)\frac{t^{n+r}}{n!}\end{aligned} \quad (3.13)$$

On equating coefficients of $(n!)^{-1}\,t^n$ on both sides of (3.13) we get (3.10).

Result 3.5. The following is a recurrence relation for the raw moments $\mu_n(1,\lambda)$ of the EHP distribution, for $n\geq 0$.

$$\mu_{n+1}(1,\lambda)=\frac{\psi_1}{\lambda}\sum_{r=0}^{n}\binom{n}{r}(\theta_1+m^{r+1}\theta_2)\mu_{n-r}(2,\lambda+1). \quad (3.14)$$

Proof. The characteristic function $\phi_W(t)$ of the EHP distribution with p. g. f. (3.1) has the following series representation. For $t\in R$,

$$\begin{aligned}\phi_W(t) &= Q\,(e^{it}) \\ &= \delta\,\phi\,(1;\lambda;\theta_1 e^{it}+\theta_2 e^{mit}) \\ &= \sum_{n=0}^{\infty}\mu_n(1,\lambda)\frac{(it)^n}{n!}\end{aligned} \tag{3.15}$$

On differentiating (3.15) with respect to t, we obtain

$$\frac{\delta}{\lambda}(\theta_1 e^{it}+m\theta_2 e^{mit})\,\phi\,(2;\lambda+1;\theta_1 e^{it}+\theta_2 e^{mit}) = \sum_{n=1}^{\infty}\mu_n(1,\lambda)\frac{(it)^{n-1}}{(n-1)!} \tag{3.16}$$

By using (3.15) with λ replaced by $\lambda+1$, we get the following from (3.16).

$$\begin{aligned}\sum_{n=0}^{\infty}\mu_{n+1}(1,\lambda)\frac{(it)^n}{n!} &= \frac{\psi_1}{\lambda}(\theta_1 e^{it}+m\theta_2 e^{mit})\sum_{n=0}^{\infty}\mu_n(2,\lambda+1)\frac{(it)^n}{n!} \\ &= \frac{\psi_1}{\lambda}\sum_{n=0}^{\infty}\left[\theta_1\sum_{k=0}^{\infty}\frac{(it)^k}{k!}+m\theta_2\sum_{k=0}^{\infty}\frac{(mit)^k}{k!}\right]\mu_n(2,\lambda+1)\frac{(it)^n}{n!}\end{aligned} \tag{3.17}$$

On equating the coefficients of $(n!)^{-1}(it)^n$ on both sides we get (3.14).

4. Estimation

In this section we consider the estimation of the parameters of the EHP distribution by the method of moments, the method of mixed moments and the method of maximum likelihood. Throughout this section we assume that m is a fixed known positive integer.

Method of Moments

Here the estimators $\overline{\lambda}$, $\overline{\theta}_1$ and $\overline{\theta}_2$ of the parameters λ, θ_1 and θ_2 of the EHP distribution are obtained by solving the following system of equations. These equations are developed by equating the first three raw moments of the EHP distribution to the corresponding sample raw moments - τ_1, τ_2 and τ_3 respectively.

$$\frac{1}{\lambda}(\theta_1+m\theta_2)\frac{\phi(2;\lambda+1;\theta_1+\theta_2)}{\phi(1;\lambda;\theta_1+\theta_2)} = \tau_1 \tag{4.1}$$

$$\frac{2}{\lambda(\lambda+1)}(\theta_1+m\theta_2)^2\frac{\phi(3;\lambda+2;\theta_1+\theta_2)}{\phi(1;\lambda;\theta_1+\theta_2)}+\frac{1}{\lambda}(\theta_1+m^2\theta_2)\frac{\phi(2;\lambda+1;\theta_1+\theta_2)}{\phi(1;\lambda;\theta_1+\theta_2)} = \tau_2 \tag{4.2}$$

$$\frac{6}{\lambda(\lambda+1)(\lambda+2)}(\theta_1+m\theta_2)^3\frac{\phi(4;\lambda+3;\theta_1+\theta_2)}{\phi(1;\lambda;\theta_1+\theta_2)}+\frac{6}{\lambda(\lambda+1)}(\theta_1+m^2\theta_2)(\theta_1+m\theta_2)\frac{\phi(3;\lambda+2;\theta_1+\theta_2)}{\phi(1;\lambda;\theta_1+\theta_2)}+$$

$$\frac{1}{\lambda}(\theta_1+m^3\theta_2)\frac{\phi(2;\lambda+1;\theta_1+\theta_2)}{\phi(1;\lambda;\theta_1+\theta_2)}=\tau_3 \quad (4.3)$$

Method of Mixed Moments

In this method, the estimators $\tilde{\lambda}$, $\tilde{\theta}_1$ and $\tilde{\theta}_2$ of the parameters λ, θ_1 and θ_2 of the EHP distribution are obtained by solving the following equation together with (4.1) and (4.2).

$$N\,\phi^{-1}(1;\lambda;\theta_1+\theta_2)=p_0, \quad (4.4)$$

where p_0 is the observed frequency of the distribution corresponding to the observation zero and N is the total observed frequency.

Method of maximum likelihood

Let A(x) be the observed frequency of x events and let k be the highest value of x observed. Then the likelihood function of the sample is

$$L=\prod_{x=o}^{k}\left[f_x(1;\lambda)\right]^{A(x)}, \quad (4.5)$$

where λ is assumed to be a known real number. Taking logarithm on both sides of (4.5) we get

$$\ell=\log L=\sum_{x=0}^{k}A(x)\log[f_x(1,\lambda)] \quad (4.6)$$

Let $\hat{\theta}_1$ and $\hat{\theta}_2$ denote the maximum likelihood estimators of θ_1 and θ_2 respectively. Now $\hat{\theta}_1$ and $\hat{\theta}_2$ are computed by solving the following equations, obtained from (4.6) on differentiation with respect to θ_1 and θ_2 respectively and equating to zero.

$$\frac{\partial\ell}{\partial\theta_1}=\sum_{x=0}^{k}T(x)\left[\frac{-1}{\lambda}\frac{\phi(2;\lambda+1;\theta_1+\theta_2)}{\phi(1;\lambda;\theta_1+\theta_2)}+\frac{\displaystyle\sum_{n=0}^{\left[\frac{r}{m}\right]}\frac{[r-(m-1)n]!\,\theta_1^{\,r-mn-1}\theta_2^{\,n}}{(\lambda)_{r-(m-1)n}(r-mn-1)!\,n!}}{\displaystyle\sum_{n=0}^{\left[\frac{r}{m}\right]}\frac{[r-(m-1)n]!\,\theta_1^{\,r-mn}\theta_2^{\,n}}{(\lambda)_{r-(m-1)n}(r-mn)!\,n!}}\right]=0 \quad (4.7)$$

$$\frac{\partial \ell}{\partial \theta_2} = \sum_{x=0}^{k} T(x) \left[\frac{-1}{\lambda} \frac{\phi(2;\lambda+1;\theta_1+\theta_2)}{\phi(1;\lambda;\theta_1+\theta_2)} + \frac{\sum_{n=0}^{\left[\frac{r}{m}\right]} \frac{[r-(m-1)n]!\theta_1^{r-mn}\theta_2^{n-1}}{(\lambda)_{r-(m-1)n}(r-mn)!(n-1)!}}{\sum_{n=0}^{\left[\frac{r}{m}\right]} \frac{[r-(m-1)n]!\theta_1^{r-mn}\theta_2^{n}}{(\lambda)_{r-(m-1)n}(r-mn)!n!}} \right] = 0 \quad (4.8)$$

Let $\bar{\lambda}$ be the moment estimate of λ and $\tilde{\lambda}$ be the mixed moment estimate of λ .Using these estimated values of λ , we can obtain $\hat{\theta}_1$ and $\hat{\theta}_2$ distinctly.

All these procedures of estimation of the parameter of the EHP distribution are demonstrated using a real life data for $m = 1, 2, 3$ and 4 by using MATHCAD software and presented in table 1. From the calculated value of χ^2 in each case, it is clear that the EHP distribution with $m = 3$ gives a better fit for the given data than the standard hyper-Poisson distribution (m = 1) as well as the distribution with $m = 2$ or $m = 4$.

Acknowledgment

The authors would like to express their sincere thanks to the Editor, Associate Editor and the anonymous referee for carefully reading the paper and for valuable suggestions. The first author is grateful to the UGC, Government of India for financial assistance under the project number F. No. 34-535/2008(SR).

References

Ahmad, M. (2007). A short note on Conway-Maxwell hyper-Poisson distribution. *Pakistan Journal of Statistics*. 23, 135–137.

Bardwell, G. E., Crow, E. L. (1964). A two parameter family of hyper-Poisson distributions. *Journal of the American Statistical Association*. 59, 133–141.

Crow, E. L. and Bardwell, G. E. (1965). Estimation of the parameters of the hyper-Poisson distributions. *Classical and Contagious Discrete Distributions*. G. P. Patil (editor), 127–140, Pergamon Press, Oxford.

Gupta, R. P. and Jain, G. C. (1974).A generalized Hermite distribution and its properties. *SIAM Journal of Applied Mathematics*. 27, 359–363.

Kemp, C. D. (2002). q-analogues of the hyper-Poisson distribution, *Journal of Statistical Planning and Inference*. 101, 179–183.

Kemp, C. D. and Kemp, A. W. (1965). Some properties of Hermite distribution. *Biometrica*. 52, 381–394.

Mathai, A. M. and Saxena, R. K. (1973). Generalized Hypergeometric Functions with Applications in Statistics and Physical Sciences, *Lecture Notes* No.348. Springer-Verlag, Heidelberg.

Morgan, M. E., Macleod, P. Anderson, E. O. and Bliss, C. I. (1951). A sequential procedure for grading milk by microscopic counts. *Storrs Agricultural Experimental Station Bulletin*. 276.

Nisida, T. (1962). On the multiple exponential channel queuing system with hyper-Poisson arrivals. *Journal of the Operations Research Society*. 5, 57–66

Roohi, A. and Ahmad, M. (2003a). Estimation of the parameter of hyper-Poisson distribution using negative moments, *Pakistan Journal of Statistics*. 19, 99–105.

Roohi, A. and Ahmad, M. (2003b). Inverse ascending factorial moments of the hyper-Poisson probability distribution. *Pakistan Journal of Statistics*. 19, 273–280.

Slater, L. J. (1960). Confluent Hypergeometric functions. Cambridge University Press, Cambridge.

Staff, P. J. (1964).The displaced Poisson distribution. *Australian Journal of Statistics*. 6, 12–20.

In: Research in Applied Statistical Science
Editor: Mohammad Ahsanullah, pp. 93-102
ISBN: 978-1-63321-818-5

Chapter 9

REPEATED MEASUREMENTS DESIGNS FOR COMPARING TWO DISJOINT SETS OF FORMULATIONS IN BIOEQUIVALENCE TRIALS

M. N. Sonawane, Cini Varghese and Seema Jaggi
Indian Agricultural Statistics Research Institute,
New Delhi, India

Abstract

Bioequivalence is the degree to which clinically important outcomes after receiving a new (test) formulation resembles a well established (reference) formulation. Bioequivalence trials form an integral part of the development programme of every drug product. Under these trials, the main interest of the experimenter is in the estimation of contrasts pertaining to test vs. test formulations and test vs. reference formulations with as high a precision as possible. Repeated measurements designs (RMDs) with two sets of formulations, one set of test and other of reference formulations, assuming the presence of residual effects are suitable for such trials. Here, two classes of RMDs for bioequivalence trials have been proposed assuming the presence of residual effects of formulations. Variances of the estimates of contrasts pertaining to direct as well as residual effects of test formulations, reference formulations and test vs. reference formulations have been worked out for these designs. Variance of test vs. reference formulations and test vs. test formulations comparisons were found to be less as compared to reference vs. reference comparisons indicating that these designs are suitable for bioequivalence trials.

Keywords: Bioequivalence, test formulation, reference formulation, repeated measurements design, residual effect, balanced design, variance balanced design

1. Introduction

Bioequivalence refers to the degree to which clinically important outcomes after receiving a new preparation known as test formulation resemble those of a previously established preparation called reference (otherwise known as standard or control) formulation. Metzler (1974) suggested that the evaluation of bioequivalence should be included in the development programme of every drug product. Evaluation of veterinary medical products is one important area where these trials are conducted. When an established drug is going off patent the company may develop a new formulation of the drug with the same active ingredients to remain in the market while other competing companies may turn up with some new formulations. Approval and marketing of these new formulations is done based on the results of the bioequivalence trials.

An important aspect in planning a bioequivalence trial is the choice of a good design. Repeated Measurements Designs (RMDs) are the most preferred designs in bioequivalence trials [Chow and Liu, 2000] as they economize on the use of experimental units or subjects (such as animals, human beings, *etc.*) and give precise comparisons among all formulations free of subject effects. In RMDs, formulations (all or some) are applied in sequences to experimental units over periods of time. RMDs are also known as change over designs or crossover designs or designs involving sequences of treatments, in literature. The presence of residual effects (carryover effect of a formulation that last beyond the period of its direct application) is the most distinguishing feature of RMDs.

As per the objective of bioequivalence trials, the interest lies on test vs. reference formulation comparisons hence the statistical problem is to obtain designs that allow estimation of these comparisons with more precision as compared to other comparisons. The problem of constructing RMDs for test vs. control comparisons was started with the introduction of control balanced RMDs balanced for residual effects by Pigeon and Raghavarao (1987). Some control balanced RMDs when number of treatments is less than the number of periods were given by Majumdar (1988) that can be constructed from existing strongly balanced uniform circular/non-circular RMDs. A two-period RMD for the comparison of two active treatments and placebo was obtained by Koch *et al.* (1989). Some methods of constructing control balanced RMDs were given by Hedayat and Zhao (1990) and Yadav (1990). Subsequently, some classes of efficient RMDs for comparing test treatments with a control were obtained by Ting (2002), Aggarwal *et al.* (2004) and Hedayat and Yang (2005). More recently, efficient RMDs for comparing test treatments with a control treatment with three periods were constructed by Yang and Park (2007).

All the designs obtained by above mentioned workers, with appropriate modifications, can be adopted in bioequivalence trials for making comparisons of a set of test treatments to a single control. However, there may occur situations wherein a set of test formulations are to be compared with another distinct set of reference formulations in bioequivalence trials. Here, the experimenter is interested in the comparisons among test vs. test and test vs. reference formulations with high precision.

In the present study, a class of complete sequence RMDs (CSRMDs) for bioequivalence trials with two sets of formulations assuming residual effects to be present has been proposed. Further, a class of incomplete sequence RMDs (ISRMDs) has also been obtained. Incomplete sequence designs are particularly advisable when more formulations are to be compared. A SAS program has been developed to calculate variances of estimates of contrasts pertaining to direct as well as residual effects among test formulations, among reference formulations and among test vs. reference formulations. A list of proposed RMDs consisting of parameters and variances has been prepared.

2. Experimental Set Up and Definitions

The experimental set up considered consists of v_1 test formulations denoted by $1,2,\ldots,v_1$ and v_2 reference formulations denoted by $v_1+1, v_1+2,\ldots,v_1+v_2$ to be compared using n experimental units in p periods.

Definition 2.1: A RMD $D(v_1+v_2,p,n)$ for comparing v_1 test formulations with v_2 reference formulations is balanced for bioequivalence trials in the presence of residual effects, if

(a) in each period each test formulation occurs ω_f times and each reference formulation occurs ω_g times ($f = 1,2,\ldots,v_1$; $g = 1,2,\ldots,v_2$),

(b) each test formulation is immediately preceded by every other test formulation equally often, say $\upsilon_{ff'}$ ($f \neq f' = 1,2,\ldots,v_1$),

(c) the reference formulation is immediately preceded by every other reference formulation equally often, say $\upsilon_{gg'}$ ($g \neq g' = 1,2,\ldots,v_2$) and

(d) the reference formulation is immediately preceded by every test formulation and vice versa equally often, say υ_{fg}.

Definition 2.2: A RMD for comparing v_1 test formulations with v_2 reference formulations $D(v_1+v_2,p,n)$ is said to be variance balanced design for bioequivalence trials in the presence of residual effects, if all elementary contrasts pertaining to

(a) direct(residual) effects among test formulations are estimated with the same variance, $V_{dff'}$ ($V_{rff'}$) where [$f \neq f' = 1,2,\ldots, v_1$],

(b) direct(residual) effects among reference formulations are estimated with the same variance, $V_{dgg'}(V_{rgg'})$ where ($g \neq g' = 1,2,\ldots, v_2$) and

(c) direct(residual) effects among test vs. reference formulations are estimated with the same variance, $V_{dfg}(V_{rfg})$.

3. CSRMDS for Bioequivalence Trials

Construct Williams (1949) Latin square(s) for v symbols (1, 2,...,v). This balanced design has one or two Latin square(s) according as v is even or odd with v rows and v or 2v columns. An easy method of constructing Williams Latin squares is given by Sharma

(1975). The different steps involved in the construction of these Latin squares are described below:

- Construct one (or two) (v × v) table(s).
- In both the squares, number the rows from 1 to v successively.
- Assign the symbols 1, 2, …, v successively to the v cells in the first column of both the squares by proceeding from top to bottom, entering only in odd-numbered cells in the first and even numbered cells in the second square, and then reversing the direction, filling in even-numbered cells in the first and odd numbered cells in the second square.
- Obtain the successive columns of the squares by adding integer 1 to each element of the previous column and reducing the elements, if necessary, by mod v.

It is to be noted that in each of the constructed squares every symbol occurs in each row and in each column precisely once. Moreover, when v is even, each level is preceded exactly once by other level in either of the two squares. Thus, in this case either of the two squares may be used. This situation occurs in neither of the two squares if v is odd. However, when both the squares are appended together, one after another horizontally, each level is preceded by every other level exactly twice. Consequently, both the squares must be used in this case.

Now, replace v_1 equi-sized distinct sets of u symbols each [$(2 \le v_1, u \le (\frac{v}{2}-1)$ for even v and $(2 \le v_1, u \le (\frac{v-1}{2}-1)$ for odd v] by test formulations $(1,2,\ldots,v_1)$ respectively, leaving at least two symbols out of v symbols without replacement. Renumber remaining symbols by reference formulations $(v_1+1, v_1+2, \ldots, v_1+v_2)$. Treating rows as periods and columns as experimental units, we get a CSRMD for bioequivalence trials with parameters v_1+v_2, p = v, n = v, $\omega_f = u$, $\omega_g = 1$, $\upsilon_{ff} = u(u-1)$, $\upsilon_{ff}' = u^2$, $\upsilon_{fg} = u$, and $\upsilon_{gg}' = 1$ for even v and v_1+v_2, p = v, n = 2v, $\omega_f = 2u$, $\omega_g = 2$, $\upsilon_{ff} = 2u(u-1)$, $\upsilon_{ff}' = 2u^2$, $\upsilon_{fg} = 2u$ and $\upsilon_{gg}' = 2$ for odd v; where υ_{ff} is the number of times each test formulation is preceded by itself.

Example 3.1. Consider the following Williams Latin Square for v = 8 symbols in 8 rows and 8 columns:

1	2	3	4	5	6	7	8
8	1	2	3	4	5	6	7
2	3	4	5	6	7	8	1
7	8	1	2	3	4	5	6
3	4	5	6	7	8	1	2
6	7	8	1	2	3	4	5
4	5	6	7	8	1	2	3
5	6	7	8	1	2	3	4

Let u = 2. Replace symbols (5, 6) by test formulation 1 and symbols (7, 8) by test formulation 2. Renumber the remaining symbols (1, 2, 3, 4) as reference formulations (3, 4, 5, 6), respectively. In the resultant arrangement, treating rows as periods and columns as experimental units we get the following CSRMD with parameters $v_1+v_2 = 2+4$, p = 8, n = 8:

		Experimental Units							
		1	**2**	**3**	**4**	**5**	**6**	**7**	**8**
Periods	1	3	4	5	6	1	1	2	2
	2	2	3	4	5	6	1	1	2
	3	4	5	6	1	1	2	2	3
	4	2	2	3	4	5	6	1	1
	5	5	6	1	1	2	2	3	4
	6	1	2	2	3	4	5	6	1
	7	6	1	1	2	2	3	4	5
	8	1	1	2	2	3	4	5	6

Example 3.2. In the Williams Latin Square for v = 8 symbols arranged in 8 rows and 8 columns as given in Example 3.1, consider u = 2 and renumber the symbols (1, 2) as reference formulations (4, 5), respectively. Now, replace symbols (3, 4) by test formulation 1, symbols (5, 6) by test formulation 2 and symbols (7, 8) by test formulation 3. Treating rows as periods and columns, the experimental units the following CSRMD with parameters v_1+v_2 = 3+2, p = 8, n = 8 is obtained:

		Experimental Units							
		1	**2**	**3**	**4**	**5**	**6**	**7**	**8**
Periods	1	4	5	1	1	2	2	3	3
	2	3	4	5	1	1	2	2	3
	3	5	1	1	2	2	3	3	4
	4	3	3	4	5	1	1	2	2
	5	1	1	2	2	3	3	4	5
	6	2	3	3	4	5	1	1	2
	7	1	2	2	3	3	4	5	1
	8	2	2	3	3	4	5	1	1

Example 3.3. Consider the Williams Latin Square for v = 8 symbols given in Example 3.1. Renumber the symbols (1, 2) as reference formulations (3, 4), respectively. Now, replace symbols (3, 4, 5) by reference formulation 1 and symbols (6, 7, 8) by reference formulation 2, taking u = 3. The resulting arrangement is a CSRMD with parameters $v_1+v_2 = 2+2$, p = 8, n = 8 (as given below) if rows as treated as periods and columns, the experimental units:

		Experimental Units							
		1	**2**	**3**	**4**	**5**	**6**	**7**	**8**
Periods	1	3	4	1	1	1	2	2	2
	2	2	3	4	1	1	1	2	2
	3	4	1	1	1	2	2	2	3
	4	2	2	3	4	1	1	1	2
	5	1	1	1	2	2	2	3	4
	6	2	2	2	3	4	1	1	1
	7	1	1	2	2	2	3	4	1
	8	1	2	2	2	3	4	1	1

The variances of estimates of contrasts pertaining to direct and residual effects among test formulations, among reference formulations and among test vs. reference formulations were calculated using a SAS program. A list of these designs has been prepared for $(v_1+v_2) \leq 10$, $p \leq 10$, $n \leq 20$ and given in Table 1 along with these variances.

From Table 1, it can be observed that the variances of estimates of contrasts pertaining to direct effects are estimated with more precision than those pertaining to residual effects for all the cases, viz., among test formulations, among test vs. reference formulations and among reference formulations. Further, the variances of estimates of contrasts pertaining to both direct effects as well as residual effects among test formulations and test vs. reference formulations are less than those among reference formulations indicating that these designs are more advantageously used for bioequivalence trials.

Table 1. List of CSRMDs

Sl. No.	v_1	v_2	p	n	ω_f	ω_g	$\upsilon_{ff'}$	υ_{fg}	$\upsilon_{gg'}$	$V_{dff'}$	V_{dfg}	$V_{dgg'}$	$V_{rff'}$	V_{rfg}	$V_{rgg'}$
1	2	2	6	6	2	1	4	2	1	0.1727	0.2590	0.3453	0.2143	0.3215	0.4286
2	2	2	8	8	3	1	9	3	1	0.0848	0.1698	0.2547	0.0988	0.1976	0.2963
3	3	2	7	14	4	2	8	4	2	0.0733	0.1099	0.1465	0.0875	0.1313	0.1750
4	3	2	9	18	4	2	8	2	2	0.0564	0.0846	0.1127	0.0643	0.0965	0.1286
5	2	3	8	8	2	1	4	2	1	0.1274	0.1910	0.2547	0.1482	0.2223	0.2963
6	4	2	8	8	2	1	4	2	1	0.1274	0.1910	0.2547	0.1482	0.2223	0.2963
7	3	3	9	18	6	2	18	3	2	0.0376	0.0752	0.1127	0.0429	0.0858	0.1286
8	4	2	10	10	3	1	9	3	1	0.0675	0.1349	0.2023	0.0758	0.1516	0.2273
9	5	2	9	18	4	2	8	2	2	0.0564	0.0846	0.1127	0.0643	0.0965	0.1286
10	4	3	10	10	2	1	4	2	2	0.1012	0.1518	0.2023	0.1137	0.1705	0.2273
11	6	2	10	10	2	1	4	2	1	0.1012	0.1518	0.2023	0.1137	0.1705	0.2273

4. ISRMDS for Bioequivalence Trials

Consider a BIB design (v_2,b,r,k,λ) with v_2 treatment symbols arranged in b blocks (columns) having k block size (rows). To this arrangement, augment an even number (odd number) v_1 (≥ 2) of rows consisting of test formulations if k is even(odd), resulting in even number of rows always. Now, renumber the (1, 2, ...,v) symbols of the original BIB design as (v_1+1, v_1+2,..., v_1+ v_2), respectively. Construct a Williams Latin Square for even number of symbols using each column contents of this arrangement. Juxtapose these Williams Latin Squares one after another horizontally. Treating rows as periods and columns as experimental units, the resultant design is ISRMD for bioequivalence trials in the presence of residual effects for comparing v_1 test formulations with v_2 reference formulations having the parameters v_1+v_2, $p = k+v_1$, $n = (k+v_1)b$, $\omega_f = b$, $\omega_g = r$, $\upsilon_{ff'} = b$, $\upsilon_{fg} = r$, and $\upsilon_{gg'} = \lambda$.

Example 4.1. Consider the blocks (in columns) of the following unreduced BIB design with parameters $v_2 = b = 4$, $r = k = 3$, $\lambda = 2$:

1	1	1	2
2	2	3	3
3	4	4	4

Renumbering symbols (1,2,3,4) in the above BIB design by reference formulations (4,5,6,7), respectively and augmenting three rows of test formulations (one row each for each test formulation denoted by 1,2,3):

4	4	4	5
5	5	6	6
6	7	7	7
1	1	1	1
2	2	2	2
3	3	3	3

Constructing Williams Latin Square from each column (of size 6) of this array, we get the following ISRMD with parameters $v_1 + v_2 = 3+4$, $p = 6$, $n = 24$:

		Experimental Units																							
		1	**2**	**3**	**4**	**5**	**6**	**7**	**8**	**9**	**10**	**11**	**12**	**13**	**14**	**15**	**16**	**17**	**18**	**19**	**20**	**21**	**22**	**23**	**24**
Periods	**1**	4	5	6	1	2	3	4	5	7	1	2	3	4	6	7	1	2	3	5	6	7	1	2	3
	2	3	4	5	6	1	2	3	4	5	7	1	2	3	4	6	7	1	2	3	5	6	7	1	2
	3	5	6	1	2	3	4	5	7	1	2	3	4	6	7	1	2	3	4	6	7	1	2	3	5
	4	2	3	4	5	6	1	2	3	4	5	7	1	2	3	4	6	7	1	2	3	5	6	7	1
	5	6	1	2	3	4	5	7	1	2	3	4	5	7	1	2	3	4	6	7	1	2	3	5	6
	6	1	2	3	4	5	6	1	2	3	4	5	7	1	2	3	4	6	7	1	2	3	5	6	7

Example 4.2. Consider the blocks (in columns) of the following BIB design ($v_2 = b = 7$, $r = k = 4$, $\lambda = 2$):

3	4	5	6	7	1	2
5	6	7	1	2	3	4
6	7	1	2	3	4	5
7	1	2	3	4	5	6

Renumbering the symbols (1,2,3,4,5,6,7) in the BIB design as reference formulations (3,4,5,6,7,8,9), respectively, and augmenting two rows of test formulations (denoted by 1,2 respectively) and we get:

5	6	7	8	9	3	4
7	8	9	3	4	5	6
8	9	3	4	5	6	7
9	3	4	5	6	7	8
1	1	1	1	1	1	1
2	2	2	2	2	2	2

Constructing Williams Latin Square from each column (of size 6) of this array, we get the following ISRMD with parameters $v_1 + v_2 = 2+7$, $p = 6$, $n = 42$:

		Experimental Units																				
		1	**2**	**3**	**4**	**5**	**6**	**7**	**8**	**9**	**10**	**11**	**12**	**13**	**14**	**15**	**16**	**17**	**18**	**19**	**20**	**21**
Periods	**1**	5	7	8	9	1	2	6	8	9	3	1	2	7	9	3	4	1	2	8	3	4
	2	2	5	7	8	9	1	2	6	8	9	3	1	2	7	9	3	4	1	2	8	3
	3	7	8	9	1	2	5	8	9	3	1	2	6	9	3	4	1	2	7	3	4	5
	4	1	2	5	7	8	9	1	2	6	8	9	3	1	2	7	9	3	4	1	2	8
	5	8	9	1	2	5	7	9	3	1	2	6	8	3	4	1	2	7	9	4	5	1
	6	9	1	2	5	7	8	3	1	2	6	8	0_7	4	1	2	7	9	3	5	1	2

		Experimental Units																				
		22	**23**	**24**	**25**	**26**	**27**	**28**	**29**	**30**	**31**	**32**	**33**	**34**	**35**	**36**	**37**	**38**	**39**	**40**	**41**	**42**
Periods	**1**	5	1	2	9	4	5	6	1	2	3	5	6	7	1	2	4	6	7	8	1	2
	2	4	5	1	2	9	4	5	6	1	2	3	5	6	7	1	2	4	6	7	8	1
	3	1	2	8	4	5	6	1	2	9	5	6	7	1	2	3	6	7	8	1	2	4
	4	3	4	5	1	2	9	4	5	6	1	2	3	5	6	7	1	2	4	6	7	8
	5	2	8	3	5	6	1	2	9	4	6	7	1	2	3	5	7	8	1	2	4	6
	6	8	3	4	6	1	2	9	4	5	7	1	2	3	5	6	8	1	2	4	6	7

The variances of estimates of contrasts pertaining to formulation effects among test formulations, among reference formulations and among test vs. reference formulation for direct as well as for residual effects of formulations were worked out. A list of these designs for $(v_1 + v_2) \leq 10$, $p \leq 10$, $n \leq 60$ is given in Table 2 along with the calculated variances.

Table 2 shows that the variances of estimates of contrasts among test formulations and among test vs. reference formulations are less than those among reference formulations for the direct effects as well as for residual effects. These designs are suitable for bioequivalence trials where test vs. test and test vs. reference comparisons are to be made with more precision, especially with regard to direct effects of formulations.

Table 2. List of ISRMDs

Sl. No.	v_1	v_2	p	n	ω_f	ω_g	$\upsilon_{ff'}$	υ_{fg}	$\upsilon_{gg'}$	$V_{dff'}$	V_{dfg}	$V_{dgg'}$	$V_{rff'}$	V_{rfg}	$V_{rgg'}$
1	2	3	4	12	3	2	3	2	1	0.1834	0.2419	0.3132	0.2667	0.3462	0.4384
2	2	4	4	24	6	3	6	3	1	0.0917	0.1509	0.2189	0.1334	0.2133	0.3019
3	3	4	6	24	4	3	4	3	2	0.0864	0.1033	0.1219	0.1072	0.1276	0.1497
4	2	5	6	30	5	4	5	4	3	0.0691	0.0792	0.0900	0.0857	0.0980	0.1110
5	2	5	4	40	10	4	10	4	1	0.0550	0.1086	0.1683	0.0800	0.1521	0.2303
6	3	5	6	60	10	6	10	6	3	0.0346	0.0482	0.0628	0.0429	0.0593	0.0767
7	3	6	8	48	6	5	6	5	4	0.0425	0.0473	0.0523	0.0494	0.0549	0.0606
8	2	7	6	42	7	4	7	4	2	0.0494	0.0712	0.0942	0.0613	0.0876	0.1150
9	3	7	6	42	7	3	7	3	1	0.0494	0.0884	0.1294	0.0613	0.1082	0.1573

Acknowledgments

The valuable suggestions received from the editor are duly acknowledged.

References

Aggarwal, M.L., Deng, L.Y. and Jha, M.K. (2004). Some new residual treatment effects designs for comparing test treatments with a control. *Journal of Applied Statistics*. 31(9), 1065-1081.

Chow, S.C. and Liu, J.P. (2000). *Design and Analysis of Bioavailability and Bioequivalence Studies*. Marcel Decker Inc., New York .

Hedayat, A. and Zhao, W. (1990). Optimal two-period repeated measurements designs. Annals of *Statistics*. 18(4), 1805-1816.

Hedayat, A.S. and Yang, M. (2005). Optimal and efficient crossover designs for comparing test treatments with a control treatment. *Annals of Statistics*. 33(2), 915-943.

Koch, G.G., Amara, I.A., Brown, B.W., Colton, T. and Gillings, D.B. (1989). A two-period crossover design for the comparison of two active treatments and placebo. *Statistics in Medicine*. 8, 487-504.

Majumdar, D. (1988). Optimal repeated measurements designs for comparing test treatments with a control. *Communication in Statistics – Theory and Methods*. 17, 3687-3703.

Metzler, C. (1974). Bioavailability - a problem in equivalence. *Biometrics*. 30, 309-317.

Pigeon, J.G. and Raghavarao, D. (1987). Crossover designs for comparing treatments with a control. *Biometrika*. 74(2), 321-328.

Sharma, V.K. (1975). An easy method of constructing Latin square designs balanced for the immediate residual and other order effects. *Canadian Journal of Statistics*. 2(1), 119-124.

Ting, C.P. (2002). Optimal and efficient repeated measurements designs for comparing test treatments with a control. *Metrika*. 56(3), 229-238.

Williams, E.J. (1949). Experimental designs balanced for the estimation of residual effects of treatment. *Australian Journal of Scientific Research*. A2, 149-168.

Yadav, P.P. (1990). *Some studies on designs involving sequence of treatments*. Unpublished Ph.D. Thesis, IARI, New Delhi.

Yang, M. and Park, M. (2007). Efficient crossover designs for comparing test treatments with a control treatment when p = 3. *Journal of Statistical Planning and Inference*. 137(6), 2056-2067.

In: Research in Applied Statistical Science
Editor: Mohammad Ahsanullah, pp. 103-117
ISBN: 978-1-63321-818-5

Chapter 10

ROLE OF CATEGORICAL VARIABLES IN MULTICOLLINEARITY IN LINEAR REGRESSION MODEL

***M. Wissmann*[1]*, Shalabh*[2] *and H. Toutenburg*[3]**
[1]WWZ Statistics and Econometrics,
University of Basel, Basel, Switzerland
[2]Department of Mathematics and Statistics,
Indian Institute of Technology, Kanpur, India
[3]University of Munich, Munich, Germany

Abstract

The present article discusses the role of categorical variable in the problem of multicollinearity in linear regression model. It exposes the diagnostic tool condition number to linear regression models with categorical explanatory variables and analyzes analytically as well as numerically how the dummy variables and choice of reference category can affect the degree of multicollinearity.

Keywords: Linear regression model, multicollinearity, dummy variable, condition number

1. Introduction

The problem of multicollinearity in linear regression analysis arises when the explanatory variables in the linear regression model are correlated. The presence of multicollinearity disturbs the optimality of statistical procedures, see Rao et al. (2008), Belsley et al. (2004) or Silvey (1969) for more details. The problem of multicollinearity has been attempted in the literature from different perspectives like as diagnostic tools, removal tools, estimation and testing of hypothesis of parameters and solutions are suggested.

The condition indices supplemented with variance decomposition method is used to detect the 'near' linear dependencies in data matrix. When explanatory variables are categorical variables then dummy variables are used and the tools of regression analysis are applied. The outcome and interpretation of dummy variable may not be exactly same as in the case of quantitative continuous explanatory variable. The role of the diagnostic measures for multicollinearity with dummy variable are still unexplored or only partially explored in the literature.

The problem whether the observations should be centered around their mean or not before applying the diagnostic tools for multicollinearity is an issue which is still not completely resolved. Belsley (1984) argues that the centering of observations around their mean is of no use when dealing with multicollinearity and demonstrates that perturbed inputs have the same influence on the estimates obtained by using the centered and uncentered observations, see also Belsley (1991). On the other hand, Marquardt (1980) states that the centering of observations removes nonessential ill conditioning.

The issue of having an intercept term in linear regression model from the multicollinearity point of view is also an unsettled issue in the literature. It is often argued that the intercept term is of no use for the interpretation of regression results. But if the linear regression model has dummy variables then the intercept term represents the mean level of study variable at the reference or baseline categories of all categorical variables, when all other variables are set to be zero. When explanatory variables are dummy variables, then the aspect of centering of observation is not meaningful because then the centered dummy variables as well as their regression coefficients loose their interpretation.

In linear regression analysis, the dummy variables can also play an important role as a possible source for multicollinearity. The choice of reference category for a categorical variable may affect the degree of multicollinearity. Such issues have not yet been addressed in the literature to the best of our knowledge. Hendrickx et al. (2004) applies Belsleys perturbation analysis to a data set using a software that can deal with categorial variables, which the original Belsleys approach cannot do and Fox (1992) considers the dummy variables in case of generalized variance inflation factors.

We have attempted in this direction and have tried to explore these issues. We have considered the condition number and variance decomposition as diagnostic tools for multicollinearity in this paper. We assume that some of the explanatory variables are categorical in nature and are represented by dummy variables. We have analytically and numerically analyzed the role of dummy variables and the choice of reference category in causing the multicollinearity.

The plan of the paper is as follows. In Section 2, we describe the model and diagnostics for multicollinearity. Several linear regression models with different combinations of qualitative and quantitative variables are considered. The tools of regression and multicollinearity analysis are exposed to these models. Their discussion and results are presented in Section 3. In the Section 4, the effect of categorical variable is explored in a real data set and its findings are reported. Some conclusions are placed in Section 5.

A more detailed version of this paper is found in Wissmann et al. (2007).

2. Model and Diagnostics

Consider the following linear regression model:

$$y = X\beta + \varepsilon \tag{1}$$

where y is a $(n \times 1)$ vector of study variable, $(n \times p)$ design matrix X is of full rank and represents n observations on each of the $(p-1)$ explanatory variables and an intercept term with $(n \times 1)$ vector of elements unity, ε is a $(n \times 1)$ vector of disturbances which is a random variables with zero mean and positive definite covariance matrix $\Sigma = \sigma_\varepsilon^2 I$, and β is a $(p \times 1)$ vector of associated regression coefficients. The ordinary least squares estimator of β from (1) is $b = (X'X)^{-1}X'y$ and its variance-covariance matrix is $Var(b) = \sigma_\varepsilon^2(X'X)^{-1}$.

Assume that some of the explanatory variables in (1) are categorical variables. A categorial variable with m categories is represented by $(m-1)$ dummy variables. The reference category or baseline category is denoted by r, which the analysts may choose freely $(r \; \varepsilon \; \{1, \ldots, m\})$, see Rao et al. (2008). A dummy variable D_k, $k \; \varepsilon \; m \backslash r$, is defined as

$$D_k = \begin{cases} 1 & \text{if in category } k \\ 0 & \text{else.} \end{cases} \tag{2}$$

Note that when dummy variables are used to represent the categorical explanatory variables, then an intercept term is needed in the model which reflects this baseline level of y.

Now we assume that the problem of multicollinearity is present in data where some of the explanatory variables are categorical in nature. We measure the multicollinearity with condition number following Belsley et al (2004). A problem with the condition number is that it has its own scaling problems, see Steward (1987). When the dimensions of the data are changed, then the condition number is also changed. Belsley et al. (2004) recommends to scale each column of the design matrix using the Euclidian norm of each column before computing the condition number. The methods of Belsley et al. (2004) are implemented in the statistical software R using its package *perturb*. This package uses the root mean square of each column for scaling as its standard procedure.

The condition number of X is defined as $\kappa(X) = \sqrt{\lambda_{max}/\lambda_{min}}$ where $\lambda_j, (j = 1, 2, \ldots, p)$ are the eigenvalues of $X'X$. If we use the singular value decompensation of X, then we can express the condition number in terms of singular values (μ) as $\kappa(X) = \mu_{max}/\mu_{min}$, where the singular values of X are the positive square root of eigenvalues of $X'X$. The values $\kappa(X) = 10$ and $\kappa(X) = 30$ indicate a medium and serious degrees of multicollinearity, respectively. The condition indices $\eta_j = \mu_{max}/\mu_j \;\; (j = 1, \ldots, p)$ provide information about the degree of involved 'near' linear dependencies. A regression coefficient variance decompensation technique can then identify the involved variables in linear dependencies. By applying the singular value decomposition, $X = UDV'$ on the covariance matrix of b, the variance of ordinary least squares estimator can be rewritten as $Var(b) = \sigma^2 VD^{-2}V'$ where U and V are $(n \times p)$ and $(p \times p)$ orthogonal matrices, respectively and D is $(p \times p)$ diagonal matrix with nonnegative diagonal elements $\mu_1, \ldots, \mu_p$. This relates the variance of jth regression coefficient with the singular values of X as

$$Var(b_j) = \sigma^2 \sum_{k=1}^{p} \frac{v_{jk}^2}{\mu_k^2}. \tag{3}$$

When a 'near' linear dependency is present then the corresponding value of μ_k is very small, in comparison to other singular values. If a high proportion of the variance of two or more regression coefficients constitute the components of (3) aligned with a small singular value, then the corresponding variables are considered to be involved in causing the 'near' linear dependency. This is the idea behind the variance-decompositions proportions. Each proportion π_{kj} is the share of one component of (3) related with one singular value μ_k relative to the total variance (3). If a proportion is high, (say, > 0.5), then more than 50% of the variance of jth coefficient is related to a small singular value μ_k. If another coefficient also has a high proportion related with μ_k and the corresponding condition index μ_{max}/μ_k is high, then a 'near' linear dependency is diagnosed. Let

$$\pi_{kj} = \frac{\phi_{jk}}{\phi_j}, \ (j, k = 1, \ldots, p); \quad \phi_{jk} = \frac{\upsilon_{jk}^2}{\mu_k^2}; \quad \phi_j = \sum_{k=1}^{p} \phi_{jk}. \tag{4}$$

These proportions may then be summarized in a matrix Π with the condition indices on the first column and remaining p columns for the proportions. The variables involved in 'near' linear dependency are identified by proportions greater then 0.5 in Π.

3. Choice of Reference Category

The choice of reference category affects the values of explanatory variable and our objective is to study the effect on the degree of multicollinearity and numerical stability of the estimates. We use the condition number of scaled design matrix to diagnose the multicollinearity.

To motivate and understand the issue, we first consider a simple situation of linear regression model $y = \beta_0 + \beta_1 D + \varepsilon$ with an intercept term (β_0), slope parameter (β_1) and a dummy variable (D) representing a categorial variable with only two categories. Note that β_1 is interpreted as the difference in the expected values of y in different categories. Out of n number of observations, we observe h times the value '1' in the sample. We use the Euclidian norm to scale the design matrix following Belsley et al. (2004). The resulting scaling factors are ($\sqrt{n}$, $\sqrt{h}$). This provides a matrix where all the columns have same length. The cross-product of the scaled matrix $X_s'X_s$ is

$$X_s'X_s = \begin{pmatrix} 1 & \sqrt{f} \\ \sqrt{f} & 1 \end{pmatrix} \tag{5}$$

with $f = h/n \neq 0$ as the share of 'ones' in the sample. If $f = 0$, then we have a column with all elements 'zero' in X and the scaled design matrix does not exist in this case. If f is near to 1, then there exists a close relationship between the intercept term and dummy variable. The condition number based on (5), which is the ratio of square root of the maximum and minimum eigenvalues of $X_s'X_s$, should also reflect the same fact. From the characteristic polynomial $(1 - \lambda)^2 - f$, we obtain the eigenvalues of $X_s'X_s$ as $\lambda_{1,2} = 1 \pm \sqrt{f}$. The condition number is then

$$\kappa(X_s) = \sqrt{\frac{\lambda_{max}}{\lambda_{min}}} = \sqrt{\frac{1+\sqrt{f}}{1-\sqrt{f}}}, \tag{6}$$

which is a function of f. The graph of (6) is shown in figure 1. We observe from the figure 1 that when f close to 0, then the condition number is close to 1 which indicates that the data is nearly orthogonal. Note that $\kappa(X) < 10$ indicates no multicollinearity at all. When $f = 0.96$, then the condition number hits the bench mark '10' and indicates the presence of

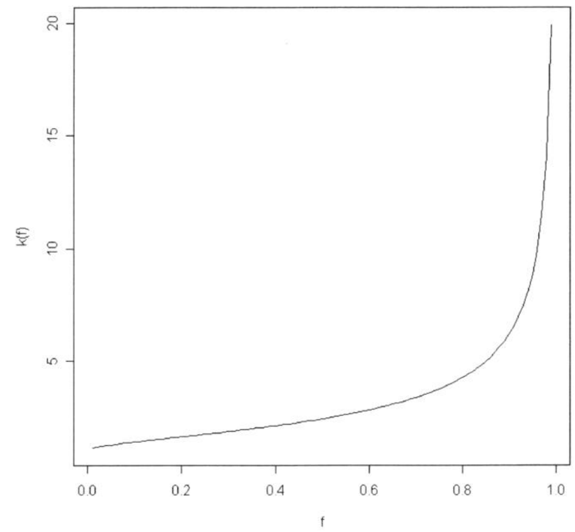

Figure 1. The condition number as a function of f.

a medium degree of multicollinearity. As f increases towards 1, the condition number tends towards infinity. Such an outcome is expected because then the dummy variable column is 'nearly' linearly depending on the column of intercept term in the design matrix.

From such illustration, we observe that the dummy variables can cause the multicollinearity problem. If only 4 percent of the observations are in the reference category, then we observe a medium degree of multicollinearity in this model. It suggests further that the problem of multicollinearity can be avoided by choosing a different reference category of dummy variable. In such a case when the reference category is changed, then 96 percent of the observations lie in the reference category. The condition number in this case is around 1.2 which indicates no multicollinearity.

It is also clear that by changing the reference category, the standard errors of the corresponding regression coefficients are not changed. The variance of least squares estimate of β_1 is

$$Var(b_1) = \frac{n\sigma_\varepsilon^2}{(n-h)h} = \frac{\sigma_\varepsilon^2}{nf(1-f)} \tag{7}$$

which is symmetric due to the role of f.

However, the variance of the least squares estimate of intercept (β_0) is affected. It is

$$Var(b_0) = \frac{\sigma_\varepsilon^2}{n-h} = \frac{\sigma_\varepsilon^2}{n(1-f)} \tag{8}$$

and is not symmetric like (7). If the share of 'ones' in the sample is large, then $(1-f)$ is small and the resulting variance in (8) is large. When $(1-f)$ is large, then the variance in (8) is small.

3.1. Interaction of One Categorical Variable and Intercept Term

We consider and simulate some models to see how the condition number behaves if the number of 'ones' are more than the number of 'zeros' in the system. We consider a model

$$y = \beta_0 + \beta_1 X_1 + \beta_2 D + \varepsilon \tag{9}$$

with an intercept term (β_0), slope parameters (β_1, β_2), dummy variable (D) and a continuous quantitative explanatory variable (X_1). The observations on D and X_1 are drawn from Binomial distribution $B(1, p)$ and exponential distribution $\exp(2.3)$, respectively. If p is small, then we expect less number of 'ones', and the share of 'ones' increases as p increases. Then we compute the condition number of the system for different values of p. We simulated 10 such designs and the mean of conditional numbers is plotted against p in figure 2.

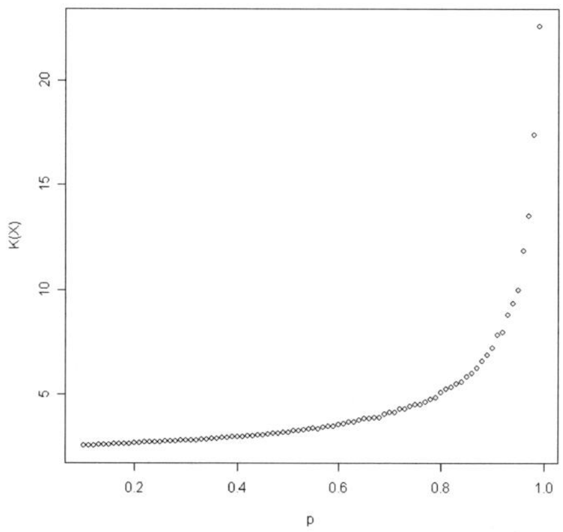

Figure 2. Conditional numbers vs. different values of p for $y = \beta_0 + \beta_1 X_1 + \beta_2 D + \varepsilon$.

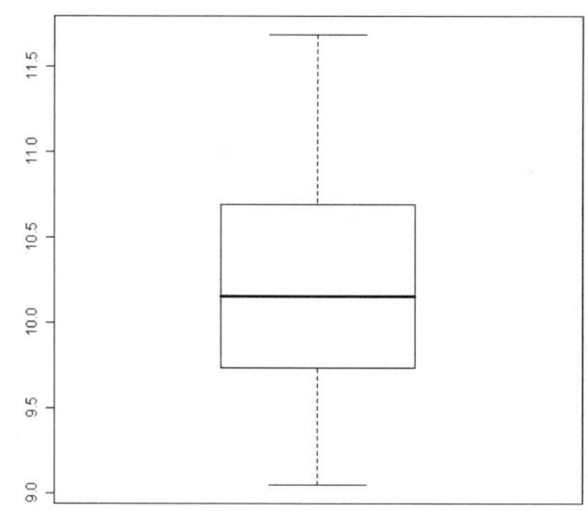

Figure 3. Boxplot of 100 condition numbers for $p = 0.95$.

We see an exponential relationship between the condition number of design matrix and p. The rate of increase of the condition number is moderate as long as the expected share of 'ones' in the sample is less then 0.8. As soon as p becomes larger than 0.8, we observe a higher rate of increment in the condition number with respect to p. We observe a medium multicollinearity problem when the value of p is around 0.95, which is quite similar to the results of the previous section.

Now we consider the simulation for the case of $p = 0.95$ in more detail and draw 100 designs matrices. We compute the condition number and present the results on its descriptive statistics in table 1 and box plot in figure 3. We observe that the condition numbers are relatively symmetrically distributed.

Table 1. Descriptive statistics of condition numbers for $y = \beta_0 + \beta_1 X_1 + \beta_2 D + \varepsilon$

Minimum	1st Quartile	Median	Mean	3^{rd}Quartile	Maximum	Standard deviation
9.048	9.738	10.15	10.22	10.69	11.69	0.643

Now we explore what happens if an additional category is added in the model (9)? To understand this issue, we add another dummy variable in (9) and have a model

$$y = \beta_0 + \beta_1 X_1 + \beta_2 D_1 + \beta_3 D_2 + \varepsilon \tag{10}$$

with two dummy variables (D_1, D_2) representing the three categories denoted by '0', '1', and '2'. The categorical variable is drawn from $B(2, p)$. The first dummy variable D_1 takes the value '1' if the categorical variable is '1' and the second dummy variable D_2 is '1' if the categorical variable is '2'. The reference category is where the categorical variable is '0'. If we choose p similar to earlier cases, then we observe no differences with the case in (10) when only one dummy variable was considered in (9). The outcomes of the simulation are very similar to the outcomes as in figure 2. This clearly shows that the nature of problem remains same even when the number of categories increase.

3.2. Interaction of More than One Explanatory Categorical Variables and Intercept Term

We now study a model with an intercept term (β_0) and two explanatory categorical variables (or factors A and B) and both have only two categories which are represented by two dummy variables D^A and D^B. More precisely, the model is now

$$y = \beta_0 + \beta_1 X_1 + \beta_2 D^A + \beta_3 D^B + \varepsilon \tag{11}$$

with $D^A \sim B(1, p_1)$, $D^B \sim B(1, p_2 = 0.5)$ and $X_1 \sim \exp(2.3)$. Then we compare the case with only one dummy variable with this one to see the sensitivity of the condition number to this addition of dummy variable in (9). We consider three values of $p_1 = (0.7, 0.9, 0.95)$ based on the idea that these values correspond to below, border line and medium multicollinearity problems for the case with only one dummy. The results of condition numbers based on 1000 simulated design matrices for the three values of p_1 are presented in figure 4.

We observe that the presence of second dummy variable increases the condition number of design matrix. The explanation is as follows. The linear combination of the two dummy variable is more similar to the intercept term than one dummy variable alone. In case of $p_1 = 0.9$, we expect a share of 90% 'ones', the second dummy variable may then fill some of the gaps with additional 'ones'. This leads to an increase in the degree of multicollinearity between the dummy variables and intercept term. As p_1 increases, the rate of increase of condition number declines. This is because if p_1 is small, then the second dummy variable has a greater chance to fill the linear combination with another 'one'. As p_1 increases, this chance become smaller because the first dummy variable contribute most of the 'ones' in linear combination.

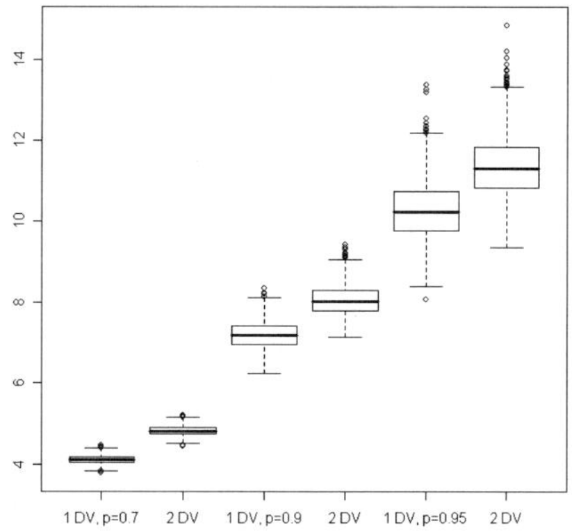

Figure 4. Results for the different values of p for $y = \beta_0 + \beta_1 X_1 + \beta_2 D^A + \beta_3 D^B + \varepsilon$.

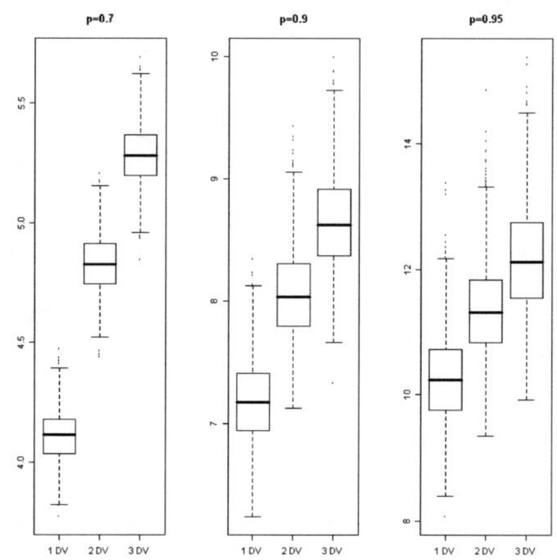

Figure 5. Results for 3 categorical variables for $y = \beta_0 + \beta_1 X_1 + \beta_2 D^A + \beta_3 D^B + \beta_4 D^C + \varepsilon$.

We now add another categorical variable D^C to the model (11) to support this argument as

$$y = \beta_0 + \beta_1 X_1 + \beta_2 D^A + \beta_3 D^B + \beta_4 D^C + \varepsilon. \tag{12}$$

The additional dummy variable (D^C) is drawn from $B(1, 0.35)$, so we expect a smaller increase in the condition number since the third dummy variable has less 'ones' than the second dummy variable. The results are presented in figure 5. It compares the condition index of the cases with one, two and three dummy variables.

Having a third dummy variable in the model increases the condition number of the design matrix. As the third dummy variable has a lower probability of success, so the increment in condition number is lower than the case of two dummy variables.

3.3. Interaction of Two Dependent Variables

Now we simulate a model with two dependent categorical variables. An issue to be explored here is if a dependent choice for the reference category can affect the stability of the design matrix?

The simulation model is $y = \beta_0 + \beta_1 X_1 + \beta_2 D^A + \beta_3 D^B + \varepsilon$ with dummy variables D^A from $B(1, 0.5)$ and D^B is dependent on D^A which is modeled as follows. If $D^A = 1$, then draw D^B from $B(1, p)$, else if $D^A = 0$, then use $B(1, (1 - p))$. Hence if p is close to 1, then we expect a positive association between the two dummy variables. If D^A has the value '1' we expect D^B to have the value '1' as well.

If we choose the reference category for D^A different from D^B, then we have a negative association as follows:

$D^A \backslash D^B$	'0'	'1'
'0'	45	462
'1'	440	53

This is then simulated by using a small value of p and we choose $p = 0.1$. We simulate 100 experiments with $p = 0.1$ and $p = 0.9$, compute the condition numbers based on the outcomes from 100 experiments and then compare them together in figure 6.

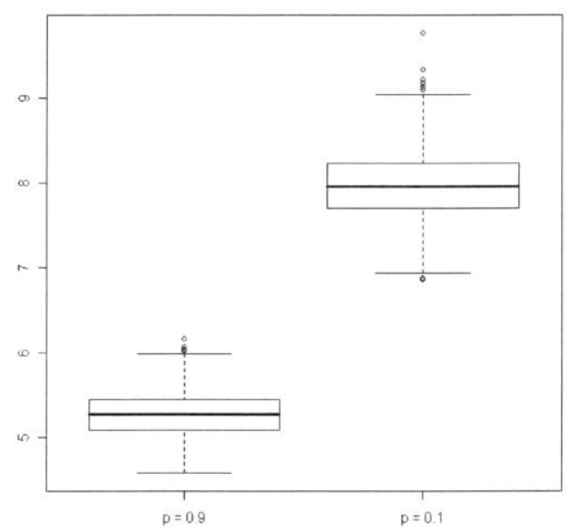

Figure 6. Dependent categorical variables under different types of coding the dummy variables for the model $y = \beta_0 + \beta_1 X_1 + \beta_2 D^A + \beta_3 D^B + \beta_4 D^C + \varepsilon$.

If we code the categorical variables such that they match in the 'ones', then we obtain smaller condition numbers compared to the case when they are coded in the other way. This means that a 'near' linear dependency with two dummy variables is less harmful in terms of the condition number than a 'near' linear dependency with two dummy variables and an intercept term.

3.4. Influence of Choice of Reference Category on Multicollinearity

Next we consider a model with two continuous quantitative variables involved in a 'near' linear dependency and one dummy variable which may have a weak reference category. A category with low frequency is termed as a weak reference category. For example, if the number of 'ones' are 95% in a category and only 5% 'zeros' in other category, then the category with '0' is a weak reference category.

The continuous quantitative variable X_1 is drawn from $\exp(2.3)$ and another continuous quantitative variable X_2 is a linear combination of X_1 and of a normal random variable $N(0, (0.085)^2 s_{x1}^2)$ where s_{x1}^2 is the sample variance of observations on X_1. The dummy variable D takes two values '0' and '1'. The study variable y is generated by

$$y_i = 23 + 1.5X_{1i} + 3X_{2i} + 0.5D_i + \varepsilon_i, \quad (i = 1, \ldots, n) \tag{13}$$

Table 2. Regression analysis output for $y = 23 + 1.5X_1 + 3X_2 + 0.5D + \varepsilon$

Coefficient:	Estimate	Standard error	$Pr(>\|t\|)$
Intercept	22.85	0.15	0
X_1	1.31	0.92	0.16
X_2	3.57	1.31	0.007
D	0.66	0.15	0

Table 3. Multicollinearity analysis output for $y = 23 + 1.5X_1 + 3X_2 + 0.5D + \varepsilon$

Condition number	Variance Decomposition Proportions Intercept	X_1	X_2	D
1	0.007	0.001	0.001	0.008
2.43	0.037	0.003	0.003	0.056
7.91	0.956	0	0	0.937
30.3	0	0.996	0.996	0

where $\varepsilon \sim N(0, 1.5^2)$. The sample size is 1000.

We then compare the situation with a weak reference, modeled by using $B(1, 0.9)$ for D, with the recoded model which is obtained by changing the reference category in (13).

The regression analysis output and Π matrix for the weak reference case are presented in tables 2 and 3, respectively. The deviations in the resulting estimates from the known β are quite large. The parameter estimate b_1 is not significant at one percent level of significance. The inflated variances of the regression estimates turn a known significant parameter into an insignificant parameter. The related condition indices and variance decomposition factors are stated in table 3. We observe from table 3 that one condition index is just below the threshold for a medium degree of multicollinearity. This is associated with the intercept term and dummy variable. Then we have a condition index that is just above the threshold for a serious multicollinearity problem, which is associated with the 'near' linear dependency of X_1 and X_2. Next we explore what happens if we choose a different reference category. The results of regression analysis obtained after reverting the reference category are presented in table 4 and the estimates are found to be more precise now. The standard error of the intercept term is lower and all other standard errors remain the same as in table 2. By changing the reference category, we change the sign of the parameter estimate associated with the dummy variable. In the earlier case, β_3 is $[E(y|D = 1) - E(y|D = 0)]$ and now the reference category is the opposite of the earlier case. Therefore the sign must be different. Note that β_3 is still $[E(y|D = 1) - E(y|D = 0)]$ but the interpretations of '0' and '1' are changed. The regression estimates associated with continuous quantitative variables are still inflated. The condition indices and variance decomposition proportions are presented in table 5. Now we have a condition index which is just below 30. The absence of the 'near' linear dependency of the dummy variable and intercept term therefore

Table 4. Regression analysis output for $y = 23 + 1.5X_1 + 3X_2 + 0.5D + \varepsilon$ with changed reference category

Coefficient:	Estimate	Standard error	$Pr(>\|t\|)$
Intercept	23.01	0.07	0
X_1	1.31	0.92	0.16
X_2	3.57	1.31	0.007
D	-0.66	0.15	0

Table 5. Multicollinearity analysis output of $y = 23 + 1.5X_1 + 3X_2 + 0.5D + \varepsilon$ with changed reference category

Condition		Variance	Decomposition	Proportions
number	Intercept	X_1	X_2	D
1	0.045	0.001	0.001	0.021
1.75	0	0	0	0.895
2.821	0.955	0.003	0.003	0.084
27.38	0	0.996	0.996	0.001

reduces the condition index. We have only problem now between X_1 and X_2. The additional dependency between D and intercept term is avoidable by the appropriate choice of the reference category.

4. Application to Real Data Set

Now we study the role of dummy variable and choice of reference category under multicollinearity in real data sets. We consider a real data set and do the similar analysis as in Section 3. We consider a data set on Infant-Mortality from Leinhardt and Wassermann (1979) which was used in Fox (1997) and is available in the R package ‘car’, R Development Core Team (2007).

It contain data on infant-mortality (y) per 1000 live births of 105 nations around the world. The influence factors are whether the country is an oil exporting country (factor A), the regions (Africa, America, Asia and Europe) (factor B, 4 categories represented by D_1^B, D_2^B, D_3^B) and the per-capita income (in U.S. Dollars) X.

First we consider the model with 2 categorical variables as

$$y = \beta_0 + \beta_1 D^A + \beta_2 D_1^B + \beta_3 D_2^B + \beta_4 D_3^B + \beta_5 X + \varepsilon \tag{14}$$

The bar plot of the two categorial variables in figures 7 and 8 indicates whether there are weak categories or not. They may give rise to multicollinearity with the intercept term if chosen as reference category. The regions seems to be well balanced and the relative frequencies of all the categories are larger then 15%. There are not much oil exporting countries in the sample as figure 8 shows. Only 8% of the countries in the sample are oil exporting countries. There is no association between the region and an oil exporting country, only Europe has no oil exporting country in the sample.

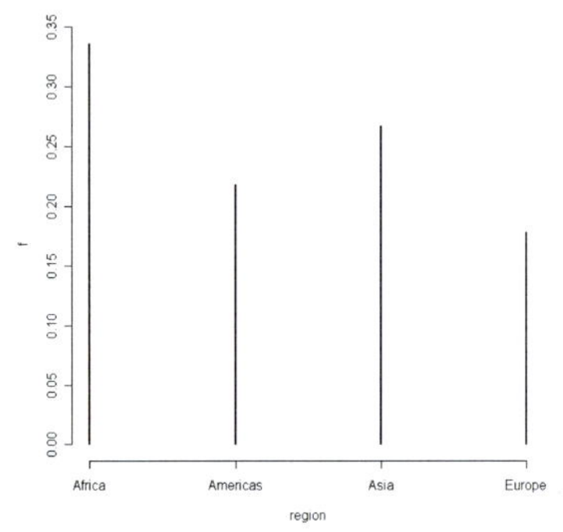

Figure 7. Barplot of the regions in infant-mortality data.

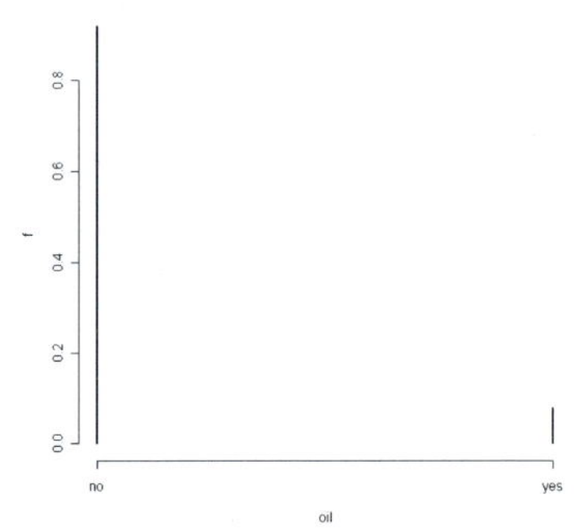

Figure 8. Barplot for the oil exporting countries in infant-mortality data.

If we choose the countries with oil export as a reference category, then we expect a 'near' dependency between the corresponding dummy variable and intercept term. Assuming homoscedastic errors, we obtain the regression results for (14) in table 6.

The intercept term in this case is the mean mortality for a non oil exporting country in Africa with income zero. It gives a reference for the interpretation of dummy variables. From the output in table 6, we observe that exporting oil lowers the infant mortality. An African state which does not export oil has a higher infant mortality than a non-African state who does export oil, as the comparison of b_1 and intercept shows. It interprets that since an African state does not export oil, so they have low income and less profit. In turn, they spend less on the welfare of the people and hence they have higher infant mortality rate.

The variance decomposition proportions shows a medium degree of multicollinearity between the dummy variable for oil and intercept term. Only the last row is printed in table 7.

Now if we choose a different reference category for the oil exporting countries, then we have no problem at all and the variance of intercept term is more precise. Since the reference category for oil is different, so the interpretation of intercept term estimate is also

Table 6. Regression analysis output of infant-mortality data

Coefficients:	Estimate	Standard Error	$Pr(>\|t\|)$
Intercept	215.2	29.7	0
oil-yes	-78.3	28.9	0.01
region-America	-83.7	21.8	0
region-Asia	-45.9	20.1	0.02
region-Europe	-101.5	30.7	0.001
income	-0.005	0.007	0.48

Table 7. Multicollinearity analysis output of infant-mortality data

Condition-number		Variance	Decomposition	Proportions		
	Intercept	oil-yes	region-America	region-Asia	region-Europe	income
9.25	0.96	0.93	0.02	0.04	0	0.02

different, see table 8. Table 8 shows that not exporting oil increases the infant mortality

Table 8. Regression analysis output of infant-mortality data with changed reference category

Coefficients:	Estimate	Standard error	$Pr(>\|t\|)$
Intercept	136.8	13.6	0
oil-no	78.3	28.9	0.01
region-America	-83.7	21.8	0
region-Asia	-45.9	20.1	0.02
region-Europe	-101.5	30.7	0.001
income	-0.005	0.007	0.48

relative to the baseline of oil exporting states, all other things equal. Note that the standard error of intercept term is less then half than the value of the standard error of intercept term in the model $y = \beta_0 + \beta_1 D + \epsilon$ with the different choice of reference category for oil as in table 6. The condition indices and variance decomposition proportions are stated in table 9. Only the last row is printed in table 9. We observe from table 9 that the choice of reference category lowers the condition index. Now there is no multicollinearity at all.

Another application of these tools to real data can be found in Wissmann et al. (2007).

Table 9. Multicollinearity analysis output of infant-mortality data with changed reference category

Condition number	Intercept	Variance oil-no	Decomposition region-America	Proportions region-Asia	region-Europe	income
3.93	0.49	0.03	0.51	0.47	0.72	0.3

5. Summary

We have considered an issue related to the problem of multicollinearity in the presence of categorical variable as explanatory variable in the context of linear regression analysis. The role of dummy variables and the choice of reference category is analyzed through different linear models to see their effect in the problem of multicollinearity. In a simple case of one dummy variable, we have demonstrated that how the choice of reference category affects the multicollinearity. A closed form of condition number is also obtained in this case as a function of a collinearity increasing factor. It is difficult to get such a closed form expression in the general case as there can be many possible combinations of dummy and quantitative variables in linear regression models. So such an issue is explored more in detail by choosing various combinations of dummy and quantitative variables. It is found that the presence of dummy variable and the choice of reference category can be a cause of multicollinearity. Also, the situation of multicollinearity can be averted by changing the reference category. We have demonstrated this issue by simulation as well as through the application on a real data set.

References

Belsley, D.A.. (1984). Demeaning Conditioning Diagnostics through Centering, *The American Statistican*, 38 (2), 7377.

Belsley, D.A.. (1991). Conditioning Diagnostics: Collinearity and Weak Data in Regression, John Wiley & Sons, New York.

Belsley, D.A., E. Kuh, and R.E. Welsch. (2004). *Regression Diagnostics: Identifying Influential Data and Sources of Collinearity*, John Wiley & Sons, New York.

Fox, J. (1992). Generalized Collinearity Diagnostics, *Journal of the American Statistical Association*, 87 (417), 178183.

Fox, J. (1997). *Applied Regression, Linear Models and Related Models,* Sage Publications.

Hendrickx, J., B. Belzer, M. Grotenhuis, and J. Lammers (2004). *Collinearity involving ordered and unordered categorical variables*, RC33 conference in Amsterdam August 2004. http://www.xs4all.nl/ jhckx/perturb/perturb.pdf.

Leinhardt, S. and S.S. Wassermann. (1979) *Explorattory data analysis: An introduction to selected methods*, in K. Schssler, ed., Sociological Methology, Jossey- Bass.

Marquardt, D. W.. (1980). A Critique on some ridge regression Methods: Comment You should standardize the predictor vairables in your regression modells, *Journal of the American Statistical Association*, 1980, 75 (369), 8791.

R Development Core Team. (2007). R: A Language and Environment for Statistical Computing R Foundation for Statistical Computing. ISBN 3-900051-07-0.

Rao, C.R., H. Toutenburg, Shalabh, and C. Heumann. (2008). *Linear Models and Generalizations - Least Squares and Alternatives,* Springer.

Silvey, S. D. (1969). Multicollinearity and Imprecise Estimation, *Journal of the Royal Statistical Society. Series B,* 31 (3), 539-552.

Steward, G.W. (1987). Collinearity and Least Squares Regression, *Statistical Science*, 1987, 2 (1), 6884.

Wissmann, M., Toutenburg, H. and Shalabh (2007). *Role of Categorical Variables in Multicollinearity in Linear Regression Model. Technical Report* No. 8, Department of Statistics, University of Munich, Munich, Germany.

In: Research in Applied Statistical Science
Editor: Mohammad Ahsanullah, pp. 119-131
ISBN: 978-1-63321-818-5

Chapter 11

PREDICTIVE INFERENCE FROM TWO-PARAMETER WEIBULL MODEL GIVEN A DOUBLY CENSORED SAMPLE OF HEPATOCELLULAR CARCINOMA OF LIVER PATIENTS

***Hafiz M.R. Khan*[1,*] *Ahmed Albatineh*[1], *Salem Tesfa*[1], *Nadine R. Jenkins*[2] and *Nasar U. Ahmed*[1]**
[1]Department of Epidemiology & Biostatistics,
Robert Stempel College of Public Health and Social Work,
Florida International University, Miami, Florida, US
[2]UMDNJ University Hospital,
University of Medicine & Dentistry of New Jersey,
Newark, New Jersey, US

Abstract

The purpose of this paper is to derive the predictive densities for future responses from a two-parameter Weibull model given a doubly censored sample. The predictive density for a single future response, bivariate future response, and a set of future responses are derived when the shape parameter α is either known or unknown. A numerical example of hepatocellular carcinoma of 69 liver patients is utilized to illustrate the predictive results for the stages of cancer. The survival days of 69 male and female patients with primary hepatocellular carcinoma of liver are analyzed. Four patients out of 69 patients are lost to follow-up and they are the first two and the last two observations of 69 patients. Furthermore, a sensitivity study of the mean and standard deviation of the predictive densities of survival days with respect to several values of the hyperparameter (τ) is performed when the shape parameter α is unknown. It is found that the rate of the males patients future survival days is lower than the females patients at the initial stage and higher at the later stage.

2000 Mathematics Subject Classification: 62F15, 62N01, 62N02, 62F15, 62M20

Keywords: Weibull model, Predictive inference, Doubly censored sample, Hepatocellular carcinoma

---

*E-mail address: hmkhan@fiu.edu

1. Introduction

Predictive inference has been playing an important role in healthcare data analysis. Healthcare researchers rely heavily on past observations to extract information and forecast treatment outcome. However, incomplete observations are inevitable in data analysis and must be accounted for. Otherwise, ignoring or failing to account for incomplete or partial unobserved data can be costly often leading to misinterpretation. Health care practitioners/researchers may reduce some amount of incomplete data by providing incentives, by calling more than once, or by using other methods. Nonetheless, the challenge of incomplete data still remains to be resolved. Therefore, appropriate methods need to be used to analyze such type of data. Data censoring may result from various factors such as loss to follow up due to migration, early termination of the study, unrelated cause of death, limited resources to conduct study up to desired length or withdrawal from study due to undesirable side effect of a drug. In clinical studies, there are various types of censored data that have been observed such as type I, type II, and doubly censored. Doubly censored data has been observed increasingly in many clinical studies in recent years. For instance, in a clinical experiment, if the first few observations and the last few observations are unavailable from a sequence of ordered observations, then the resulting sample is called a doubly censored sample. Suppose in a clinical study there are n patients diagnosed with a certain disease. These patients are given optimal doses of treatment/therapy for a period of time. Then the study follow-up is conducted and their survival times are recorded. The survival times of the patients can be ordered with $x_1 \leq \cdots \leq x_n$; it may be the case that the record of the first few observations and the last few observations are unavailable. In order to make statistical inference of this type of data, it may be reasonable to discard those unavailable observations from the statistical analysis. Such type of data constitutes doubly censored sample. For more about doubly censored samples, the reader is referred to Sarhan (1955), Khan (2003), Khan et al. (2006), Kambo (1978), Raqab (1995), Lalitha and Mishra (1996), among others.

Sarhan (1955) obtained the mean and standard deviation of certain populations from singly and doubly censored samples. Kambo (1978) derived the maximum likelihood estimators of the location and scale parameters by using a doubly censored sample. Khan et al. (2006) derived the predictive distributions of future responses on the basis of a doubly censored sample from a two-parameter exponential life testing model. One may consider a Bayesian approach to perform the predictive inference of future observations. Recently, the extensive applications of this approach has led to solutions for various healthcare related issues. Several authors have utilized the Bayesian approach for determining predictive inference in the past. For example, Aitchison and Dunsmore (1975), Aitchison and Sculthorpe (1965), and Berger (1985) considered a general Bayesian prediction problem. Chew (1968), and Geisser (1993) discussed various Bayesian prediction problems for future responses. Moreover, Geisser (1971) discussed the inferential use of predictive distributions. Prediction distributions for linear models which make use of the Bayesian approach have been determined by Dunsmore (1974). Additional Bayesian approach to predictive inference have been considered, for example, by Khan (2003), Thabane (1998), Kim et al. (2000), Khan et al. (2006, 2007, 2008), Gelman et al. (2004), Bernardo and Smith (1994), Sinha (1986), Evans and Nigm (1980a, 1980b), Jeffreys (1961) among others.

The objective of this paper is to derive the predictive densities for future responses and to draw predictive inference by making use of the Bayesian approach given a doubly censored sample of hepatocellular carcinoma of liver patients. The remainder of this paper is organized in the following order. Section 2 describes predictive inference for future responses from a two-parameter Weibull distribution when the shape parameter (α) is known. Section 3 derives the predictive inference for future responses when the shape parameter (α) is unknown. Sections 4 includes a numerical example of hepatocellular carcinoma of the liver patients to illustrate some of the results. In Section 5, conclusions are presented.

2. Predictive Inference for a Single Future Response when α Known

The probability density function of a two-parameter Weibull distribution is given by

$$f(x|\alpha,\theta) = \begin{cases} \frac{\alpha}{\theta}\, x^{\alpha-1} \exp\left\{-\frac{x^\alpha}{\theta}\right\}, & \text{for } x \geq 0;\ \beta^\alpha = \theta,\ \alpha \geq 0, \\ 0 & \text{elsewhere,} \end{cases} \tag{1}$$

where α is the shape parameter and β is the scale parameter.

Let $x_1, \ldots, x_n$ be an ordered random sample of size n from model (1), where $x_1 \leq \cdots \leq x_g$ be the g smallest ordered observations and $x_{\ell+1} \leq \cdots \leq x_n$ be the $(n-\ell)$ largest ordered observations from the sample. Only the remaining ordered observations $\mathbf{x} = (x_{g+1}, \ldots, x_\ell)$ are used for statistical analysis. It is assumed that the sample data can be modeled by the two-parameter Weibull distribution. The likelihood function of θ for a given doubly censored sample $\mathbf{x} = (x_{g+1}, \ldots, x_\ell)$ is given by

$$\mathrm{L}(\theta|\mathbf{x}) \propto \theta^{-(\ell-g)} \left[F\left(\frac{x_{g+1}^\alpha}{\theta}\right)\right]^g \left[1 - F\left(\frac{x_\ell^\alpha}{\theta}\right)\right]^{n-\ell} \left[\prod_{i=g+1}^{\ell} f\left(\frac{x_i^\alpha}{\theta}\right)\right], \quad \text{for } x_i \geq 0;\ \theta \geq 0,$$

where $F(x) \equiv F(\frac{x^\alpha}{\theta}) = 1 - \exp\left\{-\frac{x^\alpha}{\theta}\right\}$.

The prior density of θ given τ is

$$p(\theta|\tau) = \frac{(\tau)^\tau}{\Gamma(\tau)}\, \theta^{-(\tau+1)} \exp\left\{-\frac{\tau}{\theta}\right\},\ \tau,\ \theta > 0, \tag{2}$$

where τ is the hyperparameter and is in fact the shape parameter of the prior density. It is assumed that τ is non-negative. By making use of this prior, the posterior density and the predictive density are meaningfully interpretable. When $\tau = 0$, this prior reduces to Jeffreys (1961) non-informative prior. The posterior density of θ given α and $\mathbf{x}$ is

$$p(\theta|\alpha,\mathbf{x}) = \phi(\mathbf{x})\, \theta^{-(\tau+\ell+1-g)} \sum_{\eta=0}^{g} (-1)^\eta \binom{g}{\eta} \exp\left\{-\frac{\tau + \sum_{i=g+1}^{\ell} x_i^\alpha + (n-\ell)x_\ell^\alpha + \eta x_{g+1}^\alpha}{\theta}\right\},$$

where $\phi(\mathbf{x})$ is the normalizing constant and it can be evaluated as follows

$$\phi(\mathbf{x})^{-1} = \sum_{\eta=0}^{g} (-1)^\eta \binom{g}{\eta} \Gamma(\tau+\ell-g) \left(\tau + \sum_{i=g+1}^{\ell} x_i^\alpha + (n-\ell)x_\ell^\alpha + \eta x_{g+1}^\alpha\right)^{-(\tau+\ell-g)}.$$

Let y be a future response from the model specified by (1). It is assumed that y is independent of the observed data. The predictive density for y given a set of doubly censored data $\mathbf{x} = (x_{g+1}, \ldots, x_\ell)$, will be determined as follows

$$p(y|\alpha, \mathbf{x}) \propto \frac{\alpha^{(\ell-g+1)} \sum_{\eta=0}^{g}(-1)^\eta \binom{g}{\eta} y^{\alpha-1} \left(\prod_{i=g+1}^{\ell} x_i^{\alpha-1} \right)}{(\tau + \sum_{i=g+1}^{\ell} x_i^\alpha + (n-\ell)x_\ell^\alpha + \eta x_{g+1}^\alpha)^{(\tau+\ell-g+1)}} \times \left[1 + \frac{y^\alpha}{\tau + \sum_{i=g+1}^{\ell} x_i^\alpha + (n-\ell)x_\ell^\alpha + \eta x_{g+1}^\alpha} \right]^{-(\tau+\ell-g+1)},$$

which is the form of a univariate F-distribution with $(2,\ 2(\tau + \ell - g))$ degrees of freedom. When α =1, one could infer about future characteristics of the model such as the mean, standard deviation, and tolerance region in connection with the doubly censored sample. By setting $g = \eta = 0,\ \ell = r$, it is noted that the above predictive density is equivalent to the predictive density of a right censored sample.

2.1. Predictive Inference for a Bivariate Future Response

Let y_1 and y_2 be the two ordered future responses from model (1). The predictive density of y_1 and y_2 given a set of doubly censored sample **x** is given by

$$p(y_1, y_2|\alpha, \mathbf{x}) \propto \frac{\alpha^{\ell-g+2} \prod_{i=1}^{2} y_i^{\alpha-1} \sum_{\eta=0}^{g}(-1)^\eta \binom{g}{\eta} \left(\prod_{i=g+1}^{\ell} x_i^{\alpha-1} \right)}{\left(\tau + \sum_{i=g+1}^{\ell} x_i^\alpha + (n-\ell)x_\ell^\alpha + \eta x_{g+1}^\alpha \right)^{(\tau+\ell-g+2)}} \times \left[1 + \frac{\sum_{i=1}^{2} y_i^\alpha}{\tau + \sum_{i=g+1}^{\ell} x_i^\alpha + (n-\ell)x_\ell^\alpha + \eta x_{g+1}^\alpha} \right]^{-(\tau+\ell-g+2)},$$

which is the form of a bivariate F-distribution with degrees of freedom $(2,\ 2;\ 2(\tau+\ell-g))$.

2.2. Predictive Inference for a Set of Future Responses

Let $y_1, \ldots, y_m$ be a set of ordered future responses from model (1). The predictive density for m future responses $y_1, \ldots, y_m$ given a set of doubly censored sample **x** is given by

$$p(y_1, \ldots, y_m|\alpha, \mathbf{x}) \propto \alpha^{\ell-g+m} \prod_{i=1}^{m} y_i^{\alpha-1} \sum_{\eta=0}^{g}(-1)^\eta \binom{g}{\eta} \Big(\prod_{i=g+1}^{\ell} x_i^{\alpha-1} \Big) / \left(\tau + \sum_{i=g+1}^{\ell} x_i^\alpha + (n-\ell)x_\ell^\alpha + \eta x_{g+1}^\alpha \right)^{(m+\tau+\ell-g)} \times \left[1 + \frac{\sum_{i=1}^{m} y_i^\alpha}{\tau + \sum_{i=g+1}^{\ell} x_i^\alpha + (n-\ell)x_\ell^\alpha + \eta x_{g+1}^\alpha} \right]^{-(m+\tau+\ell-g)},$$

which is the form of a multivariate F- distribution with degrees of freedom $(2, 2, \ldots, 2;\ \ 2(\tau + \ell - g))$. When m=1, one could obtain the predictive density for a single future response; for m=2, a bivariate future response; and so on.

3. Predictive Inference for a Single Future Response when α Unknown

Suppose that α has a uniform prior over the interval (0, α) and θ has the prior given by (2). Furthermore, assume that α and θ are independently distributed. Then the joint prior density of α and θ is given by

$$p(\theta, \alpha|\tau) = \frac{(\tau)^\tau}{\alpha\, \Gamma(\tau)}\, \theta^{-(\tau+1)} \exp\left\{-\tfrac{\tau}{\theta}\right\},\ \tau,\ \theta > 0, \tag{3}$$

and the joint posterior of (θ, α) is

$$p(\theta, \alpha|\mathbf{x}) \ \propto\ \alpha^{(\ell-g-1)}\theta^{-(\tau+\ell+1-g)} \sum_{\eta=0}^{g}(-1)^\eta \binom{g}{\eta}\left(\prod_{i=g+1}^{\ell} x_i^\alpha\right) \times \exp\left\{-\frac{\tau + \sum_{i=g+1}^{\ell} x_i^\alpha + (n-\ell)x_\ell^\alpha + \eta x_{g+1}^\alpha}{\theta}\right\}.$$

The predictive density function of a future response y, given a set of doubly censored data $\mathbf{x}$ is

$$p(y|\mathbf{x}) \ \propto\ \sum_{\eta=0}^{g}(-1)^\eta \binom{g}{\eta} \int_{\alpha=0}^{+\infty} \frac{\alpha^{(\ell-g)}\ y^{\alpha-1}\left(\prod_{i=g+1}^{\ell} x_i^\alpha\right)}{\left(\tau + \sum_{i=g+1}^{\ell} x_i^\alpha + (n-\ell)x_\ell^\alpha + \eta x_{g+1}^\alpha\right)^{(\tau+\ell-g+1)}} \times \left[1 + \frac{y^\alpha}{\tau + \sum_{i=g+1}^{\ell} x_i^\alpha + (n-\ell)x_\ell^\alpha + \eta x_{g+1}^\alpha}\right]^{-(\tau+\ell-g+1)} d\alpha.$$

There is no closed form representation of the above density function. However, for a given doubly censored data set, it is possible to plot the density function by making use of an advanced statistical computation software package.

3.1. Predictive Inference for a Bivariate Future Response

Let y_1 and y_2 be the two ordered future responses from model (1). The predictive density for a bivariate future response y_1 and y_2 given a set of double censored sample $\mathbf{x}$ is given by

$$p(y_1, y_2|\mathbf{x}) \ \propto\ \sum_{\eta=0}^{g}(-1)^\eta \binom{g}{\eta} \int_{\alpha=0}^{+\infty} \frac{\alpha^{\ell-g+1} \prod_{i=1}^{2} y_i^{\alpha-1} \prod_{i=g+1}^{\ell} x_i^{\alpha-1}}{\left(\tau + \sum_{i=g+1}^{\ell} x_i^\alpha + (n-\ell)x_\ell^\alpha + \eta x_{g+1}^\alpha\right)^{(\tau+\ell-g+2)}} \times \left[1 + \frac{\sum_{i=1}^{2} y_i^\alpha}{\tau + \sum_{i=g+1}^{\ell} x_i^\alpha + (n-\ell)x_\ell^\alpha + \eta x_{g+1}^\alpha}\right]^{-(\tau+\ell-g+2)} d\alpha.$$

There is no closed form representation of the above density function. However, for a given doubly censored data set, it is possible to plot the density function by making use of an advanced statistical computation software package.

3.2. Predictive Inference for a Set of Future Responses

Let $y_1, ..., y_m$ be the a set of ordered future responses from model (2). The predictive density for m future responses $y_1, ..., y_m$ given a set of doubly censored sample **x** is given by

$$p(y_1, ..., y_m|\mathbf{x}) \propto \sum_{\eta=0}^{g}(-1)^{\eta}\binom{g}{\eta}\int_{\alpha=0}^{+\infty}\frac{\alpha^{\ell-g+m-1}\prod_{i=1}^{m} y_i^{\alpha-1}\prod_{i=g+1}^{\ell} x_i^{\alpha-1}}{\left(\tau+\sum_{i=g+1}^{\ell} x_i^{\alpha}+(n-\ell)x_{\ell}^{\alpha}+\eta x_{g+1}^{\alpha}\right)^{(\tau+\ell-g+m)}}$$
$$\times\left[1+\frac{\sum_{i=1}^{m} y_i^{\alpha}}{\tau+\sum_{i=g+1}^{\ell} x_i^{\alpha}+(n-\ell)x_{\ell}^{\alpha}+\eta x_{g+1}^{\alpha}}\right]^{-(\tau+\ell-g+m)} d\alpha.$$

There is no closed form representation of the above density function. However, for a given doubly censored data set, it is possible to plot the density function by making use of an advanced statistical computation software package.

4. A Numerical Example

Primary hepatocellular carcinoma (HCC), or cancer of the liver, is the most common type of malignant tumor found in the liver. HCC should not be confused with metastatic cancer of the liver, which occurs when cancer has spread to the liver from other primary cancer sites in the body. HCC is frequently seen in the presence of cirrhosis, hepatitis B or hepatitis C of the liver. Cirrhosis can be referred to as scarring of the liver and can be caused by alcoholism, hepatitis B or hepatitis C, or other diseases that cause long-term inflammation of the liver. When looking at the incidence of this tumor across the globe, Africa and Asia have a higher incidence of HCC than North America, South America and Europe (U.S. NLM, NIH, Medline, 2009). Although liver cancer is one of frequently occurring cancers in the world, liver cancer diagnoses in the United States are not that common (Mayo Clinic, 2010). HCC is diagnosed more often in men than in women, and in the United States, African American men have had the highest rate of increase in this disease since the early 1990's. This cancer typically affects adults over the age of 50, but more recently has shown an increase in those younger than age 50 (El-Serag and Mason, 1999).

The standard care for liver cancer is usually surgery, tumor ablation, tumor embolization or liver transplant for small tumor or tumors that are diagnosed early. In some instances, chemotherapy or radiation therapy may be used to decrease the size of large tumors to make it more surgically resectable (U.S. NLM, NIH, Medline, 2009). Survival rates for HCC is low, with approximately up to 10% of individuals diagnosed with HCC still being alive after 5 years (ACS, 2009). The tumors that are not completely removed, survival days can be as short as 6 months (U.S. NLM, NIH, Medline, 2009). In this paper, a survival analysis of patients diagnosed with HCC in 2002 was performed on a cancer registry data set from University Hospital located in Newark, New Jersey. This data set is composed of 69 patients of which 54 were male and 15 were female. The average age at diagnosis overall (both males and females) was 61 years. The average age for the male patients was 59 years, and the female patients averaged 67 years of age. Race/ethnicity within this data set reflected a population of the following: 55% White, 17% African American, 12% Latino, 4% Asian descent (Filipino and Chinese), and 12% were of other or unknown race/ethnicity. Forty-one percent of the cases presented with an earlier stage (0 & 1) and 58% presented with a later stage (2 & 3) of disease, where 0 = no surgery; 1 = local tumor destruction; 2 = segmental resection; and 3 = hepatectomy. There was one case (1%) that did not have a stage of disease assigned. Surgery, ablations and embolizations were the primary mode of treatment for these patients with 37% receiving tumor embolizations, 19% received a liver transplant, 16% had tumor ablation, and seven percent had segmental/partial liver resections. Twenty percent of the patients did not receive surgical treatment. There were four patients who did not follow-up beyond the study period. The observed survival days for the male: minimum = 9, maximum = 2431, and for the female: minimum = 73, maximum = 2407. The mean survival days for the male was 773.85 ($\pm$ 818.14) and for the female 1175.33 ($\pm$ 950.13).

We consider several statistical probability models to fit the data. Based on the descriptive statistics, several statistical probability plots, Kolmogorov-Smirnov, Anderson-Darling, and Chi-Squared goodness of fit tests, it is determined that the two-parameter Weibull model with parameters α = 0.74843 and β = 745.48 is appropriate fit for modeling the survival data of hepatocellular carcinoma of liver patients. We applied 'EasyFit 5.4 Professional' software from MathWave Technologies (2010) to run the goodness of fit tests and their summary results are given in Table 1.

Table 1. Goodness of fit tests based on Kolmogorov-Smirnov, Anderson-Darling and Chi-Squared methods

K-S Test Statistic = 0.10392			
Choice of probability of type I error	0.05	0.02	0.01
Critical Value	0.16204	0.18119	0.19442
Decision on Weibull Dist.	Accept	Accept	Accept
A-D Test Statistic = 1.1106			
Choice of probability of type I error	0.05	0.02	0.01
Critical Value	2.5018	3.2892	3.9074
Decision on Weibull Dist.	Accept	Accept	Accept
Chi-Squared Test Statistic = 5.2622			
Choice of probability of type I error	0.05	0.02	0.01
Critical Value	12.592	15.033	16.812
Decision on Weibull Dist.	Accept	Accept	Accept

The data set was used to estimate the survival means and standard deviations of future response. Several graphs of the predictive survival densities were given when the shape parameter α is unknown to compare the male and female future survival days.

Table 2. Mean and standard deviation of the predictive density of combined patients survival days with respect to several values of the hyperparameter (τ)

Values of hyperparameter (τ)	Male: Mean(y) ± SD(y)	Female: Mean (y) ± SD(y)
1.0	1083 ± 1071	1405 ± 1278
2.0	1063 ± 1056	1335 ± 1253
3.0	1044 ± 1041	1271 ± 1244
4.0	1025 ± 1024	1203 ± 1210
5.0	1007 ± 1008	1138 ± 1157
6.0	990 ± 992	1077 ± 1103
7.0	973 ± 976	1023 ± 1049
8.0	956 ± 961	973 ± 999
9.0	941 ± 946	927 ± 954
10.0	925 ± 931	886 ± 911

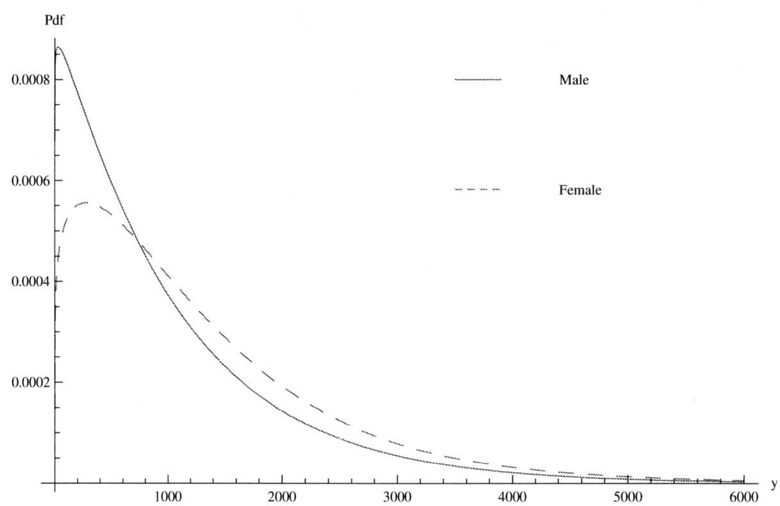

Figure 1. Predictive densities for a single future response given combined patients survival times with specific value of the hyperparameter, namely, $\tau = 1$, both for male and female.

We used the numerical integration command 'NIntegrate' in conjunction with the symbolic computational software Mathematica 7.0, Wolfram Research (2008) to evaluate the predictive survival density functions. This software package was used to perform all the calculations and to plot the graphs. The effect of hyperparamer (τ) on the predictive mean and the standard deviation can be observed from Table 2. The decreasing value of the predictive survival mean and standard deviation are shown with the increasing value of the

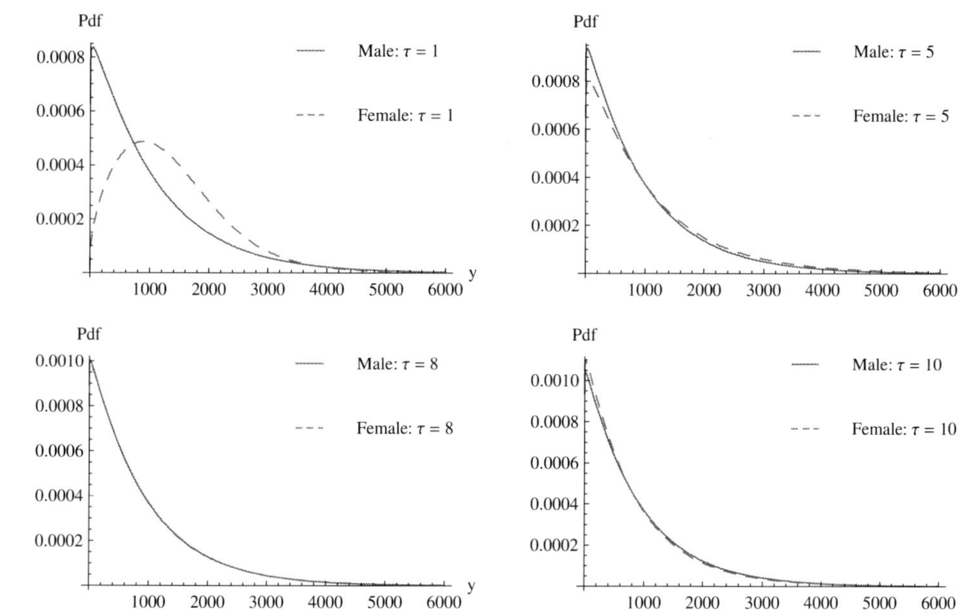

Figure 2. Predictive densities given combined patients survival times with respect to several values of the hyperparameter (τ), namely, $\tau = 1$, $\tau = 5$, $\tau = 8$, and $\tau = 10$.

Table 3. Mean and standard deviation of the predictive density with respect to specific values of the hyperparameter (τ) and stages of HCC

Stages	Hyperparameter (τ)	Male: Mean(y) $\pm$ SD(y)	Female: Mean(y) $\pm$ SD(y)
0	1	132 ± 120	325 ± 284
1	5	693 ± 692	1340 ± 1419
3	10	1086 ± 1126	985 ± 1056

hyperparameter (τ). The hyperparameter value may be considered as treatment or doses for the patients. As the treatment or dose increases the likelihood of HCC patient's future survival days decreases (Table 2). The initial survival days of female patients are higher than that of the male patients. The rate of decrease in survival days of female patients is higher than that of the male patients when comparing the hyperparameter, (τ) = 1 to the hyperparameter, (τ) = 10. The total number of 69 patients who were diagnosed in 2002, were subgrouped into four stages, namely, 0 = no surgery; 1 = local tumor destruction; 2 = segmental resection; and 3 = hepatectomy. To investigate the variability of future survival days for male and female among the stages, we obtained the sensitivity of the mean and standard deviation of the predictive densities for the subgroups 0, 1, and 3 (Table 3). Stage 0 contained 14 patients, composed of 9 male and 5 female patients. It was assumed that these patients needed primary treatment at the beginning i.e., hyperparameter, (τ) = 1. The range of the male patients survival days was between 16 and 333. The predictive

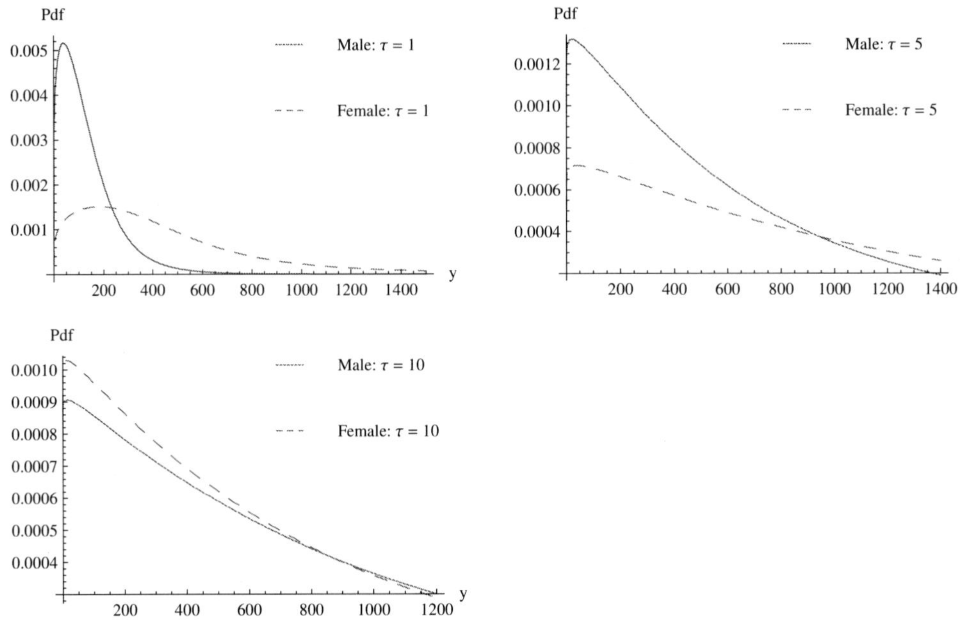

Figure 3. Predictive densities to the subgroups of male and female patients given the patients survival times with respect to certain values of the hyperparameter (τ), namely, $\tau = 1$ (stage 0), $\tau = 5$ (stage 1), and $\tau = 10$ (stage 3).

mean and standard deviation were obtained as 132 and 120, respectively. Similarly, for the five female patients' minimum survival days was 73 and the maximum survival days was 375. The predictive survival mean and the standard deviation of survival days were 325 and 284, respectively. Of the 37 patients in stage 1, 30 were male and 7 were female. The length of survival days for the 30 male patients ranged between 18 and 2352. In stage 1, it was assumed that the hyperparameter, τ = 5, due to median treatment compared to the primary treatment. The predictive mean and standard deviation obtained were 693 and 692, respectively for the male patients. In the case of the seven female patients, the minimum survival days was 166 days while the maximum survival days was 2291. The predictive mean and standard deviation of survival days for the female patients with hyperparameter, τ = 5, were obtained as 1340 and 1419, respectively. The predictive survival analysis for stage 2 was not included due to the unavailability of records for female survival days. Stage 3 on the other hand contained 3 female patients and 11 male patients. It was assumed that the value of the hyperparameter, τ = 10. The minimum and maximum survival days for the patients were male: 12 - 2431 and female: 2150 - 2407, respectively. By making use of the hyperparameter, τ = 10, due to a higher treatment compared the median treatment, the predictive mean and standard deviation of the survival days were obtained, males: 1086 and 1126; and for the females: 985 and 1056, respectively.

The predictive densities of the male and female were plotted in Figure 1. There was a total of n = 69 patients in the combined sample. Out of 69 patients, 54 were male and 15 were female. The graph of figure 1 exhibits that at the stage 0, with the primary treatment, the predictive survival days are higher for the female patients than for the male patients. In Figure 2, the predictive survival days of male and female patients were shown with several values of the hyperparameter. One would conclude that the predictive survival days are higher for the female than that of the male at the beginning of the stage with certain values of the hyperparameter. As the values of the hyperparameter increase, the predictive survival days decrease for both male and female. In Figure 3, we performed subgroup analysis for both male and female patients with respect to stages (0, 1, & 3). There were no records available for the female survival days in stage 2. In this group, only four male survival days were available. We, therefore, were unable to report a graphical comparison for stage 2 for both male and female patients. In the subgroup analysis, the predictive survival days for stage 0 is higher for the females than that of males predictive survival days. For stage 1, females predictive survival days are higher than males predictive survival days. For stage 3, the predictive survival days gradually decrease for both males and females.

5. Conclusion

We have used the Bayesian approach to derive the predictive densities for several future responses from a two-parameter Weibull model given a doubly censored sample assuming an informative prior for the parameters. In the case of known shape parameter (α), the predictive densities for the future responses have closed form representations. However, for unknown shape parameter (α), the predictive density for future responses from a two-parameter Weibull model cannot be expressed in closed form representation and are given in integral form. The graphical representation of the integral form is given by using a numerical example of hepatocellular carcinoma of the liver patients. An advanced computational software package 'Mathematica 7.0' has been used to display the integral form of the predictive density and to perform all calculations including several values of the hyperparameter. The predictive means and standard deviations of survival days are estimated and compared between males versus females. It was observed that the rate of predictive survival days of the males decreased about three times slower than the females survival days. A subgroup of predictive survival days for gender was assessed on the basis of treatments namely, primary, median, and higher treatment. It was found that under the primary and median norms, the females survival days are longer than that of the males survival days. However, the females survival days were gradually decreasing with the higher treatment norm compared to the males at stage 3. Since the observed age for males is lower than the females, it may be concluded that the males (the younger group) had longer survival days at stage 3 as compared to the females (the older group). Hence, age maybe an important factor to consider in relation to the length of survival days at stage 3 of HCC.

Acknowledgments

The authors are grateful to the Editor and the reviewers for their insightful comments and suggestions.

References

American Cancer Society (2009). How is liver cancer treated? Retrieved November 20 from web site http://www.cancer.org.

Aitchison, J. and Dunsmore, I. R. (1975). *Statistical Prediction Analysis.* Cambridge University Press, New York.

Aitchison, J. and Sculthorpe, D. (1965). Some problems of statistical prediction. *Biometrika* 52, 469-483.

Berger, J. O. (1985). *Statistical Decision Theory and Bayesian Analysis*. Springer-Verlag, New York.

Bernardo J. M. and Smith, A. F. M. (1994). *Bayesian Theory.* Wiley, New York.

Chew, V. (1968). Simultaneous prediction intervals. *Technometrics*, 323-330.

Dunsmore, I. R. (1974). The Bayesian predictive distribution in life testing models. *Technometrics*, 16, 455-460.

MathWave Technologies (2010). *EasyFit 5.4 Professional*, MathWave Technologies.

El-Serag, H. B. and Mason, A. C. (1999). Rising incidence of hepatocellular carcinoma in the United States, *New England Journal of Medicine*, 340(10):745-50. Retrieved November 20, 2009 from http://content.nejm.org.

Evans, I. G. and Nigm A. M. (1980a). Bayesian prediction for two-parameter Weibull lifetime models. *Commu. Statist.–Theor. Meth.*, A9(6), 649-658.

Evans, I. G. and Nigm A. H. M. (1980b). Bayesian prediction for the left truncated exponential distribution. *Technometrics*, 22(2), 201-204.

Geisser, S. (1971). *The Inferential Use of Predictive Distributions*. Foundation of Statistical Inference, ed., V. P. Godambe and D. A. Sprott, Holt, Toronto, 456-469.

Geiser, S. (1993). *Predictive Inference: An Introduction*. Chapman and Hall, New York.

Gelman, A., Carlin J. B., Stern H. S. and Rubin D. B. (2004). *Bayesian Data Analysis.* Chapman and Hall/CRC, New York.

Jeffreys, H. (1961). *Theory of Probability*. Third Edition, Clarendon Press, Oxford.

Kambo, N. S. (1978). Maximum likelihood estimators of the location and scale parameters of the scale and location parameters of the exponential distribution from a censored sample, *Comm. Statist. Theory Methods*, A 12, 1129-1132.

Khan, H. M. R. (2003). *Predictive Inference for Certain Life Testing Models.* Ph.D. Thesis. The University of Western Ontario, Canada.

Khan, H. M. R., Haq, M. S., and Provost, S. B. (2006). Predictive inference for future responses given a doubly censored sample from a two parameter exponential distribution. *Journal of Statistical Planning and Inference*, 136(9), 3156-3172.

Khan, H. M. R., Provost, S. B., and Hegele, R. A. (2007). Bayesian applications in cytogenetic studies. *Journal of Applied Probability and Statistics*, 2(2), 157-172.

Khan, H. M. R. and Provost, S. B. (2008). On predictive inference for future responses from the compound exponential model. *Journal of Applied Probability & Statistics*. 3(2), 263-274.

Kim, D. H., Lee, W. D., and Kang, S. G. (2000). Bayesian model selection for life time data under Type II censoring. *Communications in Statistics-Theory and Methods*, 29(12), 2865-2878.

Lalitha, S. and Mishra, A. (1996). Maximum likelihood estimation for Rayleigh distribution. *Comm. Statist. Theory Methods*, 25(2), 389-401.

Mayo Clinic (2010). Liver cancer. Mayo Foundation for Medical Education and Research. Retrieved from http://www.mayoclinic.com.

Raqab, M. Z. (1995). On the maximum likelihood prediction of the exponential distribution based on double type II censored samples. *Pakistan J. Statist.*, 11(1), 1-10.

Sarhan, A. E. (1955). Estimation of the mean and standard deviation by order statistics, Part III, *Ann. Math. Statist.*, 26, 576-592.

Sinha, S. K. (1986). *Reliability and Life Testing*. John Wiley & Sons Inc., New York.

Thabane, L. (1998). *Contribution of Bayesian Statistical Inference.* Ph.D. Thesis, The University of Western Ontario, Canada.

U.S. National Library of Medicine (2009). National Institutes of Health. Medline: Hepatocellular carcinoma. Retrieved November 20 from http://www.nlm.nih.gov.

Wolfram Research (2008). *The Mathematica Archive: Mathematica 7.0*, Wolfram Research Inc, Illinois.

In: Research in Applied Statistical Science
Editor: Mohammad Ahsanullah, pp. 133-142
ISBN: 978-1-63321-818-5

Chapter 12

Parameters Estimation of the Log-logistic Distribution Using Ranked Set Sampling

*I. Elbatal**
Institute of Statistical Studies and Research,
Department of Mathematical Statistics, Cairo University, Egypt

Abstract

The log-logistic distribution is very similar in shape to the lognormal distribution but is mathematically more tractable and hence very useful in a wide variety of applications, especially analysis of survival data, see Bennett (1983). It is also noted that the log-logistic distribution is a good alternative to the Weibull distribution, whose hazard function, while it may be increasing or decreasing, must be monotonic, whatever the values of its parameters. In this paper we derive the parameters estimator using the method of the maximum likelihood based on simple random samples, and ranked set sampling. Finally we use Montocarlo simulations to compare the results for the SRS data and the RSS data.

Keywords: Ranked set sampling, Log-logistic Distribution, Relative efficiency

1. Introduction and Motivation

In applied statistics, experimenters often encounter situations where the actual measurements of the sample observations cannot be directly measured due to cost, time and other factors but can be ranked relatively easily. Also in other subject areas such as environmental, agriculture and biomedical studies, taking measurements of subjects involves substantive expense and time consuming. Simple random sampling (SRS) is a commonly employed approach to taking measurements. Also Researchers aim to design environmental studies that optimize precision and allow for generalization of results, while keeping the costs of associated field and laboratory work at a reasonable level. McIntyre (1952) introduced an alternative to SRS, one that effectively provides more cost efficiency by utilizing information that are less expensive and / or easier to obtain.This technique was first introduced by

*E-mail address: i_elbatal@yahoo.com

McIntyre (1952) for estimating means pasture and forage yields. This method of sampling later became known as Ranked Set Sampling (RSS). Ranked Set Sampling is one method to potentially increase precision and reduce costs by using "rough but cheap" quantitative or qualitative information to obtain a more representative sample before the real, more expensive sampling is done. The method of Ranked Set Sampling is a parametric in nature. However, several statisticians have used it in estimating the parameters of various distributions and have shown their estimators to be more efficient when compared to those based on simple random samples of the same size.

The basic concept behind RSS can be briefly described as follows. Suppose X_1, X_2,..., X_n is a random sample from $F(x)$ with a mean μ and a finite variance σ^2 Then a standard nonparametric estimator of μ is $\overline{X} = \sum X_i/n$ with $Var\left(\overline{X}\right) = \frac{\sigma^2}{n}$. The RSS procedure consists of:
(i) Randomly selecting n elements from the population for which an estimator of the mean is required. The elements of this set are ranked visually or by any negligible cost method that does not require actual measurements. For this first set, the element (judgment) ranked lowest is identified.
(ii) A second set is drawn, judgment ranking is done and the element in the second lowest position is identified.

(iii) This process is continued until the element that (judgment) ranked the highest from the n_{th} chosen set have been identified.
(iv) Finally, the n identified elements, called a (judgment) ranked set, are quantified.

If we denote the values of the variable of interest, X, for this carefully chosen ranked set by $X_1; X_2; ..., X_n$, then the set $\{X_1; X_2; ..., X_n\}$ is called a Ranked set sample. The whole process can be repeated r times, if necessary, to obtain an RSS of size $n = m\,r$ In the case of no error in ranking, i.e. judgment ranks are the same as actual ranks, and the population is infinite, $X_1; X_2; ..., X_n$ are independent and X_i has the same distribution as the i_{th} order statistic, $X_{(i)}$, of a simple random sample (SRS) of size n from a common density $f(x)$.

In contrast to SRS, RSS uses only one observation, namely $X_{1.n} \equiv X_{(11)}$, the lowest observation from the first set of n units, $X_{2.n} \equiv X_{(22)}$, the second lowest from the second set of n units, and finally $X_{n.n} \equiv X_{(nn)}$, the largest observation from a last set of n unit. This process can be described in a table as follows. Display of n^2 observations in n set of n each

$X_{1(1)}$	$X_{1(2)}$	...	$X_{1(n-1)}$	$X_{1(n)}$
$X_{2(1)}$	$X_{2(2)}$	...	$X_{2(n-1)}$	$X_{2(n)}$
...	...	...	...	..
...	...	...	...	..
...	...	...	...	..
$X_{n(1)}$	$X_{n(2)}$	...	$X_{n(n-1)}$	$X_{n(n)}$

Many aspect of RSS have been studied in the literature. We refer to Takahasi and Wakimoto(1968), Stokes (1980), Dell and Clutter (1972) showed that the RSS mean remains unbiased and more efficient than the SRS mean. Even though RSS method is nonparametric in nature, several authors considered the use of RSS data to estimate the parameters of well-known distributions. Lam et al.(1994) considered the estimation of the parameters of two-parameter exponential distribution. Bhoj (1997) obtained linear unbiased estimators of the location and scale parameters of the extreme value distribution using RSS. Walid,et al (2004) have proposed different estimators for the location and scale parameters using ranked set sampling and simple random sampling . Al-Saleh M and Al-Kadiri M.(2000), Jemain et al (2007), Samawi et al (2008), Al-Saleh M and Al-Hadrami.S (2003), Bhoj.D and M. Ahsanullah (1996) and references therein.

In this paper , two estimators are proposed for the scale and shape parameters of the Log-logistic distribution. In Section 2, we derive the parameter estimator using the maximum likelihood method based on a simple random samples. The maximum likelihood method based on ranked set sampling are discussed in Section 3. Finally, the estimators are compared with each other for the cases considered via their mean square error.

2. Estimation of the Scale and Shape Parameters Using SRS

In this section we present the maximum likelihood estimators of the two unknown parameters α and β of the Log-logistic distribution. The logistic distribution has been used for growth models and used in a certain type of regression known as the logistic regression. It has also applications in modeling life data. The shape of the logistic distribution and the normal distribution are very similar. The logarithmic transformations can be applied to the logistic distribution to obtain log-logistic distribution. The log-logistic distribution is very similar in shape to the lognormal distribution but is mathematically more tractable and hence very useful in a wide variety of applications, especially analysis of survival data, see Bennett (1983). It is also noted that the the log-logistic distribution is a good alternative to the Weibull distribution, whose hazard function, while it may be increasing or decreasing , must be monotonic, whatever the values of its parameters.

Let $X_1, X_2, \ldots, X_n$ be a random sample from the Log-logistic probability distribution with scale and shape parameters α and β respectively with cumulative distribution function and probability density function are respectively given by

$$F(x) = \frac{\left(\frac{x_i}{\alpha}\right)^{\beta}}{1+\left(\frac{x_i}{\alpha}\right)^{\beta}} = \frac{x_i^{\beta}}{\alpha^{\beta}+x_i^{\beta}},\ x \geq 0, \alpha > 0, \beta > 1 \tag{2.1}$$

and

$$f(x) = \left(\frac{\beta}{\alpha}\right)\left(\frac{x_i}{\alpha}\right)^{\beta-1}\left(1+\left(\frac{x_i}{\alpha}\right)^{\beta}\right)^{-2},\ x \geq 0, \alpha > 0, \beta > 1 \tag{2.2}$$

The likelihood function is given by

$$\begin{aligned} L(X_{SRS}) &= \prod_{i=1}^{n} f(x_i) \\ &= \prod_{i=1}^{n}\left(\frac{\beta}{\alpha}\right)\left(\frac{x_i}{\alpha}\right)^{\beta-1}\left(1+\left(\frac{x_i}{\alpha}\right)^{\beta}\right)^{-2} \\ &= \left(\frac{\beta}{\alpha}\right)^{n}\prod_{i=1}^{n}\left(\frac{x_i}{\alpha}\right)^{\beta-1}\prod_{i=1}^{n}\left(1+\left(\frac{x_i}{\alpha}\right)^{\beta}\right)^{-2}. \end{aligned} \tag{2.3}$$

The the natural logarithm of the likelihood function becomes

$$\log L\left(X_{SRS}\right) = n\log\beta - n\ \log\alpha + \sum_{i=1}^{n}(\beta-1)\log\left(\frac{x_i}{\alpha}\right) -2\sum_{i=1}^{n}\log\left(1+\left(\frac{x_i}{\alpha}\right)^{\beta}\right). \tag{2.4}$$

Upon differentiating (2.4) with respect to α and β and equating each result to zero, two equations must be simultaneously satisfied to obtain MLE $\widehat{\alpha}$ and $\widehat{\beta}$. These equations are given by

$$\begin{aligned}\frac{\partial\log L\left(X_{SRS}\right)}{\partial\alpha} &= \frac{-n}{\alpha} - \sum_{i=1}^{n}\frac{\beta-1}{\alpha} - 2\sum_{i=1}^{n}\left[\frac{\beta\left(\frac{x_i}{\alpha}\right)^{\beta-1}\left(\frac{-x_i}{\alpha^2}\right)}{1+\left(\frac{x_i}{\alpha}\right)^{\beta}}\right] \\ &= \frac{-n}{\alpha} - \sum_{i=1}^{n}\frac{\beta-1}{\alpha} + \sum_{i=1}^{n}\frac{2\beta}{\alpha}\left(\frac{\left(\frac{x_i}{\alpha}\right)^{\beta}}{1+\left(\frac{x_i}{\alpha}\right)^{\beta}}\right) \\ &= 0. \end{aligned} \tag{2.5}$$

And

$$\begin{aligned}\frac{\partial\log L\left(X_{SRS}\right)}{\partial\beta} &= \frac{n}{\beta} + \sum_{i=1}^{n}\log\left(\frac{x_i}{\alpha}\right) - 2\sum_{i=1}^{n}\left[\frac{\left(\frac{x_i}{\alpha}\right)^{\beta}\log\left(\frac{x}{\alpha}\right)}{1+\left(\frac{x_i}{\alpha}\right)^{\beta}}\right] \\ &= 0. \end{aligned} \tag{2.6}$$

For solving the two equations (2.5) and (2.6) we use Newton Raphson method .

2.1. Variances and Covariances of MLE

The asymptotic variances and covariances of the MLE for the parameters α and β are given by the elements of the inverse of the Fisher information matrix

$$I_{ij} = E\left[\frac{-\partial^2 L}{\partial\alpha\partial\beta}\right];i,\ j=1,2.$$

Unfortunately, the exact mathematical expressions for the above expectations are very difficult to obtain. Therefore, we give the approximate (observed) asymptotic variance - covariance matrix for the MLE, which is obtained by dropping the expectation operator E

$$\left[\begin{array}{cc}\frac{\partial^2\log L}{\partial\alpha^2} & \frac{\partial^2\log L}{\partial\alpha\partial\beta} \\ \frac{-\partial^2\log L}{\partial\alpha\partial\beta} & \frac{-\partial^2\log L}{\partial\beta^2}\end{array}\right]^{-1}_{\widehat{\alpha},\widehat{\beta}} = \left[\begin{array}{cc} Var(\widehat{\alpha}) & Cov(\widehat{\alpha},\widehat{\beta}) \\ Cov(\widehat{\alpha},\widehat{\beta}) & Var(\widehat{\beta})\end{array}\right]$$

with

$$\frac{\partial^2 \log L}{\partial \alpha^2} = -2\sum_{i=1}^{n} \frac{\left(1+\left(\frac{x_i}{\alpha}\right)^{\beta}\right)\left[\beta\left(\frac{x_i}{\alpha}\right)^{\beta-1}\left(\frac{-x_i}{\alpha^2}\right)\right]+\left(\frac{x_i}{\alpha}\right)^{\beta}\left[\beta\left(\frac{x_i}{\alpha}\right)^{\beta-1}\left(\frac{x_i}{\alpha^2}\right)\right]}{\left(1+\left(\frac{x_i}{\alpha}\right)^{\beta}\right)^2}$$

$$= \sum_{i=1}^{n} \frac{\left(\frac{x_i}{\alpha}\right)^{\beta}}{\left(1+\left(\frac{x_i}{\alpha}\right)^{\beta}\right)^2}. \tag{2.7}$$

Also,

$$\frac{\partial^2 \log L}{\partial \beta^2} = \frac{-n}{\beta^2} - 2\sum_{i=1}^{n} \frac{\left(1+\left(\frac{x_i}{\alpha}\right)^{\beta}\left(\frac{x_i}{\alpha}\right)^{\beta}\log^2\left(\frac{x_i}{\alpha}\right)-\left(\frac{x_i}{\alpha}\right)^{2\beta}\log^2\left(\frac{x_i}{\alpha}\right)^{\beta}\right)}{\left(1+\left(\frac{x_i}{\alpha}\right)^{\beta}\right)^2}$$

$$= \frac{-n}{\beta^2} - 2\sum^{\beta}\left(\frac{x_i}{\alpha}\right)^{\beta}\log^2\left(\frac{x_i}{\alpha}\right)\left[\frac{1+\left(\frac{x_i}{\alpha}\right)^{\beta}-\left(\frac{x_i}{\alpha}\right)^{\beta}}{\left(1+\left(\frac{x_i}{\alpha}\right)^{\beta}\right)^2}\right]$$

$$= \frac{n}{\beta^2} + 2\sum^{\beta} \frac{\left(\frac{x_i}{\alpha}\right)^{\beta}\log^2\left(\frac{x_i}{\alpha}\right)}{\left(1+\left(\frac{x_i}{\alpha}\right)^{\beta}\right)^2}, \tag{2.8}$$

and

$$\frac{\partial^2 \log L}{\partial \alpha \partial \beta} = \sum_{i=1}^{n} \frac{\left(\frac{x_i}{\alpha}\right)^{\beta}\log\left(\frac{x_i}{\alpha}\right)^{\beta}}{\left(1+\left(\frac{x_i}{\alpha}\right)^{\beta}\right)^2}. \tag{2.9}$$

We used the sample sizes of $n = 20, 30, 40, 50$ to study the bias and mean square error (MSE) of the estimators. Table (2.1) gives the values of MSE and bias for the MLE estimators for different values of α and β and different sample sizes using SRS data when both α and β are unknown.

We conclude from table(2.1) the bias is small. The MSE is decreasing in n for fixed α and β and increasing in α and β when of them is fixed for fixed n.

3. Maximum Likelihood Estimation of Parameters under RSS

In this section, we find the maximum likelihood method based on ranked set sampling . Let $X_1, X_2, \ldots, X_n$ be a random sample from the Log-logistic probability distribution function with scale and shape parameters α and β respectively with probability density function

$$f(x) = \left(\frac{\beta}{\alpha}\right)\left(\frac{x_i}{\alpha}\right)^{\beta-1}\left(1+\left(\frac{x_i}{\alpha}\right)^{\beta}\right)^{-2}.$$

Table 2.1.

n		$\alpha=1\,,\beta=2$	$\alpha=1\,,\beta=2.5$	$\alpha=3\,,\beta=3$	$\alpha=3\,,\beta=5$
	$\widehat{\alpha}$	1.0559	1.0507	3.0205	3.0062
	bias($\widehat{\alpha}$)	-0.0559	-0.0507	-0.0205	-0.0062
	MSE($\widehat{\alpha}$)	0.1398	0.1149	0.1563	0.0554
20	$\widehat{\beta}$	1.0010	1.6338	3.2023	5.3372
	bias($\widehat{\beta}$)	-0.1403	0.2239	0.3525	0.5874
	MSE($\widehat{\beta}$)	0.0247	0.1586	0.4614	1.2816
	$\widehat{\alpha}$	1.0569	1.0366	3.0136	3.0040
	bias $\widehat{\alpha}$	-0.0569	-0.0366	-0.0136	-0.0040
	MSE $\widehat{\alpha}$	0.0687	0.0654	0.1042	0.0371
30	$\widehat{\beta}$	0.9911	1.5770	3.1234	5.2056
	bias($\widehat{\beta}$)	-0.0674	-0.1691	-0.2832	-0.4720
	MSE($\widehat{\beta}$)	0.0101	0.0701	0.2617	0.7268
	$\widehat{\alpha}$	1.1240	1.0163	3.0098	3.0029
	bias $\widehat{\alpha}$	-0.1240	-0.0163	-0.0098	-0.0029
	MSE $\widehat{\alpha}$	0.0924	0.0389	0.0762	0.0272
40	$\widehat{\beta}$	1.0252	1.5440	3.0905	5.1509
	bias($\widehat{\beta}$)	0.0701	0.0203	-0.3655	-0.6092
	MSE($\widehat{\beta}$)	0.0278	0.0447	0.1868	0.5190
	$\widehat{\alpha}$	1.0287	1.0122	3.0081	3.0024
	bias $\widehat{\alpha}$	-0.0287	-0.0122	-0.0081	-0.0024
	MSE $\widehat{\alpha}$	0.1024	0.0284	0.0617	0.0220
50	$\widehat{\beta}$	1.0161	1.5356	3.0714	5.1190
	bias($\widehat{\beta}$)	0.0221	-0.1172	-0.2345	-0.3908
	MSE($\widehat{\beta}$)	0.0076	0.0357	0.1429	0.3970

The likelihood function is given by

$$\begin{aligned} L\left(X_{RSS}\right) &= A\prod_{i=1}^{n}\left(F\left(x_i\right)\right)^{i-1}\left(1-F\left(x_i\right)\right)^{n-i}f\left(x_i\right) \\ &= \prod_{i=1}^{n}\left[\frac{\left(\frac{x_i}{\alpha}\right)^{\beta}}{1+\left(\frac{x_i}{\alpha}\right)^{\beta}}\right]^{i-1}\left[1-\frac{\left(\frac{x_i}{\alpha}\right)^{\beta}}{1+\left(\frac{x_i}{\alpha}\right)^{\beta}}\right]^{n-i}\left(\frac{\beta}{\alpha}\right)^{n}\left(\frac{x_i}{\alpha}\right)^{\beta-1}\left(1+\left(\frac{x_i}{\alpha}\right)^{\beta}\right)^{-2} \end{aligned} \tag{3.1}$$

where A is constant.

The log-likelihood function becomes

$$\begin{aligned}
\log L\left(X_{RSS}\right) &= \sum_{i=1}^{n}(i-1)\left[\beta\log\left(\frac{x_i}{\alpha}\right)-\log\left(1+\left(\frac{x_i}{\alpha}\right)^{\beta}\right)\right] \\
&\quad -\sum_{i=1}^{n}(n-i)\log\left(1+\left(\frac{x_i}{\alpha}\right)^{\beta}\right)+n\log\beta-n\log\alpha \\
&\quad +\sum_{i=1}^{n}(\beta-1)\log\left(\frac{x_i}{\alpha}\right)-2\sum_{i=1}^{n}\log\left(1+\left(\frac{x_i}{\alpha}\right)^{\beta}\right). \qquad (3.2)
\end{aligned}$$

Upon differentiating (3.2) with respect to α and β and equating each result to zero, two equations must be simultaneously satisfied to obtain MLE $\widehat{\alpha}$ and $\widehat{\beta}$. These equations are given by

$$\begin{aligned}
\frac{\partial\log L\left(X_{RSS}\right)}{\partial\alpha} &= \sum\nolimits_{i=1}^{n}(i-1)\left[\frac{-\beta}{\alpha}-\frac{\beta\left(\frac{x_i}{\alpha}\right)^{\beta-1}\left(\frac{-x_i}{\alpha^2}\right)}{1+\left(\frac{x_i}{\alpha}\right)^{\beta}}\right]-\frac{n}{\alpha} \\
&\quad -\sum\nolimits_{i=1}^{n}(n-i)\left[\frac{\beta\left(\frac{x_i}{\alpha}\right)^{\beta-1}\left(\frac{-x_i}{\alpha^2}\right)}{1+\left(\frac{x_i}{\alpha}\right)^{\beta}}\right]-\sum\nolimits_{i=1}^{n}\left(\frac{\beta-1}{\alpha}\right) \\
&\quad -2\sum\nolimits_{i=1}^{n}\left[\frac{\beta\left(\frac{x_i}{\alpha}\right)^{\beta-1}\left(\frac{-x_i}{\alpha^2}\right)}{1+\left(\frac{x_i}{\alpha}\right)^{\beta}}\right] \\
&= \sum\nolimits_{i=1}^{n}(n-i+2)\left[\frac{\left(\frac{\beta}{\alpha}\right)\left(\frac{x_i}{\alpha}\right)^{\beta}}{1+\left(\frac{x_i}{\alpha}\right)^{\beta}}\right]-\sum\nolimits_{i=1}^{n}(i-1)\left[\frac{\left(\frac{\beta}{\alpha}\right)}{1+\left(\frac{x_i}{\alpha}\right)^{\beta}}\right]-\frac{n\beta}{\alpha} \\
&= 0. \qquad (3.3)
\end{aligned}$$

And

$$\begin{aligned}
\frac{\partial\log L\left(X_{RSS}\right)}{\partial\beta} &= \sum_{i=1}^{n}(i-1)\left[\log\left(\frac{x_i}{\alpha}\right)-\frac{\left(\frac{x_i}{\alpha}\right)^{\beta}\log\left(\frac{x_i}{\alpha}\right)}{1+\left(\frac{x_i}{\alpha}\right)^{\beta}}\right]+\frac{n}{\beta} \\
&\quad -\sum_{i=1}^{n}(n-i)\left[\frac{\left(\frac{x_i}{\alpha}\right)^{\beta}\log\left(\frac{x_i}{\alpha}\right)}{1+\left(\frac{x_i}{\alpha}\right)^{\beta}}\right]+\sum_{i=1}^{n}\log\left(\frac{x_i}{\alpha}\right) \\
&\quad -2\sum_{i=1}^{n}\left[\frac{\left(\frac{x_i}{\alpha}\right)^{\beta}\log\left(\frac{x_i}{\alpha}\right)}{1+\left(\frac{x_i}{\alpha}\right)^{\beta}}\right]
\end{aligned}$$

$$\begin{aligned}
&= \sum_{i=1}^{n}(i-1)\left[\frac{\log\left(\frac{x_i}{\alpha}\right)}{1+\left(\frac{x_i}{\alpha}\right)^{\beta}}\right]+\frac{n}{\beta} \\
&\quad -\sum_{i=1}^{n}(n-i+2)\left[\frac{\left(\frac{x_i}{\alpha}\right)^{\beta}\log\left(\frac{x_i}{\alpha}\right)}{1+\left(\frac{x_i}{\alpha}\right)^{\beta}}\right]+\sum_{i=1}^{n}\log\left(\frac{x_i}{\alpha}\right) \\
&= 0. \qquad (3.4)
\end{aligned}$$

For solving the two equations (3.3) and (3.4) we use Newton Raphson method. We need the following, second derivatives

$$\begin{aligned}
\frac{\partial^2 \log L\left(X_{RSS}\right)}{\partial \alpha^2} &= \sum\nolimits_{i=1}^{n} (n-i+2) \left\{ \frac{\left(\frac{x_i}{\alpha}\right)^{\beta} \left[1+\frac{2}{\beta}\left(\frac{x_i}{\alpha}\right)^{\beta}\right]}{\left[1+\left(\frac{x_i}{\alpha}\right)^{\beta}\right]^2} \right\} \\
&\quad + \sum\nolimits_{i=1}^{n} (i-1) \left\{ \frac{-\frac{\beta}{\alpha^2}}{\left[1+\left(\frac{x_i}{\alpha}\right)^{\beta}\right]} + \frac{\left(\frac{x_i}{\alpha}\right)^{\beta}\left(\frac{\beta}{\alpha}\right)^2}{\left[1+\left(\frac{x_i}{\alpha}\right)^{\beta}\right]^2} \right\} - \frac{n\beta}{\alpha^2} \\
&= 0.
\end{aligned}$$

also

$$\begin{aligned}
\frac{\partial^2 \log L\left(X_{RSS}\right)}{\partial \beta^2} &= -\sum_{i=1}^{n} (i-1) \frac{\left(\frac{x_i}{\alpha}\right)^{\beta}\left[\log\left(\frac{x_i}{\alpha}\right)\right]^2}{\left(1+\left(\frac{x_i}{\alpha}\right)^{\beta}\right)^2} - \frac{n}{\beta^2} \\
&\quad - \sum_{i=1}^{n} (n-i+2) \frac{\left(\frac{x_i}{\alpha}\right)^{\beta}\left[\log\left(\frac{x_i}{\alpha}\right)\right]^2}{\left(1+\left(\frac{x_i}{\alpha}\right)^{\beta}\right)^2} \\
&= 0.
\end{aligned}$$

and

$$\begin{aligned}
\frac{\partial^2 \log L\left(X_{RSS}\right)}{\partial \beta \partial \alpha} &= \sum (n-i+2) \left\{ \frac{\left(\frac{x_i}{\alpha}\right)^{\beta}\left(1+\beta\log\left(\frac{x_i}{\alpha}\right)\right)}{\alpha\left(1+\left(\frac{x_i}{\alpha}\right)^{\beta}\right)} \right. \\
&\quad \left. - \frac{\left(\frac{\beta}{\alpha}\right)\left(\frac{x_i}{\alpha}\right)^{2\beta}\log\left(\frac{x_i}{\alpha}\right)}{\left(1+\left(\frac{x_i}{\alpha}\right)^{\beta}\right)^2} \right\} - \frac{n}{\alpha} \\
&\quad - \sum (i-1) \frac{1}{\alpha\left(1+\left(\frac{x_i}{\alpha}\right)^{\beta}\right)} - \frac{\left(\frac{\beta}{\alpha}\right)\left(\frac{x_i}{\alpha}\right)^{\beta}\log\left(\frac{x_i}{\alpha}\right)}{\left(1+\left(\frac{x_i}{\alpha}\right)^{\beta}\right)^2} \\
&= 0.
\end{aligned}$$

Table (3.1) gives the values of MSE and bias for the MLE estimators for different values of α and β, and set sizes m and for $r = 10$ using RSS data when both α and β are unknown.

Finally comparing the results in Table (2.1) and Table (3.1) for the SRS and RSS we notice that

(i) The RSS estimators dominate the SRS estimators for both parameters α and β.

(ii) The efficiency of RSS estimators with respect to SRS estimator is greater than one and increases when the sample size increases.

Table 3.1.

m		$\alpha = 1\,, \beta = 2$	$\alpha = 1\,, \beta = 2.5$	$\alpha = 3\,, \beta = 3$	$\alpha = 3\,, \beta = 5$
	$\widehat{\alpha}$	1.0073	1.0045	3.0021	3.0001
	bias $\widehat{\alpha}$	-0.0073	-0.0045	-0.0021	-0.0001
	MSE $\widehat{\alpha}$	0.0265	0.0125	0.0278	0.0100
2	$\widehat{\beta}$	1.0941	1.6329	3.2659	5.4431
	bias($\widehat{\beta}$)	-0.2314	-0.0257	-0.0514	-0.0857
	MSE($\widehat{\beta}$)	0.0309	0.0673	0.2692	0.7479
	$\widehat{\alpha}$	1.0006	1.0018	2.9996	2.9990
	bias $\widehat{\alpha}$	-0.0006	-0.0018	0.0004	0.0010
	MSE $\widehat{\alpha}$	0.0188	0.0083	0.0185	0.0067
3	$\widehat{\beta}$	1.0603	1.5914	3.1828	5.3047
	bias($\widehat{\beta}$)	-0.1223	-0.1096	-0.2193	-0.3655
	MSE($\widehat{\beta}$)	0.0173	0.0385	0.1540	0.4278
	$\widehat{\alpha}$	1.0045	1.0014	2.9999	2.9994
	bias $\widehat{\alpha}$	-0.0045	-0.0014	0.0001	-0.0006
	MSE $\widehat{\alpha}$	0.0140	0.0062	0.0139	0.0050
4	$\widehat{\beta}$	1.0468	1.5702	3.1403	5.2339
	bias($\widehat{\beta}$)	-0.1461	-0.2192	-0.4383	-0.7306
	MSE($\widehat{\beta}$)	0.0118	0.0265	0.1059	0.2941
	$\widehat{\alpha}$	1.0026	1.0026	2.9988	2.9989
	bias $\widehat{\alpha}$	-0.0026	-0.0026	0.0012	0.0011
	MSE $\widehat{\alpha}$	0.0114	0.0114	0.0113	0.0041
5	$\widehat{\beta}$	1.0380	1.0380	3.1141	5.1902
	bias($\widehat{\beta}$)	0.0843	0.0843	0.2529	0.4216
	MSE($\widehat{\beta}$)	0.0091	0.0091	0.0815	0.2263.

4. Conclusion

In this paper two estimators are proposed for the scale and shape parameters of the Log-logistic distribution. The maximum likelihood method based on a simple random samples and ranked set sampling are discussed. The estimators are compared with each other for the cases considered via their mean square error. The numerical results show that the estimators based on RSS are more efficiently than that from SRS at the same sample size.

References

Al-Saleh MF, Al-Hadrami S. (2003). Estimation of the mean of the exponential distribution using moving extremes ranked set sampling. *Statistical Papers*, 44, 3, 367-382.

Al-Saleh MF, Al-Kadiri M.(2000). Double ranked set sampling. *Statistics and Probability Letters* 48: 205–212.

Bhoj.D.S (1997) Estimation of parameters of the extreme value distribution using ranked set sampling. *Communication in Statistics-Theory and Methods* 26 (3), 653- 667.

Bhoj.D and M.Ahsanullah (1996). Estimations of the generalized geometric distribution using ranked set sampling. *Biometrics*. 52, 685- 694.

Bennett, S. (1983). Log-logistic regression models for survival data. *Applied Statistics.* 32, (2), 165-171.

Dell.D.R, Clutter.J.L. (1972). Ranked set sampling theory with order statistics background, *Biometrics*. 28, 545 –555.

Jemain, A, Al-Omari, A. and Ibrahim, K.(2007). Multistage Median Ranked Set sampling for estimating the population median. *Journal of Mathematics and Statistics.* 2, 58-64.

Lam.K,Sinha.B.K., Z.Wu.(1994). Estimation of parameters in two-parameter exponential distribution using ranked set sample., *Annals of the Institute of statistical Mathematics* 46, 723–736.

McIntyre.G.A., A.(1952). Method of unbiased selective sampling, using ranked sets,*Australian Journal of Agricultural Research* 3, 385–390.

Samawi, M., Al-Saleh,M. and Al_Saidy,O (2008). On the matched pairs sign test using bivariate ranked set sampling: an application to environmental issues. *African journal of Environmental science and Technology.*, 2, 001-009.

Stokes.S.L.(1980). Estimation of variance using judgment ordered ranked set samples, *Biometrics* 36, 35–42.

Takahasi.K,Wakimoto.K. (1968). On unbiased estimates of the population mean based on the sample stratified by means of ordering, *Annals of the Institute of Statistical Mathematics* 21 249–255.

Walid, A.M, Sameer, A. and Mottlak, H (2004). Logistic parameters estimation using simple random sampling and ranked set sampling data. *Applied Mathematics and Computation.* 150, 543-554.

In: Research in Applied Statistical Science
Editor: Mohammad Ahsanullah, pp. 143-154
ISBN: 978-1-63321-818-5

Chapter 13

Characterization of Some Discrete Families Based on Doubly Truncated Mean and Covariance

M. A. H. Khorasadizadeh[1], Rezaei Roknabadi[2] and G. R. Borzadaran[2]
[1]Department of Statistics, University of Birjand, Iran
[2]Department of Statistics, Ferdowsi University of Mashhad, Mashhad, Iran

Abstract

In the present paper, we generalize the characterizations of discrete lifetime distributions based on conditional doubly truncated mean and covariance which are related to the general failure rates, doubly truncated mean and doubly truncated variance of a lifetime random variable. In addition, we obtained the results for some general families and well-known distributions such as, modified power series, Ord, Katz and Telescopic form of distributions.

Keywords: Doubly truncated random variable, General failure rates, Katz family, Mean residual life, Modified power series family, Ord family, Reversed mean residual life, Reversed variance residual life, Telescopic form of distribution, Variance residual life

1. Introduction

In the reliability literature and analyzing survival data for the components of a system or a device, there have been defined several measures in various condition and situation of the system. Sometimes, we only have information about between two lifetime points, so we should study the reliability measures under the condition of doubly truncated random variables. If the random variable X denotes the lifetime of a unit, then the random variable ${}_xX_y = (X|x \leq X \leq y)$ where $(x, y) \in \{(x, y) : F(x) < F(y)\}$ is called the doubly truncated lifetime.

Suppose that, T be a non-negative discrete random variable with cumulative distribution function (cdf), $F(t)$ and probability mass function (pmf), $p(t)$. Studying the doubly truncation of a lifetime random variable has an important role in reliability. Navarro and Ruiz

(1996) generalized the failure rate function for discrete doubly truncated random variables as follows,

$$h_i(t_1, t_2) = P_r(T = t_i | t_1 \leq T \leq t_2) = \frac{p(t_i)}{F(t_2) - F(t_1 - 1)}; \quad i = 1, 2, \tag{1}$$

where $(t_1, t_2) \in D = \{(t_1, t_2) : F(t_1^-) < F(t_2)\}$. Later, Sankaran and Sunoj (2004) defined and obtained some properties of the mean of the doubly truncated lifetime distributions. Recently, many authors such as Su and Huang (2000), Ahmad (2001), Betensky and Martin (2003), Navarro and Ruiz (2004), Bairamov and Gebizlioglu (2005), Poursaeed and Nematollahi (2008) and Sunoj et al.(2009), studied the properties of the conditional expectations of doubly truncated random variables in various areas like order statistics and k-out-of-n systems. Khorashadizadeh et al. (2012) characterized some of lifetime distributions based on doubly truncated mean residual life and mean past to failure.

Finding properties of discrete failure data came relatively late in comparison to its continuous analogue. Discrete failure data arise in several common situations such as below,

- Reports on field failures are collected weekly, monthly, and the observations are the number of failures, without specification of the failure times.
- A piece of equipment operates in cycles and the experimenter observes the number of cycles successfully completed prior to failure. For example the life length of many devices in industry such as switches and mechanical devices depend on the number of times that the devices are turned on or off.
- An experimenter often discretizes or groups continuous data.

The subject of characterization lifetime distributions based on mean and variance, is recently has gathered the attention of many researchers in the reliability studying.

Characterizing most of the discrete families of distributions based on left and right truncated mean and its relation with hazard rate and reversed hazard rate, has been done by many authors like, Osaki and Li (1988), Ahmed (1991), Nair and Sankaran (1991), Glanzel (1991), Sindhu (2002) and Nair and Sudheesh (2008).

The role and properties of the variance residual life in reliability have been discussed considerably for continuous lifetime random variables by various authors such as Gupta and Kirmani (1998, 2000, 2004), El-Arishi (2005), Al-Zahrani and Stoyanov (2008) and Abu-Youssef (2004, 2007, 2009).

Recently, Khorashadizadeh et al. (2010) defined and studied the properties of variance of $T_k = (T - k | T \geq k)$ where T is a discrete random variable. Sudheesh and Nair (2010) discuss characterization of a class of life distributions by properties of right and left truncated conditional mean and variance.

In this paper, we extend the characterizations of some discrete family of distributions by properties of mean and covariance based on doubly truncated lifetime random variable. Various characterizations and results in literature are special case of our results. Many families of distributions which are consider in these characterizations are, modified power series, Ord, Katz and Telescopic form of distributions. Finally, for some well known distributions that belongs to the above families, we illustrate the results.

2. Main Results

In this section, first we characterize a general family of discrete distribution by properties of conditional mean of doubly truncated and general failure rate functions.

Let $c(.)$ be any real valued function, so that,

$$m_c(t_1,t_2) \quad = \quad E(c(T)|t_1 \leq T \leq t_2) = \frac{\sum_{i=t_1}^{t_2} c(i)p(i)}{F(t_2)-F(t_1-1)};\,(t_1,t_2)\in D, \qquad (2)$$

is the conditional mean of doubly (interval) truncated random variable.

Theorem 1. *Let T be a non-negative discrete random variable with pmf, $p(t)$ and cdf, $F(t)$. Also suppose that for any real valued function $c(.)$, $\mu = E(c(T))$ and $\sigma^2 = Var(c(T))$, then for $(t_1,t_2)\in D$, T follows the family of distributions satisfying,*

$$\frac{p(t+1)}{p(t)} = \frac{\sigma g(t)}{\sigma g(t+1)-\mu+c(t+1)}, \qquad t=0,1,2,..., \qquad (3)$$

with $\sigma g(t) = \sum_{i=0}^{t}\frac{p(i)}{p(t)}[\mu - c(i)]$ if and only if,

$$m_c(t_1,t_2) = \mu + \sigma g(t_1-1)\frac{h_1(t_1-1,t_2)}{1-h_1(t_1-1,t_2)} - \sigma g(t_2)h_2(t_1,t_2). \qquad (4)$$

Proof. Suppose (3) holds, then,

$$c(t)p(t) = \mu p(t) + \sigma p(t-1)g(t-1) - \sigma p(t)g(t). \qquad (5)$$

Summation the both side of (5) from t_1 to t_2 leads to,

$$\sum_{i=t_1}^{t_2}[c(i)-\mu]p(i) = \sigma[p(t_1-1)g(t_1-1) - p(t_2)g(t_2)]. \qquad (6)$$

Dividing the both sides of (6) by $F(t_2) - F(t_1-1)$ and using (1) and (2), we get (4). Conversely, let (4) is satisfied or equivalently (6) is true. So, changing t_1 to t_1+1 and subtracting from (6) and using notation t instead of t_1, leads to (5), which is the required result.□

Remark 1. It can be seen that, in special cases, when $t_1 \to 0$ or $t_2 \to \infty$, Theorem 1. is identical with those of Nair and Sudheesh (2008).

In the next theorem, we characterize the family of the form (3) based on doubly truncated conditional covariance and mean.

Theorem 2. *The distribution function of the non-negative discrete random variable T, belongs to the family of the form (3) if and only if for all $(t_1,t_2)\in D$,*

$$\begin{aligned} Cov(s(T),c(T)|t_1\leq T\leq t_2) \quad &= \quad \sigma E(\Delta s(T).g(T)|t_1\leq T\leq t_2) + [\mu - m_c(t_1,t_2)][m_s(t_1,t_2) - s(t_2+1)] \\ &- \quad \sigma g(t_1-1)\frac{h_1(t_1-1,t_2)}{1-h_1(t_1-1,t_2)}[s(t_2+1)-s(t_1)], \end{aligned} \qquad (7)$$

where $c(.)$ and $s(.)$ are any real valued functions such that $E(s^2(T)) < \infty$, $E(\Delta s(T).g(T)) < \infty$, $\Delta s(T) = s(T+1) - s(T) \neq 0$ and $m_s(t_1,t_2) = E(s(T)|t_1 \leq T \leq t_2)$.

Proof. First, we know that,

$$\begin{aligned}
E(s(T)(c(T) - \mu)|t_1 \leq T \leq t_2) &= \frac{1}{F(t_2)-F(t_1-1)} \sum_{i=t_1}^{t_2} s(i)[c(i)-\mu]p(i) \\
&= \frac{1}{F(t_2)-F(t_1-1)} \sum_{j=0}^{t_2-t_1} s(t_1+j) \left[\sum_{i=t_1+j}^{t_2} (c(i)-\mu)p(i) - \sum_{i=t_1+j+1}^{t_2} (c(i)-\mu)p(i) \right] \\
&= \frac{1}{F(t_2)-F(t_1-1)} \left[\sum_{j=t_1}^{t_2} \Delta s(j) \sum_{i=j+1}^{t_2} (c(i)-\mu)p(i) \right] + s(t_1)(m_c(t_1,t_2)-\mu). \quad (8)
\end{aligned}$$

Now, suppose that T has a distribution of form (3), hence, on using Theorem 1., we have,

$$\sum_{i=j+1}^{t_2} (c(i)-\mu)p(i) = \sigma[p(j)g(j) - p(t_2)g(t_2)].$$

Therefore,

$$\begin{aligned}
E(s(T)(c(T)-\mu)|t_1 \leq T \leq t_2) &= \frac{1}{F(t_2)-F(t_1-1)} \left[\sigma \sum_{j=t_1}^{t_2} \Delta s(j)g(j)p(j) - \sigma g(t_2)p(t_2) \sum_{j=t_1}^{t_2} \Delta s(j) \right] \\
&+ s(t_1)(m_c(t_1,t_2)-\mu) \quad (9) \\
&= \sigma E(\Delta s(T).g(T)|t_1 \leq T \leq t_2) + \sigma s(t_1)g(t_1-1)\frac{h_1(t_1-1,t_2)}{1-h_1(t_1-1,t_2)} \\
&+ s(t_2+1)[m_c(t_1,t_2) - \mu - \sigma g(t_1-1)\frac{h_1(t_1-1,t_2)}{1-h_1(t_1-1,t_2)}],
\end{aligned}$$

and after simplifying we get,

$$\begin{aligned}
E(s(T).c(T)|t_1 \leq T \leq t_2) &= \sigma E(\Delta s(T).g(T)|t_1 \leq T \leq t_2) + s(t_2+1)(m_c(t_1,t_2)-\mu) \\
&- \sigma g(t_1-1)\frac{h_1(t_1-1,t_2)}{1-h_1(t_1-1,t_2)}[s(t_2+1)-s(t_1)] + \mu m_s(t_1,t_2),
\end{aligned}$$

which easily leads to (7).

Conversely, let (7) is true, then, by retracing the above steps we have (8). Using (8) and (9), the following equation results,

$$\sum_{j=t_1}^{t_2} \Delta s(j) \sum_{i=j+1}^{t_2} (c(i)-\mu)p(i) = \sigma \sum_{j=t_1}^{t_2} \Delta s(j)g(j)p(j) - \sigma g(t_2)p(t_2) \sum_{j=t_1}^{t_2} \Delta s(j). \quad (10)$$

Changing t_1 to $t_1 - 1$ and subtracting from (10) leads to (6), which is equivalent to the distribution with the form (3). □

Remark 2. In special case of Theorem 2., when $s(t) = c(t)$, we can characterize the distributions of the form (3) with doubly truncated conditional variance by,

$$\begin{aligned}
Var(c(T)|t_1 \leq T \leq t_2) &= \sigma E(\Delta c(T).g(T)|t_1 \leq T \leq t_2) + [\mu - m_c(t_1,t_2)][m_c(t_1,t_2) - c(t_2+1)] \\
&- \sigma g(t_1-1)\frac{h_1(t_1-1,t_2)}{1-h_1(t_1-1,t_2)}[c(t_2+1)-c(t_1)],
\end{aligned}$$

Also, when $c(T) = T$, we have,

$$\begin{aligned}
Var(T|t_1 \leq T \leq t_2) &= \sigma E(g(T)|t_1 \leq T \leq t_2) + [\mu - m(t_1,t_2)][m(t_1,t_2) - t_2 - 1]. \\
&- \sigma g(t_1-1)\frac{h_1(t_1-1,t_2)}{1-h_1(t_1-1,t_2)}[t_2 - t_1 + 1]. \quad (11)
\end{aligned}$$

where $m(t_1, t_2) = E(T|t_1 \leq T \leq t_2)$ is the doubly truncated mean time to failure function (DMTTF) and $\mu = E(T)$ and $\sigma^2 = Var(T)$.

Remark 3. Theorem 1 and Remark 1 of Sudheesh and Nair (2010) are respectively the special cases of Theorem 2. and Remark 2., when $t_2 \to \infty$.

In the following subsection, we verify our results in some well- known families of distributions.

2.1. Modified Power Series Distribution (MPSD)

Let T be a discrete random variable from modified power series distribution (MPSD) with pmf,

$$p(t) = P(T = t) = \frac{a(t)u(\theta)^t}{A(\theta)}, \quad t = 0, 1, 2, ..., \tag{12}$$

where $u(\theta)$ and $A(\theta)$ are positive, finite and differentiable and $\mu = E(T) = \frac{u(\theta)A'(\theta)}{u'(\theta)A(\theta)}$. So,

$$A(\theta)[F(t_2) - F(t_1 - 1)] = \sum_{i=t_1}^{t_2} a(i)u(\theta)^i,$$

differentiation w.r.t θ, implies that,

$$m(t_1, t_2) = \mu \left[1 + \frac{A(\theta)}{A'(\theta)} . \frac{F'(t_2) - F'(t_1 - 1)}{F(t_2) - F(t_1 - 1)}\right].$$

Hence, using (4), we have,

$$\sigma[g(t_1 - 1)p(t_1 - 1) - g(t_2)p(t_2)] = \frac{u(\theta)}{u'(\theta)} \frac{\partial}{\partial\theta}[F(t_2) - F(t_1 - 1)],$$

which leads to,

$$g(t) = -\frac{u(\theta)}{u'(\theta)} \frac{1}{\sigma p(t)} \frac{\partial}{\partial\theta} F(t). \tag{13}$$

Theorem 3. *The distribution of the non-negative discrete random variable T belongs to the modified power series distribution (MPSD) of the form (12), if and only if,*

$$Var(T|t_1 \leq T \leq t_2) = \frac{u(\theta)}{u'(\theta)} \frac{\partial}{\partial\theta} m(t_1, t_2). \tag{14}$$

Proof. When $c(T) = T$, using (13) we have,

$$\begin{aligned}
\sigma E(\Delta T.g(T)|t_1 \leq T \leq t_2) &= \frac{u(\theta)}{u'(\theta)[F(t_2) - F(t_1 - 1)]} \sum_{i=t_1}^{t_2} [i - (i + 1)] \frac{\partial F(i)}{\partial\theta} \\
&= \frac{u(\theta)}{u'(\theta)[F(t_2) - F(t_1 - 1)]} \frac{\partial}{\partial\theta} [t_1 F(t_1 - 1) - (t_2 + 1)F(t_2) + \sum_{i=t_1}^{t_2} ip(i)] \\
&= \frac{u(\theta)}{u'(\theta)[F(t_2) - F(t_1 - 1)]} \frac{\partial}{\partial\theta} [t_1 F(t_1 - 1) - (t_2 + 1)F(t_2) \\
&+ \ m(t_1, t_2)(F(t_2) - F(t_1 - 1))].
\end{aligned}$$

On the other hand,

$$\begin{aligned}\frac{\partial}{\partial\theta}[t_1F(t_1-1) &- (t_2+1)F(t_2)+m(t_1,t_2)(F(t_2)-F(t_1-1))]\\ &= \frac{\partial m(t_1,t_2)}{\partial\theta}(F(t_2)-F(t_1-1))+(t_1-m(t_1,t_2))\frac{\partial F(t_1-1)}{\partial\theta}\\ &+ (m(t_1,t_2)-t_2-1)\frac{\partial F(t_2)}{\partial\theta}.\end{aligned}$$

Now, using (4) and (13) and after some computations,

$$\begin{aligned}\sigma E(g(T)|t_1\le T\le t_2) &= \frac{u(\theta)}{u'(\theta)}\frac{\partial}{\partial\theta}m(t_1,t_2)+(t_1-m(t_1,t_2))\,(\mu-m(t_1,t_2)-\sigma g(t_2)h_2(t_1,t_2))\\ &+ (m(t_1,t_2)-t_2-1)\left(m(t_1,t_2)-\mu-\sigma g(t_1-1)\frac{h_1(t_1-1,t_2)}{1-h_1(t_1-1,t_2)}\right).\end{aligned}$$

Hence, by comparing with equation (11) and some computation the rest terms are vanishes and (14) is obtained. □

We can consider distributions such as the binomial, Poisson, negative binomial, and logarithmic as the well-known member of this class.

2.2. Ord Family

Let T belongs to the Ord family defined by,

$$\frac{p(t+1)-p(t)}{p(t)} = -\frac{t+d}{b_0+b_1t+b_2t^2};t=0,1,\ldots. \qquad (15)$$

Theorem 4. *T belongs to the family (15), if and only if,*

$$m(t_1,t_2)=\mu+(a_0+a_1(t_1-1)+a_2(t_1-1)^2)\frac{h_1(t_1-1,t_2)}{1-h_1(t_1-1,t_2)}-(a_0+a_1t_2+a_2t_2^2)h_2(t_1,t_2), \quad (16)$$

where, $\mu=E(T)$, $d=\frac{a_1-a_2-\mu}{2a_2+1}$ and $b_i=\frac{a_i}{2a_2+1}$; $i=0,1,2$.

Proof. First suppose (16) is hold, then, by replacing t_1 with t_1-1 in (16) and subtracting it from (16), we get the distribution of the form (15). The converse is followed by reacting the previous steps. □

In the next corollary, characterization of Ord family based on doubly truncated variance is considered.

Corollary 1. *The discrete random variable T belongs to the Ord family of the form (15), if and only if,*

$$\begin{aligned}V(T|t_1\le T\le t_2) &= E(a_0+a_1T+a_2T^2|t_1\le T\le t_2)+[\mu-m(t_1,t_2)][m(t_1,t_2)-t_2-1]\\ &- (a_0+a_1(t_1-1)+a_2(t_1-1)^2)\frac{h_1(t_1-1,t_2)}{1-h_1(t_1-1,t_2)}[t_2-t_1+1].\end{aligned}$$

Proof. According to Theorems 1. and 4., in this family we have,

$$\sigma g(t) = a_0 + a_1 t + a_2 t^2.$$

So, we get the results via (11). □

Note that, Katz family as a spacial case of Ord family is defined by,

$$\frac{p(t+1)}{p(t)} = \frac{\alpha + \beta t}{1+t}, \quad t = 0, 1, 2, ...,$$

where $\alpha > 0$ and $\beta < 1$. So, in this family we have the following corollary.

Corollary 2. *The non-negative discrete random variable T belongs to the Katz family, if and only if,*

$$\begin{aligned} V(T|t_1 \le T \le t_2) &= \frac{\alpha + \beta m(t_1, t_2)}{1-\beta} + [\mu - m(t_1, t_2)][m(t_1, t_2) - t_2 - 1] \\ &- \frac{\alpha + \beta t_1}{1-\beta} \frac{h_1(t_1 - 1, t_2)}{1 - h_1(t_1 - 1, t_2)} [t_2 - t_1 + 1]. \end{aligned}$$

Proof. From Johnson et al. (2005), in Katz family we have, $\mu = E(T) = \frac{\alpha}{1-\beta}$. So from,

$$\frac{\alpha + \beta t}{1+t} = \frac{\sigma g(t)}{\sigma g(t+1) - \mu + t + 1},$$

it can be verified that,

$$\sigma g(t) = \frac{\alpha + \beta t}{1-\beta},$$

which is proved the required result. □

Note that, binomial, Poisson, negative binomial, discrete student's t, hypergeometric and negative hyper geometric distributions are the some of well-known distributions that belongs to the Ord family.

2.3. Telescopic Form of Discrete Distributions

Rezaei Roknabadi et al., (2009) defined the telescopic form of distribution as follow:

A discrete non-negative random variable T has a Telescopic form and is denoted by $T \sim T(q, k_\theta)$, if its probability mass function is of the form,

$$p(t) = q^{k_\theta(t)} - q^{k_\theta(t+1)}, \quad t = 0, 1, 2, ... \tag{17}$$

where $0 < q < 1$ and $k_\theta(t)$ is a strictly increasing function of t with $k_\theta(0) = 0$ and $k_\theta(t) \to \infty$ as $t \to \infty$. Some of discrete distributions in this form are as follow, when,

- $k_\theta(t) = t$, we have geometric distribution,
- $k_\theta(t) = t^2/2$, we have discrete Rayleigh distribution,

- $k_\theta(t) = t^\theta$, we have discrete Weibull distribution,
- $k_\theta(t) = e^{\theta t} - 1$, we have discrete Gomperts distribution,
- $\theta = (\beta, \gamma)$ and $k_\theta(t) = t^{2\beta} e^{-\frac{\gamma}{t^2}}$, we have discrete Brittle-fracture distribution.

In this form of presenting the distributions, it can be shown that $\overline{F}(t) = 1 - F(t) = q^{k_\theta(t+1)}$, $\mu = E(T) = \sum_{i=0}^{\infty} q^{k_\theta(i+1)}$ and by using (3), when $c(T) = T$, after some computation, we have,

$$\begin{aligned} \sigma g(t) &= \frac{\sum_{i=t}^{\infty} q^{k_\theta(i+1)} + q^{k_\theta(t+1)}[t - \sum_{i=0}^{\infty} q^{k_\theta(i+1)}]}{q^{k_\theta(t)} - q^{k_\theta(t+1)}} \\ &= \frac{\sum_{i=t}^{\infty} \overline{F}(i) + \overline{F}(t)(t-\mu)}{p(t)}, \qquad (18) \\ &= \frac{\mu}{h_2(0,t)} + \frac{t}{h_1(t,\infty)} - \frac{\sum_{i=0}^{t-1} \overline{F}(i-1)}{p(t)}. \end{aligned}$$

Theorem 5. *The non-negative discrete random variable T belongs to the Telescopic form of distributions (17), if and only if,*

$$m(t_1, t_2) = \frac{\sum_{i=t_1}^{t_2-1} \overline{F}(i) + t_1 \overline{F}(t_1 - 1) - t_2 \overline{F}(t_2)}{\overline{F}(t_1 - 1) - \overline{F}(t_2)}, \qquad (19)$$

or

$$\begin{aligned} Var(T|t_1 \le T \le t_2) &= \frac{(t_1 - t_2)(t_1 + t_2 - 1)\overline{F}(t_2) + 2\sum_{i=t_1}^{t_2-1} i\overline{F}(i)}{\overline{F}(t_1 - 1) - \overline{F}(t_2)} \\ &- (2t_1 - 1)(m(t_1,t_2) - t_1) - (m(t_1,t_2) - t_1)^2. \qquad (20) \end{aligned}$$

Proof. Via Theorem 1. and equation (11) and by using (18), the results are obtained. □

2.4. Discrete Rayleigh and Discrete Weibull Distributions

For these two important distributions, based on their Telescopic forms, we have, for discrete Rayleigh distribution,

$$\sigma g(t) = \frac{\sum_{i=t}^{\infty} q^{i+i^2/2}}{q^{t^2/2}(q^{-1/2} - q^t)} + \frac{t - \mu}{q^{-(t+1/2)} - 1},$$

and for discrete Weibull distribution,

$$\sigma g(t) = \frac{\sum_{i=t}^{\infty} q^{(i+1)^\theta} - q^{(t+1)^\theta}(t-\mu)}{q^{t^\theta} - q^{(t+1)^\theta}}.$$

Hence we can characterize these distributions by computing $Var(T|t_1 \le T \le t_2)$ via Theorem 1. and equation (11).

3. Examples

In this section, for some specific distributions that belong to the above families and have an interesting form, we find out our results.

3.1. Binomial Distribution

Katz (1965) showed that in Katz family when $\beta < 0$, the outcome is the binomial distribution with $\alpha = nq/p$ and $\beta = -q/p$, that is, $n = -\alpha/\beta$ and $p = \beta/(\beta - 1)$. Also Consul (1990) has shown that the binomial distribution belongs to the modified power series distribution with $u(\theta) = \theta$ and $A(\theta) = (1 + \theta)^n$, where $p = \theta/(\theta + 1)$. Hence, in binomial distribution $\sigma g(t) = (n - t)p$. So, the binomial distribution is characterized by,

$$\begin{aligned} Var(T|t_1 \leq T \leq t_2) &= p\left[m(t_1, t_2)(n - 1) - nt_2 - \frac{h_1(t_1 - 1, t_2)}{1 - h_1(t_1 - 1, t_2)}(t_2 - t_1 + 1)(n - t_1 + 1)\right] \\ &- m(t_1, t_2)(m(t_1, t_2) - t_2 - 1). \end{aligned}$$

3.2. Poisson Distribution

From Johnson et al. (2005), in Katz family, when $\beta = 0$, the outcome is the Poisson distribution with $\alpha = \theta$. Also, by assuming $u(\theta) = \theta$ and $A(\theta) = e^{\theta}$ in MPSD, we have Poisson distribution. It can be seen that $\sigma g(t) = \theta$, so this distribution is characterized by,

$$\begin{aligned} Var(T|t_1 \leq T \leq t_2) &= \theta\left[m(t_1, t_2) - \frac{t_2}{1 - h_1(t_1 - 1, t_2)} + (t_1 - 1)\frac{h_1(t_1 - 1, t_2)}{1 - h_1(t_1 - 1, t_2)}\right] \\ &- m(t_1, t_2)(m(t_1, t_2) - t_2 - 1). \end{aligned}$$

3.3. Negative Binomial Distribution

Johnson et al. (2005) showed that the negative binomial distribution with parameters θ and k and pmf,

$$p(t) = \binom{k + t - 1}{k - 1}\theta^k(1 - \theta)^t; \quad t = 0, 1, 2, ...; \quad 0 \leq \theta \leq 1,$$

belongs to the Ord and MPS families of distributions. In this distribution, $\sigma g(t) = \frac{1-\theta}{\theta}(t + k)$, so, it can be characterized by,

$$\begin{aligned} Var(T|t_1 \leq T \leq t_2) &= \frac{1}{\theta}[(t_1 + k - 1)(t_2 - t_1 + 1)(\theta - 1)\frac{h_1(t_1 - 1, t_2)}{1 - h_1(t_1 - 1, t_2)} \\ &+ (m(t_1, t_2) - t_2)k + m(t_1, t_2)] - (m(t_1, t_2) + k)(m(t_1, t_2) - t_2). \end{aligned}$$

In particular case, when $k = 1$, we have geometric distribution, and,

$$Var(T|t_1 \leq T \leq t_2) = \frac{t_1(t_2 - t_1 + 1)}{1 - (1 - \theta)^{t_2 - t_1 + 1}} + \frac{2m(t_1, t_2) - t_2}{\theta} - (m(t_1, t_2) + 1)(m(t_1, t_2) - t_2).$$

Remark 4. In special case, when $t_2 \to \infty$, we get the characterization results of Nair and Sudheesh (2008) for the binomial, Poisson, negative binomial and geometric distributions ,based on $Var(T|T \geq t_1)$. Furthermore, when, $t_1 = 0$, we have new characterizations based on reversed variance residual life of form $Var(t_2 - T|T \leq t_2) = Var(T|T \leq t_2)$ as follow:

- In binomial distribution,

$$Var(T|T \leq t_2) = p\left[m(t_2)(n-1) - nt_2\right] - m(t_2)(m(t_2) - t_2 - 1).$$

- In Poisson distribution,

$$Var(T|T \leq t_2) = \theta\left[m(t_2) - t_2\right] - m(t_2)(m(t-2) - t_2 - 1).$$

- In negative binomial distribution,

$$Var(T|T \leq t_2) = \frac{1}{\theta}[(m(t_2) - t_2)k + m(t_2)] - (m(t_2) + k)(m(t_2) - t_2).$$

- In geometric distribution,

$$Var(T|T \leq t_2) = \frac{2m(t_2) - t_2}{\theta} - (m(t_2) + 1)(m(t_2) - t_2).$$

Where $m(t_2) = m(0, t_2) = E(T|T \leq t_2)$.

4. Summary

In this paper, we generalized the characterization of some discrete families by using doubly truncated conditional mean and covariance. Furthermore, for some well-known families of distributions we obtained the characterizations based on variance of one sided (left and right) and doubly truncated random variables.

References

Abu-Youssef, S. E. (2004). Nonparametric test for monotone variance residual life class of life distributions with hypothesis testing applications.*Applied Mathematics and Computations,* 158, 817-826.

Abu-Youssef, S. E. (2007). Testing decreasing (increasing) variance residual class of life distributions using kernel method. *Applied Mathematical Sciences,* 1, 1915-1927.

Abu-Youssef, S. E. (2009). A Goodness of fit approach to monotone variance residual life class of life distributions.*Applied Mathematical Sciences,* 3, 15, 715-724.

Al-Zahrani, B. and Stoyanov, J. (2008). On some properties of life distributions with increasing elasticity and log-concavity.*Applied Mathematical Sciences,* 2, 48, 2349-2361.

Ahmed, A. N. (1991). Characterization of beta, binomial and Poisson distribution.*IEEE Trans. Reliab.*, 40, 290-293.

Ahmad, A. A. (2001). Moments of order statistics from doubly truncated continuous distributions and characterizations. *Statistics*, 35, 4, 479-494.

Bairamov, I., Gebizlioglu, O. (2005). On the characterizations of distributions through the properties of conditional expectations of order statistics. In: Balakrishnan, N., Bairamov, I., Gebizlioglu, O. L., eds. *Advances on Models, Characterizations and Applications.* Boca Raton, FL: Chapman and Hall/CRC Press.

Betensky, R. A. and Martin, E. C. (2003). Commentary: Failure-rate functions for doubly truncated random variables. *IEEE Trans. Reliab.*, 52(1), 7-8.

Consul, P. C. (1990). New class of location-parameter discrete probability distributions and their characterizations. *Communications in StatisticsTheory and Methods*, 19, 4653-4666.

El-Arishi, S. (2005). A conditional variance characterization of some discrete probability distributions.*Statistical Papers*, 46, 31-45.

Glanzel, W. (1991). Characterization through some conditional moments of Pearson-type distributions and discrete analogues.*Sankhya-B*, 53, 17-24.

Gupta, R. C. and Kirmani, S. N. U. A. (1998). Residual life function in reliability studies. In: *Fronteriors of Reliability*, A. P. Basu, S. K. Basu and S. Mukhopadhyay (eds.), World Scientific, New Jersey.

Gupta, R. C. and Kirmani, S. N. U. A. (2000). Residual coefficient of variation and some characterization results. *Journal of Statistical Planning and Inference*, 91, 23-31.

Gupta, R. C. and Kirmani, S. N. U. A. (2004). Moments of residual life and some characterization results. *Journal of Applied Statistical Science*, 13, 2, 155-167.

Johnson, N. L., Kemp, A. W. and Kotz, S. (2005). *Univariate Discrete Distributions, (Third Edition).* John Wiley and Sons, Inc.

Katz, L. (1965). *Unified treatment of a broad class of discrete probability distributions, Classical and Contagious Discrete Distributions.* G. P. Patil (editor), 175182. Calcutta: Statistical Publishing Society; Oxford: Pergamon.

Khorashadizadeh, M., Rezaei Roknabadi, A. H. and Mohtashami Borzadaran, G. R. (2010). Variance residual life function in discrete random ageing.*METRON-International Journal of Statistics.* LXVIII, 1, 67-75.

Khorashadizadeh, M., Rezaei Roknabadi, A.H. and Mohtashami Borzadaran, G.R. (2012). Characterization of lifetime distributions based on doubly truncated mean residual life and mean past to failure. *Communications in Statistics-Theory and Methods*, 41, 6, 1105-1115.

Nair, N. U. and Sankaran, P. G. (1991). Characterization of Pearson family of distribution. *IEEE Trans. Reliab.*, 40(1), 75-77.

Nair, N. U. and Sudheesh, K. K. (2008). Some results on lower variance bounds useful in reliability modeling and estimation. *Ann. Inst. Stat. Math.*, 60, 591-603.

Navarro, J. and Ruiz, J. M. (1996). Failure-rate function for doubly-truncated random variables. *IEEE Transactions on Reliability,* 45(4), 685-690.

Navarro, J. and Ruiz, J. M. (2004). Characterization from relationships between failure rate functions and conditional moments.*Commun. Statist. Theor. Meth.*, 33(12), 3159-3171.

Osaki, S. and Li, X. (1988). Characterization of gamma and negative binomial distributions. *IEEE Trans. Reliab.*, 37(4), 379-382.

Poursaeed, M. H. and Nematollahi, A. R. (2008). On the mean past and the mean residual life under double monitoring. *Communications in Statistics - Theory and Methods,* 37, 7, 1119-1133.

Rezaei Roknabadi, A. H., Mohtashami Borzadaran, G. R. and Khorashadizadeh, M. (2009). Some aspects of discrete hazard rate function in Telescopic families.*Economic Quality Control,* Vol. 24, No. 1, 35-42.

Sankaran, P. G. and Sunoj, S. M. (2004). Identification of models using failure rate and mean residual life of doubly truncated random variables. *Statistical Papers,* 45, 97-109.

Su, J. C. and Huang, W. J. (2000). Characterizations based on conditional expectations. *Statistical Papers,* 41, 423-435.

Sindhu, T. K. (2002). *An extended Pearson system useful in reliability analysis.* Unpublished Ph. D thesis, Cochin University.

Sudheesh, K. K. and Nair, N.U. (2010). Characterization of discrete distributions by conditional variance. *METRON - International Journal of Statistics*, LXVIII, 1, 77-85.

Sunoj, S. M., Sankaran, P. G. and Maya, S. S. (2009). Characterizations of life distributions using conditional expectations of doubly (interval) truncated random variables. *Communications in Statistics-Theory and Methods,* 38, 9, 1441-1452.

In: Research in Applied Statistical Science
Editor: Mohammad Ahsanullah, pp. 155-171
ISBN: 978-1-63321-818-5

Chapter 14

Different Methods of Estimation for Marshall Olkin Exponential Distribution

Omar M. Bdair*
Department of Applied Sciences, Faculty of Engineering Technology,
Balqa Applied University, Amman, Jordan

Abstract

Marshall Olkin exponential distribution (MOE) has been recently studied quite extensively by several authors. In this paper, we examine different methods of estimation for different parameters of the Marshall Olkin exponential distribution function. An extensive numerical simulation is used to compare the performances of these procedures.

Keywords: Maximum likelihood estimators, Method of moment estimators, Least squares estimators, Weighted least squares estimators, Percentiles estimators, L-estimators, Asymptotic distribution, Simulations

AMS Subject Classification: 60E05, 62E10

1. Introduction

Marshall and Olkin (1997) introduced an interesting method for adding a new parameter to an existing distribution. The resulting new distribution, known as the Marshall-Olkin extended distribution. Suppose we have a given distribution with survival function $\overline{F}(x)$, $-\infty < x < \infty$. The Marshall-Olkin extended distribution is defined in the form given by

$$\overline{G}(x) = \frac{\alpha\overline{F}(x)}{1-(1-\alpha)\overline{F}(x)}, \quad -\infty < x < \infty, \ \alpha > 0. \tag{1}$$

*E-mail address: bdairmb@yahoo.com

If we take the survival function of the exponential distribution, i.e. $\overline{F}(x) = e^{-\lambda x}$, we obtain the survival function

$$\overline{G}(x;\alpha,\lambda) = \frac{1}{e^{\lambda x} - (1-\alpha)}, \; x > 0, \; \alpha > 0, \; \lambda > 0. \tag{2}$$

The distribution with survival function (2) is called Marshall-Olkin exponential (MOE) distribution with parameters α and λ. The probability density function (p.d.f.) of the MOE distribution with survival function (2) is given by

$$f(x;\alpha,\lambda) = \frac{\alpha\lambda e^{\lambda x}}{(e^{\lambda x} - (1-\alpha))^2}, \; x > 0, \; \alpha > 0, \; \lambda > 0. \tag{3}$$

Here α and λ are the shape and scale parameters, respectively. The two-parameter MOE distribution will be denoted by $MO(\alpha,\lambda)$. For $\alpha = 1$ the $MO(\alpha,\lambda)$ is the exponential distribution. Marshall and Olkin (1997) noted that the hazard rate $r(x;\alpha,\lambda) = \lambda e^{\lambda x}/(e^{\lambda x} - (1-\alpha))$, $\alpha, \; \lambda > 0$ for $MO(\alpha,\lambda)$ is decreasing in x for $0 < \alpha \leq 1$ and increasing in x for $\alpha \geq 1$. The pdf $f(x;\alpha,\lambda)$ is log-convex for $0 < \alpha \leq 1$ and log-concave for $\alpha \geq 1$. This means that for $\alpha \leq 1$, $f(x;\alpha,\lambda)$ is decreasing and for $\alpha \geq 1$, $f(x;\alpha,\lambda)$ is unimodal. The mean and variance of $MO(\alpha,\lambda)$ according to Marshall and Olkin (1997) can be written as

$$E(X) = -\frac{\alpha \log \alpha}{\lambda(1-\alpha)}$$

and

$$Var(X) = \frac{1}{\lambda^2}\left(-2 \; \alpha \; A(\alpha) - \left(\frac{\alpha \ln \alpha}{1-\alpha}\right)^2\right),$$

where

$$A(\alpha) = \int_0^1 \frac{\ln u}{1-(1-\alpha)u} du.$$

The main aim of this paper is to introduce different estimators and study how the estimators of the unknown parameter(s) behave for different sample sizes and for different parameter(s) values. We mainly compare, the maximum likelihood estimators, the modified moment estimators, estimators based on percentiles, least-squares estimators, weighted least-squares estimators and the L-moment estimators by using extensive simulation techniques.

The rest of the paper is organized as follows. In section 2, we describe the MLE of parameter(s). In sections 3-6, we describe other methods of estimation. Numerical results and discussions are provided in section 7.

2. Maximum Likelihood Estimators

In this section, we consider the maximum likelihood estimators (MLEs) of $MO(\alpha,\lambda)$. Let $x_1, x_2, ..., x_n$ be a random sample of size n from $MO(\alpha,\lambda)$, then the log-likelohood function $L(\alpha,\lambda)$ can be written as

$$L(\alpha,\lambda) = n \ln \alpha + n \ln \lambda + \lambda \sum_{i=1}^{n} x_i - 2 \sum_{i=1}^{n} \ln(\alpha - 1 + e^{\lambda x_i}). \tag{4}$$

By differentiating (4) with respect to α and λ, respectively. We obtain the following normal equations

$$\frac{\partial L}{\partial \alpha} = \frac{n}{\alpha} - 2\sum_{i=1}^{n} \frac{1}{\alpha - 1 + e^{\lambda x_i}} = 0, \tag{5}$$

and

$$\frac{\partial L}{\partial \lambda} = \frac{n}{\lambda} + \sum_{i=1}^{n} x_i - 2\sum_{i=1}^{n} \frac{x_i e^{\lambda x_i}}{\alpha - 1 + e^{\lambda x_i}} = 0. \tag{6}$$

The two normal equations (5) and (6) are nonlinear and we can not solve them explicitly. Since the parameters of the two normal equations are impossible to be solved in an explicit form, we use numerical method to find the nonlinear solutions ($\widehat{\alpha}_{ML}$ and $\widehat{\lambda}_{ML}$). Further, the asymptotic normality results of MLE's of $\theta = (\alpha, \lambda)$ is used since it is impossible to obtain the variances of $\widehat{\alpha}_{ML}$ and $\widehat{\lambda}_{ML}$, which can be stated as follows:

$$\sqrt{n}\left(\widehat{\theta}_{ML} - \theta\right) \rightarrow N_2\left(0, I^{-1}(\theta)\right),$$

where $I(\theta)$ is the Fisher information matrix, i.e.

$$\begin{aligned} I(\theta) &= -\frac{1}{n}\begin{bmatrix} E\left(\frac{\partial^2 L}{\partial \alpha^2}\right) & E\left(\frac{\partial^2 L}{\partial \alpha \partial \lambda}\right) \\ E\left(\frac{\partial^2 L}{\partial \lambda \partial \alpha}\right) & E\left(\frac{\partial^2 L}{\partial \lambda^2}\right) \end{bmatrix} \\ &= -\begin{bmatrix} I_{11} & I_{12} \\ I_{21} & I_{22} \end{bmatrix}, \end{aligned}$$

such that

$$\begin{aligned} I_{11} &= \frac{5}{3\alpha^2}, \\ I_{12} &= \frac{1}{3}\left(\frac{1}{(1-\alpha)\alpha\lambda} + \frac{\alpha \ln \alpha}{(\alpha-1)^2 \lambda}\right), \\ I_{22} &= -\frac{1}{\lambda^2} - \frac{1}{3}(5 - 2\alpha) - \frac{1}{\alpha^2}(2 - \alpha). \end{aligned}$$

The asymptotic variances of $\widehat{\alpha}_{ML}$ and $\widehat{\lambda}_{ML}$ are

$$Var(\widehat{\alpha}_{ML}) \approx \frac{I_{22}}{n(I_{11}I_{22} - I_{12}^2)},$$

and

$$Var(\widehat{\lambda}_{ML}) \approx \frac{I_{11}}{n(I_{11}I_{22} - I_{12}^2)}.$$

Now, without loss of generality, we assume that $\lambda = 1$. The MLE of α, when the scale parameter λ is known, say $\widehat{\alpha}_{ML}^{sc}$, is obtained by solving the non-linear equation

$$\frac{n}{\widehat{\alpha}_{ML}^{sc}} = 2\sum_{i=1}^{n} \frac{1}{\widehat{\alpha}_{ML}^{sc} - 1 + e^{x_i}}. \tag{7}$$

Let us consider the MLE of λ, say $\widehat{\lambda}_{ML}^{sh}$ when the shape parameter α is known. For known α, $\widehat{\lambda}_{ML}^{sh}$ can be obtained by maximizing

$$w(\lambda) = n \ln \lambda + \lambda \sum_{i=1}^{n} x_i - 2 \sum_{i=1}^{n} \ln(\alpha - 1 + e^{\lambda x_i}), \tag{8}$$

with respect to λ. Therefore, $\widehat{\lambda}_{ML}^{sh}$ can be obtained by a numerical solution using the Newton Raphson method of the non-linear equation

$$\frac{n}{\lambda} + \sum_{i=1}^{n} x_i = 2 \sum_{i=1}^{n} \frac{x_i e^{\lambda x_i}}{\alpha - 1 + e^{\lambda x_i}}.$$

3. Method of Moments Estimators

First of all, we consider the case when both parameters are unknown. If X follows $MO(\alpha, \lambda)$, then the mean and variance of X can be written as

$$\mu = \mathbb{E}(X) = \frac{\alpha \ln \alpha}{\lambda(\alpha - 1)}, \tag{9}$$

and

$$\sigma^2 = \frac{1}{\lambda^2} \left(-2\ \alpha\ A(\alpha) - \left(\frac{\alpha \ln \alpha}{1 - \alpha} \right)^2 \right), \tag{10}$$

where

$$A(\alpha) = \int_0^1 \frac{\ln u}{1 - (1 - \alpha) u} du.$$

Therefore, the coefficient of variation (C.V.) is given by

$$\frac{\sigma}{\mu} = C.V. = \frac{(\alpha - 1) \sqrt{-2\ \alpha\ A(\alpha) - \left(\frac{\alpha \ln \alpha}{1 - \alpha} \right)^2}}{\alpha \ln \alpha}.$$

The method of moments estimator (MME) of α ($\widehat{\alpha}_{MM}$) can be obtained by equating the sample C.V. by the population C.V.. So, the MME of α is a solution of the non-linear equaion

$$\frac{S}{\overline{X}} = \frac{(\alpha - 1) \sqrt{-2\ \alpha\ A(\alpha) - \left(\frac{\alpha \ln \alpha}{1 - \alpha} \right)^2}}{\alpha \ln \alpha}, \tag{11}$$

where $S^2 = \frac{\sum_{i=1}^{n} (X_i - \overline{X})^2}{n-1}$ is the sample variance. First, we estimate α ($\widehat{\alpha}_{MM}$) by solving (11) numerically. Then MME of λ is readily obtained by

$$\widehat{\lambda}_{MM} = \frac{\widehat{\alpha}_{MM} \ln \widehat{\alpha}_{MM}}{\overline{X}(\widehat{\alpha}_{MM} - 1)}. \tag{12}$$

The asymptotic distribution of the MME's of α and λ can be obtained as follows. Let's define

$$f_1(\alpha,\lambda) = \overline{X} - \frac{\alpha\ln\alpha}{\lambda(\alpha-1)},$$

and

$$f_2(\alpha,\lambda) = S^2 - \frac{1}{\lambda^2}\left(-2\ \alpha\ A(\alpha) - \left(\frac{\alpha\ln\alpha}{1-\alpha}\right)^2\right).$$

By considering $f(\alpha,\lambda) = (f_1(\alpha,\lambda), f_2(\alpha,\lambda))$ and expanding $f(\widehat{\alpha}_{MM},\widehat{\lambda}_{MM})$ using Taylor series expansion about the true value of (α,λ), we obtain

$$f(\widehat{\alpha}_{MM},\widehat{\lambda}_{MM}) - f(\alpha,\lambda) = [\widehat{\alpha}_{MM}-\alpha, \widehat{\lambda}_{MM}-\lambda]\left[\begin{pmatrix}\frac{\partial f_1}{\partial\alpha}\\ \frac{\partial f_1}{\partial\lambda}\end{pmatrix}\ \begin{pmatrix}\frac{\partial f_2}{\partial\alpha}\\ \frac{\partial f_2}{\partial\lambda}\end{pmatrix}\right]_{(\alpha,\lambda)=(\overline{\alpha},\overline{\lambda})},$$

where $(\overline{\alpha},\overline{\lambda})$ is a point between $(\widehat{\alpha}_{MM},\widehat{\lambda}_{MM})$ and (α,λ). Note that $f(\widehat{\alpha}_{MM},\widehat{\lambda}_{MM}) = 0$ and as $n\to\infty$, $(\widehat{\alpha}_{MM},\widehat{\lambda}_{MM}) \to (\alpha,\lambda)$, so $(\overline{\alpha},\overline{\lambda}) \to (\alpha,\lambda)$. Hence, as $n\to\infty$, the distribution of $[\sqrt{n}(\widehat{\alpha}_{MM}-\alpha), \sqrt{n}(\widehat{\lambda}_{MM}-\lambda)]$ is the same as

$$-[\sqrt{n}(\overline{X}-E(\overline{X})), \sqrt{n}(S^2-E(S^2))]\left[\begin{pmatrix}\frac{\partial f_1}{\partial\alpha}\\ \frac{\partial f_1}{\partial\lambda}\end{pmatrix}\ \begin{pmatrix}\frac{\partial f_2}{\partial\alpha}\\ \frac{\partial f_2}{\partial\lambda}\end{pmatrix}\right]^{-1}.$$

Now using the central limit theorem we obtain

$$\sqrt{n}(\overline{X}-E(\overline{X})) \to N\left(0, \frac{\left(-2\ \alpha\ A(\alpha) - \left(\frac{\alpha\ln\alpha}{1-\alpha}\right)^2\right)}{\lambda^2}\right),$$

and

$$\sqrt{n}(S^2-E(S^2)) \to N\left(0, \frac{\alpha(n-1)^2B_4(\alpha) - \alpha^2(n-1)(n-3)B_2^2(\alpha)}{n^2\lambda^4}\right),$$

where

$$B_r(\alpha) = \int_0^1 \frac{(\ln u)^r}{(1-(1-\alpha u))^2}du.$$

Therefore

$$[\sqrt{n}(\overline{X}-E(\overline{X})), \sqrt{n}(S^2-E(S^2))] \to N_2\left(\mathbf{0}, \begin{bmatrix}\sigma^2 & \mu_3\\ \mu_3 & \mu_4-\sigma^4\end{bmatrix}\right),$$

where

$$\mu_3 = \frac{-\alpha}{\lambda^3(\alpha-1)^3}((\alpha-1)^3B_3(\alpha) + 3\alpha(\alpha-1)^2\ln\alpha B_2(\alpha) - 2\alpha^2(\ln\alpha)^3),$$

and

$$\mu_4 = \frac{\alpha}{\lambda^4}\left(B_4(\alpha) + 4\frac{\alpha\ln\alpha}{(\alpha-1)} + 6\frac{\alpha^2(\ln\alpha)^2}{(\alpha-1)^2} - 3\frac{\alpha^3(\ln\alpha)^4}{(\alpha-1)^4}\right).$$

If the scale parameter is known, then the MME of α ($\widehat{\alpha}^{sc}_{MM}$) can be obtained by solving the non-linear equation

$$\lambda\overline{X} = \frac{\alpha \ln \alpha}{(\alpha - 1)}. \tag{13}$$

Eq. (13) can be solved numerically. If the shape parameter is known, the MME of λ is

$$\widehat{\lambda}^{sh}_{MM} = \frac{\alpha \ln \alpha}{\overline{X}(\alpha - 1)}. \tag{14}$$

Eq. (14) follows directly from (9). Note that $\widehat{\lambda}^{sh}_{MM}$ is not unbiased for λ but $1/\widehat{\lambda}^{sh}_{MM}$ is an unbiased estimator for $1/\lambda$ and then

$$Var\left(\frac{\lambda}{\widehat{\lambda}^{sh}_{MM}}\right) = \frac{(\alpha - 1)^2}{n\alpha^2(\ln \alpha)^2}\left(-2\ \alpha\ A(\alpha) - \left(\frac{\alpha \ln \alpha}{1 - \alpha}\right)^2\right). \tag{15}$$

4. Estimators Based on Percentiles

If the data come from a distribution function which has a closed form, then it is natural to estimate the unknown parameters by fitting straight line to the theoretical percentile points obtained from the distribution function and the sample percentile points. This method was originally proposed by Kao (1958, 1959) and it has been used quite successfully for Weibull distribution and for the generalized exponential distribution. Here, we apply the same technique for the MOE distribution. First let us consider the case when both parameters are unknown. Since

$$F(x; \alpha, \lambda) = 1 - \frac{\alpha e^{-\lambda x}}{1 - (1 - \alpha)e^{-\lambda x}},$$

therefore

$$\frac{1}{\lambda} \ln\left(\frac{\alpha}{1 - F(x; \alpha, \lambda)} + 1 - \alpha\right) = x.$$

Let us denote $X_{(i)}$ $(1 \le i \le n)$ as the i-th order statistics from a sample of size n. If p_i denotes some estimate of $F(x_{(i)}, \alpha, \lambda)$, then the estimate of α and λ can be obtained by minimizing

$$\sum_{i=1}^{n}\left(x_{(i)} - \frac{1}{\lambda}\ln\left(\frac{\alpha}{1 - p_i} + 1 - \alpha\right)\right)^2, \tag{16}$$

with respect to α and λ. It is clear that (16) is non-linear function and it has to be minimized using non-linear optimization technique. We call the corresponding estimators as the percentile estimators (PCE's). Several estimators of p_i can be used. In this paper, we mainly consider $p_i = \frac{i}{n+1}$, which is in fact the expected value of $F(X_{(i)})$. When α is known, then the PCE of λ, say $\widehat{\lambda}_{PCE}$ is as follows

$$\widehat{\lambda}_{PCE} = \frac{\sum_{i=1}^{n}\left(\ln\left(\frac{\alpha}{1-p_i} + 1 - \alpha\right)\right)^2}{\sum_{i=1}^{n} x_{(i)} \ln\left(\frac{\alpha}{1-p_i} + 1 - \alpha\right)}. \tag{17}$$

To obtain the asymptotic distribution of the PCE of λ, let us consider

$$z_n = \frac{1}{n}\sum_{i=1}^{n} J(p_i)X_{(i)},$$

where $J(u) = n\ln\left(\frac{\alpha}{1-u} + 1 - \alpha\right)$. From Stigler (1974), it follows that for $n \to \infty$, $\sqrt{n}(z_n - \mu(F,J)) \to N(0, \sigma^2(F,J))$, where

$$\mu(F,J) = \int_{-\infty}^{\infty} xJ(F(x))f(x)dx,$$

and

$$\sigma^2(F,J) = 2\int_{-\infty}^{\infty}\int_{-\infty}^{y} J(F(x))J(F(y))F(x)(1-F(y))dxdy.$$

The PCE of λ can be written as

$$\widehat{\lambda}_{PC}^{sh} = \frac{\frac{1}{n^2}\sum_{i=1}^{n} J^2(p_i)}{\frac{1}{n}\sum_{i=1}^{n} J(p_i)X_{(i)}}.$$

Taking the transformation $g(u) = c/u$, where $c = \sum_{i=1}^{n} J^2(p_i)/n^2$, and apply the Slutsky's Theorem, we immediately obtain

$$\sqrt{n}\left(\widehat{\lambda}_{PC}^{sh} - \frac{c}{\mu(F,J)}\right) \to N\left(0, \frac{c^2\sigma^2(F,J)}{\mu^4(F,J)}\right).$$

Therefore, the asymptotic variance of $\widehat{\lambda}_{PC}^{sh}$ is

$$Var(\widehat{\lambda}_{PC}^{sh}) \approx \frac{\left[\sum_{i=1}^{n} \ln^2\left(\frac{\alpha}{1-p_i} + 1 - \alpha\right)\right]^2 \sigma^2(F,J)}{n\mu^4(F,J)},$$

where $\mu(F,J)$ and $\sigma^2(F,J)$ can be computed numerically based on (2) and (3). Now, let us consider the case when the scale parameter λ is known. Without loss of generality, we can assume $\lambda = 1$. If we denote $F(x;\alpha) = F(x;\alpha,1)$ then

$$\frac{\alpha}{1-F(x)} = e^x - 1 + \alpha,$$

or equivalently

$$\alpha = \frac{(e^x - 1)(1-F(x))}{F(x)}.$$

Therefore, the PCE of α say $\widehat{\alpha}_{PCE}$, can be obtained by minimizing

$$\sum_{i=1}^{n}\left(\alpha - \frac{(e^{x_{(i)}} - 1)(1-p_i)}{p_i}\right)^2$$

with respect to α. In this case the PCE of α is found to be

$$\widehat{\alpha}_{PCE}^{sc} = \frac{\sum_{i=1}^{n}\frac{(e^{x_{(i)}}-1)(1-p_i)}{p_i}}{n}. \tag{18}$$

One can also note that

$$\widehat{\alpha}^{sc}_{PCE} = \frac{\sum_{i=1}^{n} J^*(p_i)(W_{(i)})}{n},$$

where $J^*(u) = (1/u) - 1$. Therefore the asymptotic variance for $\widehat{\alpha}^{sc}_{PC}$ is

$$Var(\widehat{\alpha}^{sc}_{PC}) \approx \frac{\sigma^2(F, J^*)}{n^2}.$$

5. Least Squares and Weighted Least Squares Estimators

In this section we provide the regression based method estimators of the unknown parameters, which was originally suggested by Swain,Venkatraman and Wilson (1988) to estimate the parameters of beta distributions.The method can be described as follows: Suppose $Y_1, Y_2, ..., Y_n$ is a random sample of size n from a distribution function $G(.)$ and $Y_{(1)}, Y_{(2)}, ..., Y_{(n)}$ denote the order statistics of the observed sample. It is well-known that $G(Y_{(i)})$ behaves like the ith order statistics of a sample of size n from $U(0, 1)$, therefore we have

$$E(G(Y_{(j)})) = \frac{j}{n+1}, \qquad Var(G(Y_{(j)})) = \frac{j(n-j+1)}{(n+1)^2(n+2)}, \tag{19}$$

and

$$Cov(G(Y_{(i)}), G(Y_{(j)})) = \frac{i(n-j+1)}{(n+1)^2(n+2)}, \quad for \quad i < j, \tag{20}$$

see Johnson, Kotz and Balakrishnan (1995). The proposed method use (19) and obtain estimates of the unknown parameters. The least squares estimators of the unknown parameters can be obtained by minimizing

$$\sum_{i=1}^{n} \left(G(Y_{(i)}) - \frac{j}{n+1} \right)^2,$$

with respect to the unknown parameters. Therefore the least squares estimators of α and λ say $\widehat{\alpha}_{LSE}$ and $\widehat{\lambda}_{LSE}$ can be obtained by minimizing

$$\sum_{i=1}^{n} \left(\frac{e^{\lambda Y_{(i)}} - 2 + \alpha}{e^{\lambda Y_{(i)}} - 1 + \alpha} - \frac{j}{n+1} \right)^2, \tag{21}$$

with respect to α and λ. On the other hand, the weighted least squares estimators of the unknown parameters can be obtained by minimizing

$$\sum_{i=1}^{n} \frac{1}{Var(G(Y_{(i)}))} \left(G(Y_{(i)}) - \frac{j}{n+1} \right)^2,$$

with respect to the unknown parameters. Therefore, the weighted least squares estimators of α and λ say $\hat{\alpha}_{WLSE}$ and $\widehat{\lambda}_{WLSE}$ can be obtained by minimizing

$$\sum_{i=1}^{n} \frac{(n+1)^2(n+2)}{j(n-j+1)} \left(\frac{e^{\lambda Y_{(i)}} - 2 + \alpha}{e^{\lambda Y_{(i)}} - 1 + \alpha} - \frac{j}{n+1} \right)^2, \tag{22}$$

with respect to α and λ only.

6. L-moment Estimators

In this section we propose a method of estimating the unknown parameters of a MOE distribution based on the linear combination of order statistics, see for example David and Nagaraja (2003) or Hosking (1990). The estimators obtained by this method are popularly known as L-moment estimators (LME's). The LME's are analogous to the conventional moment estimators but can be estimated by linear combinations of order statistics. The LME's have certain advantages over conventional moment estimators. It is observed that LME's are less subject to bias in estimation and sometimes more accurate in small samples than even the MLE's. First we discuss the case when both parameters of a MOE distribution are unknown. Here, we equate the first two L-moments with the respective population L-moments. From Hosking (1990), we obtain the first and second sample L-moments as

$$L_1 = \frac{1}{n}\sum_{i=1}^{n} x_{(i)}, \quad \text{and} \quad L_2 = \frac{2}{n(n-1)}\sum_{i=1}^{n}(i-1)x_{(i)} - L_1.$$

Similarly, the first two population L-moments of the MOE distribution are

$$\phi_1 = \frac{-\alpha \ln \alpha}{(1-\alpha)\lambda},$$

and

$$\phi_2 = \frac{\alpha(\alpha-1) - \alpha \ln \alpha}{(\alpha-1)^2 \lambda}.$$

Therefor, the LME's can be obtained by solving the following equations:

$$L_1 = \frac{-\alpha \ln \alpha}{(1-\alpha)\lambda}, \quad \text{and} \quad L_2 = \frac{\alpha(\alpha-1) - \alpha \ln \alpha}{(\alpha-1)^2 \lambda}.$$

First we compute the LME of α, say $\widehat{\alpha}_{LM}$, as a solution of non-linear equation:

$$\frac{L_2}{L_1} = \frac{\alpha(\alpha-1) - \alpha \ln \alpha}{\alpha(\alpha-1)\ln \alpha}. \tag{23}$$

Once $\widehat{\alpha}_{LM}$ is computed, the LME of λ, say $\widehat{\lambda}_{LM}$, can be computed from (23) as

$$\widehat{\lambda}_{LM} = \frac{-\widehat{\alpha}_{LM} \ln \widehat{\alpha}_{LM}}{(1-\widehat{\alpha}_{LM})L_1}. \tag{24}$$

The asymptotic normality of L-estimates of α and λ can be computed as follows. As $n \to \infty$,

$$\left[\sqrt{n}(L_1 - \phi_1), \sqrt{n}(L_2 - \phi_2)\right] \to N(\mathbf{0}, \Lambda),$$

where

$$\Lambda = \begin{bmatrix} \Lambda_{11} & \Lambda_{12} \\ \Lambda_{21} & \Lambda_{22} \end{bmatrix}.$$

From Theorem 3 of Hosking (1990), $\sqrt{n}(L_i - \phi_i),\ i = 1, 2$ converges to $N(\mathbf{0}, \Lambda)$ with $\Lambda = \Lambda_{rs},\ r = 1, 2; s = 1, 2$.The elements Λ_{rs}'s are defined as

$$\Lambda_{rs} = \int\int_{x<y} [P_{r-1}(F(x))P_{s-1}(F(y)) + P_{s-1}(F(x))P_{r-1}(F(y))]F(x)(1-F(y))dxdy,$$

where $P_r(x)$ is the rth shifted polynomial defined by

$$P_r(x) = \sum_{k=0}^{r}(-1)^{r-k}\left(\begin{array}{c} r \\ k \end{array}\right)\left(\begin{array}{c} r+k \\ k \end{array}\right)x^k.$$

This yields that $P_0(x) = 1$ and $P_1(x) = -1 + 2x$.

Let us first define the polylogarithmic function ($Polylog[s, z]$)

$$Polylog[s, z] = \frac{1}{\Gamma(s)}\int_0^\infty \frac{t^{s-1}}{(e^t/z)^{-1}},$$

which is also equivalent to

$$Polylog[s, z] = \sum_{k=1}^{\infty}\frac{z^k}{k^s}.$$

The dilogarithmic function can be evaluated when $s = 2$ which is denoted by

$$Li_2(z) = -\int_0^\infty \frac{\ln(1-t)}{t}dt = -\int_0^1 \frac{\ln(1-zt)}{t}dt.$$

The dilogarithmic function has many valuable properties,see for example Abramowitz, M., Stegun, I.A. (1972) for more details.

This yields that

$$\begin{aligned}
\Lambda_{11} &= \frac{-2\alpha\left(\alpha Li_2(1-\alpha) - (\ln\alpha)^2\right)}{(\alpha-1)^2\lambda^2} \\
&- \frac{-\pi^2/3 + 2\ln(\alpha-1)\ln\alpha + \ln(\alpha/\alpha-1)^2 + 2Li_2(1/1-\alpha)}{(\alpha-1)^2\lambda^2},
\end{aligned}$$

$$\begin{aligned}
\Lambda_{12} &= \frac{2}{6(\alpha-1)^3\lambda^2}[-6\alpha^2 - \alpha^3\pi^2 + 6\alpha^3 - 6\alpha^2\ln(\alpha-1)(\alpha\ln\alpha + 1 - \alpha) \\
&- 3\alpha^2\ln(\alpha/\alpha - 1)(2 - 2\alpha + \alpha\ln(\alpha/\alpha - 1)) - 6\alpha^2 Li_2(1/1-\alpha) \\
&+ (2\alpha^2(\alpha-1) - \alpha)(\pi^2 + 6\ln(\alpha-1)\ln\alpha + 3\ln(\alpha/\alpha-1)^2 + 6Li_2(1/1-\alpha)) \\
&+ \alpha^2(\pi^2 - 6\ln\alpha + 6\ln(\alpha-1)\ln\alpha - 3\ln\alpha^2 - 6Li_2(1/\alpha)) \\
&+ (6 + 6\alpha^2\ln\alpha)(1 - \alpha + \alpha\ln\alpha) - 6\alpha(1 - 2\alpha(\alpha-1))Li_2(1-\alpha) \\
&+ 6\ln\alpha(1 - \alpha + \alpha\ln\alpha(3 - 2\alpha))],
\end{aligned}$$

and

$$\begin{aligned}\Lambda_{22} &= \frac{1}{(\alpha-1)^4\lambda^2}[2\alpha(3-3\alpha+\ln\alpha+2\alpha\ln\alpha)(2-2\alpha+\ln\alpha-3\alpha(\alpha-1)\ln\alpha)\\ &+ \frac{\alpha}{3}(2-2\alpha^2+\alpha)(-\pi^2(2\alpha+1)+3\ln\alpha(6\alpha-(2\alpha+1)(\ln\alpha-2\ln(\alpha-1))))\\ &+ 6(2\alpha+1)Li_2(1/\alpha)+\frac{\alpha}{3}(3\alpha(\alpha-1)-1)(-6(\alpha-1)Li_2(1-\alpha)\\ &+ (2\alpha+1)(6+\pi^2-6\alpha+6\ln(\alpha-1)\ln\alpha-3\ln\alpha^2-6Li_2(1/\alpha)))\\ &+ 1-8\alpha+7\alpha^2-2\alpha(2\alpha+1)\ln\alpha].\end{aligned}$$

If the scale parameter λ is known, then the LME of α (say $\widehat{\alpha}_{LM}^{sc}$) can be obtained by solving the equation

$$\lambda(\alpha-1)L_1 = \alpha\ln\alpha.$$

When the shape parameter α is known, then the LME of λ (say$\widehat{\lambda}_{LM}^{sh}$) is

$$\widehat{\lambda}_{LM}^{sh} = \frac{\alpha\ln\alpha}{(\alpha-1)L_1}.$$

It is noted here that if α (or λ) is known, then the LME of α (or λ) is the same as the method of moment estimator given in Section 3. It turns out that the asymptotic variances of LME's of α and λ are given in (2) and (15).

The asymptotic distribution of L-moment estimators can be shown by setting $V(\alpha,\lambda) = (V_1(\alpha,\lambda), V_2(\alpha,\lambda))$, where

$$V_1(\alpha,\lambda) = L_1 + \frac{\alpha\ln\alpha}{(1-\alpha)\lambda},$$

and

$$V_2(\alpha,\lambda) = L_2 - \frac{\alpha(\alpha-1)-\alpha\ln\alpha}{(\alpha-1)^2\lambda}.$$

Now, for large n, $[\sqrt{n}(\widehat{\alpha}_{LM}-\alpha), \sqrt{n}(\widehat{\lambda}_{LM}-\lambda)]$ is distributed as

$$-[\sqrt{n}(L_1-E(L_1)), \sqrt{n}(L_2-E(L_2))]\begin{bmatrix}\frac{\partial V_1}{\partial\alpha} & \frac{\partial V_2}{\partial\alpha}\\ \frac{\partial V_1}{\partial\lambda} & \frac{\partial V_2}{\partial\lambda}\end{bmatrix}^{-1}.$$

7. Numerical Computations and Discussions

In this section, we perform some numerical computations to compare the performances of the different estimators proposed in the previous sections. We perform an extensive simulation study to compare the performances of the different methods in the sense of bias and mean square error (MSE) for different sample sizes and for different parameter values. The generation of the $MO(\alpha,\lambda)$ can be easily obtained through the transformation $X = (1/\lambda)\ln(\alpha(1-U)^{-1}+1-\alpha)$, where U is a uniform distribution deviate on $(0,1)$. Mathematica 7 codes are used for generating the MOE random variables and for solving the non-linear equations as well as for computing the minimization or maximization of the related functions.

Since λ is the scale parameter and all the estimators are scale invariant, we take $\lambda = 1$ in all cases considered. We consider various choices of the shape parameter α and sample sizes $n = 10, 20, 30, 50$ and 100.

We compute the average relative biases and average relative MSE's over 10,000 runs. This number of runs will give the accuracy in the order $\pm(10,000)^{-5} = \pm 0.01$ (see Karian and Dudewicz (1999)). Therefore, we report all the results up to three decimal places.

First we consider the estimation of α when λ is known. If λ is known, the MLE's and PCE's of α can be obtained directly from (7) and (18) respectively. The MME's of α can be obtained by solving the non-linear equation (13). The LSE's and WLSE's can be obtained by minimizing (21) and (22), respectively, with respect to α only. If $\widehat{\alpha}$ is an estimate, then we present the average value of $(\widehat{\alpha}/\alpha)$ and the average MSE of $(\widehat{\alpha}/\alpha)$. The relative average bias and relative average MSE of $(\widehat{\theta}/\theta)$, where $\widehat{\theta}$ is an estimate of θ, are defined, respectively, as follows:

$$RelativeBias(\widehat{\theta}/\theta) = (1/k)\sum_{i=1}^{k}(\widehat{\theta}_i/\theta)$$

and

$$RelativeMSE(\widehat{\theta}/\theta) = (1/k\theta^2)\sum_{i=1}^{k}(\widehat{\theta}_i - \rho)^2,$$

where $\rho = \theta RelativeBias(\widehat{\theta}/\theta)$ and k is the number of iterations. We present the results for $\alpha = 0.5, 1.0, 1.5, 2.0, 2.5, 3.0, 3.5$ and for $n = 10, 20, 30, 50$ and 100. The results are presented in Table 1. For each method, the average value of $(\widehat{\alpha}/\alpha)$ is given in each box and the corresponding MSE is reported within parenthesis.

It is observed from Table 1 that all of the estimators usually overestimate α for small values of α, $(\alpha \leq 1)$. For $\alpha > 1$, most of the estimators tends to be underestimates α. One can also observe that for each estimation method, the average relative MSE's decreases as the sample size increases and also as the value of α increases. According to the average relative biases, it is observed that it decreases as the value of α increases. For $(\alpha > 1)$, the average relative biases increase as the sample size increases.

The MLE's provide the best results for all sample sizes. The WLSE's work better than the LSE's for all sample sizes and all values of α. For large values of α $(\alpha > 3.0)$, the PCE's provide the best results but for other values of α, it is noticed that MLE's are the best. In the context of computational issues, the MLE's, MME's and PCE's could not be easily implemented since they involve non-linear equations. The LSE's and WLSE's involve non-linear functions that should be minimized.

Now consider the estimation of λ when α is known. The MLE's of λ can be obtained by maximizing (8). The MME's and PCE's can be obtained directly from (14) and (17), respectively. The LSE and WLSE can be immediately obtained by minimizing (21) and (22), respectively with respect to λ only. Since λ is the scale parameter, we take $\lambda = 1$ throughout without loss of generality. We present the average biases and average MSE's of $(\widehat{\lambda}/\lambda)$ over 10,000 runs for various choices of sample sizes $n = 10, 20, 30, 50, 100$ and

Table 1. Average relative estimators and average relative MSE's of α when λ is known

n	*method*	$\alpha = 0.5$	$\alpha = 1.0$	$\alpha = 1.5$	$\alpha = 2.0$	$\alpha = 2.5$	$\alpha = 3.0$	$\alpha = 3.5$
10	MLE	1.244(0.499)	1.171(0.311)	1.118(0.232)	1.061(0.149)	0.937(0.088)	0.873(0.055)	0.857(0.053)
	MME	1.780(0.198)	1.038(0.090)	0.882(0.184)	0.880(0.194)	0.793(0.134)	0.719(0.106)	0.695(0.109)
	PCE	1.513(1.193)	1.325(0.531)	1.205(0.306)	1.105(0.193)	0.939(0.113)	0.849(0.073)	0.813(0.073)
	LSE	2.461(0.099)	1.031(0.035)	0.613(0.018)	0.420(0.013)	0.341(0.004)	0.354(0.001)	0.232(0.002)
	WLSE	2.343(0.061)	1.018(0.024)	0.621(0.015)	0.433(0.011)	0.352(0.003)	0.349(0.001)	0.240(0.002)
20	MLE	1.089(0.174)	1.093(0.181)	1.064(0.149)	1.040(0.112)	0.975(0.074)	0.896(0.051)	0.882(0.044)
	MME	1.629(0.257)	1.048(0.059)	0.938(0.143)	0.958(0.150)	0.906(0.108)	0.840(0.076)	0.820(0.072)
	PCE	1.341(0.589)	1.271(0.363)	1.193(0.221)	1.120(0.143)	1.010(0.086)	0.917(0.053)	0.902(0.047)
	LSE	2.419(0.048)	1.022(0.013)	0.611(0.007)	0.424(0.005)	0.326(0.002)	0.346(0.001)	0.219(0.001)
	WLSE	2.255(0.019)	1.013(0.006)	0.631(0.004)	0.450(0.003)	0.349(0.002)	0.341(0.001)	0.237(0.001)
30	MLE	1.052(0.108)	1.079(0.141)	1.074(0.132)	1.041(0.087)	1.011(0.068)	0.926(0.044)	0.929(0.038)
	MME	1.543(0.283)	1.055(0.048)	0.975(0.135)	0.981(0.117)	0.970(0.085)	0.890(0.058)	0.886(0.051)
	PCE	1.231(0.267)	1.250(0.291)	1.204(0.197)	1.124(0.109)	1.059(0.069)	0.941(0.042)	0.945(0.041)
	LSE	2.404(0.031)	1.016(0.008)	0.610(0.004)	0.424(0.003)	0.322(0.002)	0.339(0.000)	0.216(0.001)
	WLSE	2.217(0.010)	1.009(0.003)	0.635(0.002)	0.456(0.002)	0.353(0.001)	0.339(0.001)	0.241(0.001)
50	MLE	1.028(0.064)	1.026(0.065)	1.045(0.066)	1.025(0.060)	1.004(0.051)	0.955(0.033)	0.962(0.030)
	MME	1.493(0.299)	1.064(0.018)	1.011(0.085)	1.008(0.081)	0.987(0.062)	0.937(0.037)	0.941(0.035)
	PCE	1.159(0.184)	1.143(0.122)	1.154(0.114)	1.116(0.088)	1.057(0.052)	0.990(0.031)	0.980(0.030)
	LSE	2.386(0.018)	1.010(0.004)	0.607(0.002)	0.425(0.001)	0.321(0.001)	0.256(0.001)	0.214(0.001)
	WLSE	2.180(0.004)	1.005(0.001)	0.638(0.001)	0.464(0.001)	0.361(0.001)	0.294(0.001)	0.247(0.001)
100	MLE	1.029(0.032)	1.024(0.031)	1.015(0.031)	1.015(0.030)	1.015(0.028)	0.983(0.023)	0.984(0.021)
	MME	1.390(0.275)	1.072(0.012)	1.008(0.042)	1.017(0.040)	1.010(0.037)	0.977(0.026)	0.975(0.024)
	PCE	1.099(0.080)	1.098(0.066)	1.084(0.056)	1.078(0.043)	1.064(0.035)	1.013(0.023)	1.004(0.022)
	LSE	2.380(0.009)	1.005(0.002)	0.604(0.001)	0.422(0.001)	0.320(0.001)	0.256(0.001)	0.212(0.001)
	WLSE	2.148(0.002)	1.002(0.001)	0.643(0.001)	0.469(0.001)	0.367(0.001)	0.301(0.001)	0.255(0.001)

different values of the shape parameter, $\alpha = 0.5, 1.0, 1.5, 2.0, 2.5, 3.0, 3.5$. The results are presented in Table 2. Similarly as in Table 1, the average MSE's of $\widehat{\lambda}$ are presented within brackets.

Clearly, as sample size increases, the MLE's and MME's decrease for all values of α. It could be noticed that the PCE's increase as the sample size increases but this is not true for the small sample size ($n = 10$). According to LSE's and WLSE's, it could be noticed that both of them increase in most cases as the sample size increases. When $\alpha = 1$, the MME's equal to one by taking the limit The MSE's decrease for all methods as the sample size increase. As a consequence of that, all estimators are asymptotically unbiased and consistent of λ when α is known. It is observed from Table 2 that MLE's and MME's usually overestimate λ while PCE's underestimates λ regardless of the sample size or the value of α taken. It is also observed that the average biases and MSE's of these estimators depend on α. For all methods, as α increases, the average relative MSE's of $\widehat{\lambda}$ decrease. The same thing is valid for the average relative biases, for most of the methods.

Comparing the biases, it is observed that PCE's perform well for small and moderate sample sizes ($n = 10, 20$) and for all values of α considered. For moderate or large sample sizes ($n \geq 30$), MLE's works well. In the sense of MSE's, it is observed that LSE's and WLSE's perform well for very large sample sizes and for all choices of α's considered. For large sample size and large values of α, MSE's of all estimators performs well. The MSE's of the WLSE's are usually smaller than that of the LSE's. Considering all the points and taking into the account that PCE's and MME's are easiest to compute, we recommend the

Table 2. Average relative estimators and average relative MSE's of λ when α is known

n	*method*	$\alpha = 0.5$	$\alpha = 1.0$	$\alpha = 1.5$	$\alpha = 2.0$	$\alpha = 2.5$	$\alpha = 3.0$	$\alpha = 3.5$
10	MLE	1.086(0.135)	1.096(0.118)	1.088(0.105)	1.065(0.080)	1.068(0.077)	1.065(0.071)	1.061(0.062)
	MME	1.098(0.137)	1.076(0.109)	1.085(0.107)	1.060(0.081)	1.062(0.077)	1.058(0.072)	1.056(0.064)
	PCE	0.974(0.204)	0.979(0.128)	0.966(0.087)	0.963(0.075)	0.975(0.072)	0.976(0.063)	0.974(0.053)
	LSE	2.620(0.579)	1.058(0.172)	0.566(0.033)	0.364(0.014)	0.202(0.008)	0.110(0.005)	0.084(0.003)
	WLSE	2.758(0.609)	1.052(0.161)	0.592(0.035)	0.430(0.018)	0.305(0.011)	0.206(0.008)	0.143(0.006)
20	MLE	1.056(0.078)	1.047(0.056)	1.036(0.044)	1.040(0.040)	1.037(0.040)	1.033(0.033)	1.026(0.028)
	MME	1.064(0.081)	1.047(0.055)	1.034(0.044)	1.038(0.040)	1.036(0.041)	1.032(0.034)	1.024(0.029)
	PCE	0.943(0.081)	0.954(0.052)	0.957(0.040)	0.969(0.036)	0.971(0.037)	0.972(0.030)	0.971(0.026)
	LSE	2.931(0.415)	1.049(0.085)	0.570(0.015)	0.371(0.007)	0.231(0.004)	0.114(0.004)	0.075(0.003)
	WLSE	3.119(0.354)	1.045(0.078)	0.606(0.017)	0.455(0.009)	0.364(0.006)	0.289(0.004)	0.218(0.005)
30	MLE	1.046(0.057)	1.031(0.037)	1.024(0.031)	1.028(0.027)	1.027(0.023)	1.025(0.021)	1.021(0.018)
	MME	1.053(0.059)	1.035(0.039)	1.022(0.030)	1.027(0.027)	1.026(0.023)	1.023(0.022)	1.020(0.019)
	PCE	0.953(0.059)	0.960(0.036)	0.962(0.030)	0.971(0.025)	0.975(0.021)	0.977(0.020)	0.975(0.017)
	LSE	3.054(0.301)	1.028(0.051)	0.571(0.011)	0.376(0.004)	0.243(0.003)	0.116(0.003)	0.071(0.002)
	WLSE	3.389(0.198)	1.025(0.04)	0.612(0.012)	0.470(0.006)	0.387(0.004)	0.326(0.003)	0.267(0.003)
50	MLE	1.022(0.032)	1.016(0.020)	1.018(0.017)	1.014(0.015)	1.018(0.013)	1.012(0.011)	1.014(0.011)
	MME	1.024(0.033)	1.019(0.021)	1.018(0.017)	1.013(0.015)	1.016(0.013)	1.010(0.011)	1.013(0.011)
	PCE	0.950(0.036)	0.961(0.020)	0.972(0.017)	0.974(0.015)	0.981(0.013)	0.978(0.011)	0.981(0.011)
	LSE	3.220(0.235)	1.011(0.028)	0.569(0.007)	0.380(0.002)	0.250(0.002)	0.121(0.002)	0.056(0.002)
	WLSE	3.556(0.108)	1.010(0.025)	0.615(0.006)	0.482(0.004)	0.408(0.003)	0.353(0.002)	0.309(0.002)
100	MLE	1.011(0.013)	1.012(0.010)	1.007(0.008)	1.004(0.007)	1.007(0.006)	1.005(0.006)	1.006(0.005)
	MME	1.013(0.014)	1.013(0.010)	1.006(0.008)	1.004(0.007)	1.006(0.006)	1.005(0.006)	1.005(0.005)
	PCE	0.966(0.016)	0.977(0.011)	0.977(0.008)	0.978(0.007)	0.983(0.006)	0.983(0.006)	0.986(0.005)
	LSE	3.375(0.147)	1.015(0.014)	0.573(0.003)	0.383(0.001)	0.256(0.001)	0.128(0.001)	0.045(0.001)
	WLSE	3.073(0.105)	1.014(0.013)	0.623(0.003)	0.493(0.002)	0.424(0.001)	0.378(0.001)	0.342(0.001)

PCE as a good estimate of λ.

Finally, we consider the estimation of α and λ when both of them are unknown. The $\widehat{\lambda}_{ML}$ and $\widehat{\alpha}_{ML}$ can be obtained by solving the system of two non-linear equations (5) and (6) using the generalized Newton Raphson method. The $\widehat{\alpha}_{MM}$ and $\widehat{\alpha}_{LM}$ can be obtained by solving the non-linear equations (11) and (23), and then $\widehat{\lambda}_{MM}$ and $\widehat{\lambda}_{LM}$ can be computed from (12) and (24), respectively. The PCE's, LSE's and WLSE's can be computed by minimizing (16), (21) and (22), respectively, with respect to α and λ. As for this case, we consider $n = 10, 20, 30, 50, 100$ and $\alpha = 0.5, 1.0, 1.5, 2.0, 2.5, 3.0, 3.5$. Throughout, we take $\lambda = 1$. The results are presented in Tables 3 and 4. In Table 3, we present the average values of $(\widehat{\alpha}/\alpha)$ for each method and the corresponding MSE's are presented within brackets. Similar results for $(\widehat{\lambda}/\lambda)$ are presented in Table 4.

From Tables 3 and 4, the average biases and the average MSE's decrease as sample size increases. Therefore all estimators are asymptotically unbiased and consistent. It is observed that the average biases and the average MSE's of $(\widehat{\alpha}/\alpha)$ and $\widehat{\lambda}$ depend on α. For all methods, as α increases, the average biases and average MSE's of $(\widehat{\alpha}/\alpha)$ decrease for most of the methods. On the other hand there is no pattern observed for the average biases and MSE's of $\widehat{\lambda}$. For LSE's and WLSE's, the MSE's are smaller than other methods and tends to be near zero for $n \geq 30$ for both $(\widehat{\alpha}/\alpha)$ and $\widehat{\lambda}/\lambda$. One could notice that the MSE's of LM's is smaller than the MSE's of other estimators.

Comparing the performance of all the estimators, it is observed that as far as biases are concerned, PCE's work the best for $\alpha \leq 2.0$ while MLE's work better for other cases

Table 3. Average relative estimators and average relative MSE's of α when λ is unknown

n	*method*	$\alpha = 0.5$	$\alpha = 1.0$	$\alpha = 1.5$	$\alpha = 2.0$	$\alpha = 2.5$	$\alpha = 3.0$	$\alpha = 3.5$
10	MLE	2.218(1.944)	1.061(0.075)	1.041(0.071)	1.043(0.070)	1.042(0.069)	1.044(0.069)	1.046(0.070)
	MME	2.340(0.050)	1.293(0.106)	0.852(0.037)	0.656(0.026)	0.692(0.027)	0.791(0.001)	1.269(0.056)
	PCE	1.386(1.658)	0.946(0.080)	0.972(0.080)	0.954(0.074)	0.534(0.196)	0.534(0.189)	0.484(0.180)
	LSE	2.136(0.072)	1.029(0.018)	0.728(0.007)	0.543(0.001)	0.377(0.005)	0.311(0.003)	0.260(0.003)
	WLSE	2.097(0.054)	1.016(0.014)	0.719(0.006)	0.534(0.001)	0.376(0.004)	0.310(0.003)	0.259(0.003)
	LM	1.366(1.686)	0.914(0.078)	0.915(0.077)	0.918(0.079)	0.665(0.145)	0.669(0.143)	0.631(0.140)
20	MLE	1.897(1.397)	1.050(0.064)	1.035(0.066)	1.030(0.068)	1.023(0.069)	1.035(0.065)	1.028(0.070)
	MME	2.290(0.028)	2.769(0.359)	0.891(0.020)	0.633(0.018)	0.730(0.024)	0.747(0.001)	0.813(0.052)
	PCE	1.253(1.117)	0.937(0.081)	0.911(0.078)	0.903(0.067)	0.506(0.174)	0.404(0.160)	0.341(0.147)
	LSE	2.106(0.026)	1.017(0.007)	0.712(0.003)	0.525(0.000)	0.379(0.001)	0.310(0.001)	0.261(0.001)
	WLSE	2.066(0.014)	1.007(0.005)	0.710(0.002)	0.518(0.000)	0.382(0.001)	0.314(0.001)	0.266(0.001)
	LM	1.363(1.212)	0.933(0.081)	0.936(0.074)	0.908(0.074)	0.783(0.127)	0.758(0.123)	0.777(0.125)
30	MLE	1.668(0.941)	1.046(0.064)	1.038(0.065)	1.030(0.066)	1.027(0.064)	1.030(0.065)	1.023(0.065)
	MME	2.226(0.011)	1.202(0.100)	0.768(0.025)	0.640(0.025)	0.827(0.014)	0.751(0.001)	0.759(0.051)
	PCE	0.892(0.079)	0.913(0.078)	0.904(0.072)	0.909(0.071)	0.911(0.075)	0.924(0.078)	0.916(0.068)
	LSE	1.277(0.004)	1.012(0.004)	0.655(0.001)	0.514(0.000)	0.379(0.001)	0.311(0.001)	0.272(0.001)
	WLSE	1.264(0.001)	1.006(0.002)	0.659(0.001)	0.507(0.001)	0.385(0.001)	0.318(0.001)	0.274(0.001)
	LM	0.946(0.076)	0.945(0.076)	0.943(0.072)	0.955(0.071)	0.931(0.075)	0.941(0.078)	0.938(0.074)
50	MLE	1.404(0.468)	1.039(0.061)	1.042(0.064)	1.043(0.060)	1.045(0.060)	1.043(0.061)	1.032(0.062)
	MME	2.449(0.335)	1.294(0.111)	0.908(0.077)	0.666(0.046)	0.900(0.015)	0.753(0.010)	0.658(0.009)
	PCE	0.923(0.075)	0.914(0.074)	0.904(0.072)	0.909(0.071)	0.921(0.073)	0.916(0.069)	0.930(0.067)
	LSE	1.277(0.004)	1.008(0.002)	0.655(0.001)	0.514(0.001)	0.376(0.001)	0.310(0.001)	0.267(0.001)
	WLSE	1.264(0.002)	1.004(0.001)	0.659(0.001)	0.507(0.001)	0.386(0.001)	0.319(0.001)	0.273(0.001)
	LM	0.964(0.072)	0.963(0.071)	0.955(0.071)	0.955(0.071)	0.968(0.070)	0.944(0.069)	0.957(0.068)
100	MLE	1.142(0.186)	1.042(0.054)	1.041(0.050)	0.998(0.059)	1.001(0.060)	1.002(0.060)	0.996(0.061)
	MME	2.722(0.463)	1.260(0.113)	0.948(0.076)	0.706(0.049)	0.859(0.003)	0.754(0.011)	0.640(0.006)
	PCE	0.993(0.243)	0.925(0.069)	0.924(0.066)	0.922(0.065)	0.865(0.087)	0.876(0.084)	0.871(.082)
	LSE	2.083(0.004)	1.004(0.002)	0.651(0.001)	0.508(0.001)	0.376(0.001)	0.308(0.001)	0.260(0.001)
	WLSE	2.035(0.002)	1.002(0.002)	0.659(0.001)	0.504(0.001)	0.388(0.001)	0.321(0.001)	0.273(0.001)
	LM	1.072(0.169)	0.983(0.060)	0.976(0.061)	0.980(0.057)	0.967(0.064)	0.971(0.063)	0.967(0.061)

considered for estimating α. According to λ, PCE's work the best for all cases considered. For all sample sizes, the MLE performs well when compared with LSE and WLSE for estimating α and λ. The performances of the LM's show that the bias of LM's are better than other estimations except MLE's and PCE's for estimating $\widehat{\alpha}$. The MSE's of the WLSE's are usually smaller than that of the LSE's whereas the MSE's of the MME's are getting larger in most of the cases considered especially for estimating λ in Table 4. Now if we consider the computational complexities, it is observed that MME's and LM's involve one dimensional minimization, whereas MLE's, PCE's, LSE's and WLSE's involve a two dimensional minimization technique. Summing up, we recommend to use the PCE as the best method for estimating α and λ for most of the cases considered.

Acknowledgment

The author would like to thank Professor Mohammad Raqab for suggesting the problem and for his valuable comments that lead to an improved version of the paper.

Table 4. Average relative estimators and average relative MSE's of λ when α is unknown

n	*method*	$\alpha = 0.5$	$\alpha = 1.0$	$\alpha = 1.5$	$\alpha = 2.0$	$\alpha = 2.5$	$\alpha = 3.0$	$\alpha = 3.5$
10	MLE	1.533(0.429)	1.064(0.065)	1.060(0.063)	1.071(0.062)	1.071(0.061)	1.065(0.060)	1.080(0.058)
	MME	1.605(0.256)	1.192(0.120)	0.826(0.268)	0.621(0.508)	0.578(1.355)	0.572(3.270)	0.675(4.212)
	PCE	0.976(0.494)	0.946(0.075)	0.964(0.074)	0.972(0.062)	0.992(0.059)	0.988(0.061)	0.958(0.057)
	LSE	1.525(0.312)	0.934(0.061)	0.831(0.047)	0.730(0.034)	0.685(0.025)	0.658(0.021)	0.639(0.018)
	WLSE	1.526(0.300)	0.942(0.059)	0.839(0.047)	0.735(0.034)	0.691(0.025)	0.660(0.020)	0.641(0.018)
	LM	1.750(0.276)	0.964(0.056)	0.863(0.047)	0.759(0.034)	0.648(0.017)	0.627(0.014)	0.605(0.010)
20	MLE	1.379(0.309)	1.065(0.058)	1.069(0.054)	1.074(0.055)	1.064(0.053)	1.070(0.050)	1.067(0.050)
	MME	1.703(0.354)	1.658(1.367)	0.807(0.308)	0.586(0.477)	0.572(1.577)	0.542(2.896)	0.527(3.961)
	PCE	0.972(0.348)	0.922(0.070)	0.930(0.065)	0.954(0.060)	0.971(0.052)	1.024(0.059)	1.061(0.054)
	LSE	1.577(0.200)	0.977(0.045)	0.829(0.033)	0.704(0.020)	0.65(0.013)	0.612(0.008)	0.595(0.006)
	WLSE	1.582(0.193)	0.987(0.044)	0.838(0.032)	0.712(0.019)	0.660(0.013)	0.616(0.008)	0.599(0.006)
	LM	1.762(0.195)	0.987(0.039)	0.855(0.031)	0.719(0.019)	0.634(0.010)	0.594(0.006)	0.584(0.005)
30	MLE	1.297(0.243)	1.069(0.057)	1.065(0.051)	1.071(0.048)	1.070(0.046)	1.067(0.045)	1.070(0.044)
	MME	1.742(0.430)	1.129(0.048)	0.742(0.183)	0.592(0.484)	0.605(2.150)	0.535(2.958)	0.503(4.640)
	PCE	0.917(0.070)	0.916(0.064)	0.916(0.050)	0.941(0.051)	0.950(0.052)	0.985(0.049)	1.010(0.047)
	LSE	1.255(0.028)	0.994(0.036)	0.793(0.013)	0.687(0.008)	0.630(0.008)	0.596(0.005)	0.575(0.003)
	WLSE	1.266(0.027)	1.002(0.034)	0.804(0.012)	0.696(0.008)	0.638(0.008)	0.602(0.005)	0.578(0.003)
	LM	1.250(0.027)	0.999(0.031)	0.810(0.011)	0.717(0.008)	0.651(0.008)	0.616(0.006)	0.588(0.004)
50	MLE	1.188(0.147)	1.059(0.049)	1.058(0.043)	1.064(0.039)	1.065(0.036)	1.061(0.034)	1.058(0.034)
	MME	1.933(0.537)	1.151(0.046)	0.801(0.326)	0.590(0.555)	0.630(2.632)	0.528(3.002)	0.461(3.392)
	PCE	0.927(0.065)	0.916(0.057)	0.916(0.052)	0.941(0.051)	0.942(0.044)	0.946(0.039)	0.969(0.037)
	LSE	1.315(0.028)	0.998(0.024)	0.796(0.013)	0.687(0.008)	0.620(0.005)	0.581(0.003)	0.558(0.002)
	WLSE	1.325(0.027)	1.003(0.022)	0.804(0.012)	0.696(0.008)	0.628(0.005)	0.588(0.003)	0.564(0.002)
	LM	1.114(0.020)	0.997(0.019)	0.810(0.011)	0.717(0.008)	0.652(0.005)	0.604(0.003)	0.580(0.002)
100	MLE	1.084(0.071)	1.045(0.034)	1.044(0.027)	1.025(0.025)	1.028(0.022)	1.032(0.021)	1.038(0.021)
	MME	2.025(0.455)	1.127(0.031)	0.807(0.385)	0.607(0.651)	0.610(2.368)	0.529(3.003)	0.455(3.190)
	PCE	0.941(0.080)	0.933(0.044)	0.940(0.034)	0.941(0.033)	0.944(0.030)	0.953(0.027)	0.954(0.024)
	LSE	1.592(0.049)	1.004(0.012)	0.795(0.006)	0.682(0.004)	0.615(0.003)	0.569(0.002)	0.543(0.001)
	WLSE	1.595(0.047)	1.007(0.011)	0.801(0.006)	0.691(0.004)	0.624(0.003)	0.577(0.002)	0.549(0.001)
	LM	1.520(0.032)	1.004(0.009)	0.820(0.005)	0.719(0.004)	0.650(0.003)	0.602(0.002)	0.568(0.001)

References

Abramowitz, M. and Stegun, I.A. (1972). Handbook of Mathematical Functions with Formulas, Graphs and Mathematical Tables, New York: Dover Publications.

David, H. A. and Nagaraja, H. N. (2003). Order Statistics, John Wiley and Sons, Inc., New York.

Hosking, J. R. M. (1990). L-Moment: Analysis and Estimation of Distribution Using Linear Combinations of Order Statistics, *Journal of the Royal Statistical Society, Series B* 52(1), 105-124.

Johnson, N.L., Kotz, S. and Balakrishnan, N. (1995). *Continuous Univariate Distribution*, Vol. 1, 2nd Ed., New York, Wiley.

Kao, J.H.K. (1958). Computer Methods for Estimating Weibull Parameters in Reliability Studies, *Transactions of IRE-Reliability and Quality Control* 13, 15-22.

Kao, J.H.K. (1959). A Graphical Estimation of Mixed Weibull Parameters in Life Testing Electron Tube, *Technometrics* 1, 389-407.

Karian, Z. A. and Dudewicz, E. J. (1999). *Modern Statistical Systems and GPSS Simulations*, 2nd edition, CRC Press, Florida.

Marshall, A. W. and Olkin, I. (1997). A New Method for Adding a Parameter to a Family of Distributions with Application to the Exponential and Weibull Families, *Biometrika* 84(3), 641-652.

Nadarajah S. (2008). Marshall and Olkins Distributions, *Acta Appl Math* 103, 87-100.

Stigler, S. M. (1974). Linear Functions of Order Statistics with Smooth Weight Functions, *The Annals of Statistics* 2, 676-693.

Swain, J., Venkatraman, S. and Wilson, J. (1988). Least Squares Estimation of Distribution Function in Johnson's Translation System, *Journal of Statistical Computation and Simulation* 29, 271-297.

In: Research in Applied Statistical Science
Editor: Mohammad Ahsanullah, pp. 173-181
ISBN: 978-1-63321-818-5

Chapter 15

CHARACTERIZATION OF BETA AND GAMMA DISTRIBUTION BASED ON THE DOUBLY TRUNCATED MEAN FUNCTION

Zohdy M. Nofal
Department of Statistics, Faculty of Commerce, Benha University, Egypt

Abstract

In this paper we introduce a new characterization to beta and gamma distribution using conditional expectation in terms of failure rate and reversed failure rate. A new result is presented in this paper.

Keywords: characterization, failure rate, reversed failure rate, conditional expectation, Beta distributions, gamma distributions, doubly truncated

1. Introduction

Characterizations of distributions have always played an important role in statistical theory and are widely published in the literature, e.g., Arnold (1980), Strivastava (1981), Sreehari (1983), Zijlstra (1983), and Dallas (1987). The survey book by Kagan, Linnik, Rao (1973) and the recent notes by Galambos &Kotz (1978) cover many results. A good discussion of the major directions of characterization is in the Kotz survey (1973). Ahmed (1991) characterized beta, binomial, and Poisson distributions by connecting conditional expectation with hazard rate. Osaki & Li (1988) characterized the gamma and negative binomial distributions. Kagan, Linnik and Rao (1973). Yehia and Ahmed (1993) introduced two characterizations of Geometric distributions. EL-Arishi (2005) presented the A conditional variance characterization of some discrete probability distributions. Shanbhag (1970) has characterized the exponential distribution using the linearity of the conditional expectation. Talwalker (1977) has characterized various distributions like Pareto, Power and Burr distributions in terms of conditional expectation of a function of the absolutely continuous random variable,

Unnikrishnan and Sudheesh (2010).studied the characterization of continuous distributions by properties of conditional variance. Shanbhag (1970) has characterized the exponential distribution using the linearity of the conditional expectation. Talwalker (1977) has characterized various distributions like Pareto, Power and Burr distributions in terms of conditional expectation of a function of the absolutely continuous random variable. Also, characterizations based on the properties of the failure rate function have been considered by many authors. Xekalaki (1983) has identified the Pareto distribution through decreasing failure rate, Roy and Mukherjee (1986) have characterized the Weibull distribution via increasing failure rate. Let X be a random variable (r.v.), usually representing the life length for a certain unit (where this unit can have multiple interpretations), then r.v.$(X - x \mid X \geq x)$, represents the residual life of a unit with age x.

Several functions are defined related to the residual life. The failure rate function, defined by:

$$r(x) = f(x)/(1 - F(x))$$

represents the failure rate of X (or F) at age x, where $F(x) = P(X \leq x)$, and f(x) is the density function when X is absolutely continuous, or $f(x) = P(X = x)$ when X is discrete.

Another interesting function is the mean residual life function, defined by

$$E(X - x \mid X \geq x),$$

and it represents the expected additional life length for a unit which is alive at age x. This function is equivalent to the left censored mean function, also called vitality function (see Gupta (1975)), defined by $E(X \mid X \geq x)$.Applications of hazard functions are quite well known in the statistical literature.

Another interesting function is the mean inactivity time, defined by

$$E(x\text{-}X \mid X \leq x)$$

and it is equivalent to the left censored lifetimes .it become quite popular among the statisticians, see for example Gupta and Han (2001). Anderson et al. (1993) show that the reversed hazard function plays the same role in the analysis of left-censored data as the hazard function plays in the analysis of right-censored data. Interestingly, The properties of the mean inactivity time have been considered by many authors, see, eg., Kayid and Ahmad (2004), and Ahmad, Kayid and Pellery (2005). Several characterizations of probability models have been obtained in the last 30 years based on the univariate failure rate or mean residual life functions. The problems of characterization of distributions are today a substantial part of probability theory and mathematical statistics. The mean inactivity time and mean residual life are applicable in biostatistics and many other actuarial science, engineering, economics, biometry and applied probability areas. They also are useful in survival analysis studies when we take are faced with left or right censored data.

2. Definitions

Note that

$$h(x) = \frac{f(x)}{\bar{F}(x)} \qquad r(x) = \frac{f(x)}{F(x)}$$

$$\frac{h(x)}{r(x)} = \frac{F(x)}{\bar{F}(x)} \qquad \frac{r(x)}{h(x)} = \frac{\bar{F}(x)}{F(x)}$$

3. Characterization Theorems

3.1. Beta Distribution

Theorem 3.1. Let X is a non negative continuous random variable with cdf F(x), pdf f(x); then X has beta distribution with

$$f(x) = \frac{(x-a)^{m-1}(b-x)^{n-1}}{\beta(m,n)(b-a)^{m+n-1}} \; a < x < b, a \geq 0, b > 0, m > 0, n > 0$$

If and only if

$$E[X|x \leq X \leq y] = \mu + \frac{1}{[h(y)r(x)-h(x)r(y)]}[-A_1h(y)r(x)r(y)$$
$$-A_1h(y)h(x)r(y) + A_2h(x)r(x)r(y) + A_2h(x)h(y)r(x)]$$

where

$$A_1 = \frac{(y-a)(b-y)}{n+m}$$

$$A_2 = \frac{(x-a)(b-x)}{n+m}$$

Lemmas (3.2, 3.3, 3.4 and 3.5) presume the conditions of theorem3.1. All lemmas are proved in the appendix.

Lemma 3.2.

$$F(y) - F(x) = F(y)\bar{F}(x) - F(x)\bar{F}(y)$$

Lemma 3.3.

$$A_1f(x) = A_1f(x)\bar{F}(x) + A_1f(x)F(x)$$

Lemma 3.4.

$$\frac{-A_1f(y)+A_2f(x)}{\bar{F}(x)\bar{F}(y)} =$$

$$\frac{-A_1h(y)r(x)r(y)-A_1h(y)h(x)r(y)+A_2h(x)r(x)r(y)+A_2h(x)h(y)r(x)}{r(x)r(y)}$$

Lemma3.5.

$$\frac{F(y)-F(x)}{\bar{F}(y)\bar{F}(x)}=\frac{[h(y)r(x)-r(y)h(x)]}{r(x)r(y)}$$

Proof of Theorem 3.1.

a. **Necessity**. One knows that

$$E[X|x\le X\le y]=\int_x^y uf(u)du/[F(y)-F(x)]$$
$$=E((X-a)|x\le X\le y)+a,$$

$$E[(X-a)|x\le X\le y]=\int_x^y\frac{(u-a)^m(b-u)^{n-1}}{\beta(m,n)(b-a)^{m+n-1}}du$$

Integrating by parts

$$\therefore E[(X-a)|x\le X\le y]=\frac{1}{F(y)-F(x)}\left[\frac{-(y-a)^m(b-y)^n}{n\beta(m,n)(b-a)^{m+n-1}}\right.$$
$$\left.+\frac{(x-a)^m(b-x)^n}{n\beta(m,n)(b-a)^{m+n-1}}+\frac{m}{n}E[(b-x)|x\le X\le y]\right]$$

Then,

$$E[X|x\le X\le y]=$$
$$=\frac{an+mb}{n+m}-\frac{(y-a)(b-y)}{n+m}\frac{f(y)}{F(y)-F(x)}+\frac{(x-a)(b-x)}{n+m}\frac{f(x)}{F(y)-F(x)}$$
$$=\mu-A_1\frac{f(y)}{F(y)-F(x)}+A_2\frac{f(x)}{F(y)-F(x)}$$

Using lemmas 1-3, then

$$E[X|x\le X\le y]=\mu+\frac{-A_1f(y)\bar{F}(x)-A_1f(y)F(x)+A_2f(x)\bar{F}(y)+A_2f(x)F(y)}{F(y)-F(x)}$$

By dividing all of (numerator and denominator) on $\bar{F}(y)\bar{F}(x)$and using lemmas (3.4 and 3.5) then

$$E[X|x\le X\le y]=\mu+$$
$$\frac{A_2h(x)r(x)r(y)+A_2h(x)h(y)r(x)-A_1h(y)r(x)r(y)-A_1h(x)h(x)r(y)}{[h(y)r(x)-h(x)r(y)]} \quad (1)$$

b. Sufficiency

Equation (1) can be written as

$$\int_x^y uf(u)du=\frac{an+mb}{n+m}[\bar{F}(y)-\bar{F}(x)-A_1f(y)++A_2f(x)]$$

By differentiating both sides with respect to y, then

$$yf(y) = \frac{an+mb}{n+m} f(y) - \frac{(y-a)(b-y)}{n+m} \acute{f}(y) + \frac{(y-a)}{n+m} f(y) - \frac{(b-y)}{n+m} f(y)$$

Gathering the similar terms, then

$$\frac{(y-a)(b-y)}{n+m} \acute{f}(y) = \left[\frac{an+mb}{n+m} - y + \frac{(y-a)}{n+m} - \frac{(b-y)}{n+m}\right] f(y) \tag{2}$$

One can write (2) as

$$\frac{f(y)}{\acute{f}(y)} = -\frac{n-1}{b-y} + \frac{m-1}{y-a}$$

Integrating both sides with respect to y, then

$$f(x) = \frac{(x-a)^{m-1}(b-x)^{n-1}}{\beta(m,n)(b-a)^{m+n-1}}$$

which is Beta distribution.

3.2. Gamma Distribution

Theorem 3.6. Let X is a non negative continuous r.v. X has gamma distribution with:

$$f(x) = \frac{1}{\Gamma\beta} \alpha(\alpha x)^{\beta-1} e^{-\alpha x} \; x > 0 \; \alpha, \beta > 0$$

If and only if

$$E[X|x \leq X \leq y] =$$
$$= \frac{1}{\alpha}\left[\beta + \frac{xh(x)r(x)r(y) + xh(x)h(y)r(x) - yh(y)r(x)r(y) - yh(x)h(x)r(y)}{[h(y)r(x) - h(y)r(x)]}\right]$$

Lemma 3.7.

$$\frac{xf(x) - yf(y)}{\bar{F}(y)\bar{F}(x)}$$
$$= \frac{xh(x)r(x)r(y) + xh(x)h(y)r(x) - yh(y)r(x)r(y) - yh(x)h(x)r(y)}{r(x)r(y)}$$

Proof of Theorem 3.6.

a. Necessity. One knows that

$$E[X|x \leq X \leq y] = \int_x^y uf(u)du / [F(y) - F(x)]$$
$$= \frac{1}{F(y) - F(x)}\left[\int_x^y u \frac{1}{\Gamma\beta} \alpha(\alpha u)^{\beta-1} e^{-\alpha u} du\right]$$

Integrating by parts, then

$$E[X|x \leq X \leq y] = \frac{1}{\alpha[F(y)-F(x)]}\left[[-yf(y) + xf(x)] + \beta[F(y) - F(x)]\right] \quad (3)$$

Using lemma (3.2, 3.3 and 3.7), then

$$E[X|x \leq X \leq y]=$$
$$= \frac{1}{\alpha}\left[\beta + \frac{xh(x)r(x)r(y)+xh(x)h(y)r(x)-yh(y)r(x)r(y)-yh(x)h(x)r(y)}{[h(y)r(x)-h(y)r(x)]}\right]$$

b. Sufficiency

Equation (3) can be written as

$$\int_x^y uf(u)du = \left[-\frac{y}{\alpha}f(y) + \frac{x}{\alpha}f(x) + \frac{\beta}{\alpha}F(y) - \frac{\beta}{\alpha}F(x)\right]$$

By differentiating both sides with respect to y, then

$$yf(y) = -\frac{y}{\alpha}\acute{f}(y) - \frac{1}{\alpha}f(y) - \frac{\beta}{\alpha}f(y),$$

one can write it as

$$y\acute{f}(y) = f(y)[\beta - \alpha y - 1],$$

then

$$\frac{\acute{f}(y)}{f(y)} = \left[\frac{\beta-\alpha y-1}{y}\right]$$
$$= -\alpha + \frac{\beta-1}{y}$$

By integrating both sides with respect to y, then

$$lnf(y) = -\alpha y + (\beta - 1)\text{lny, then}$$
$$= lny^{\beta-1}.e^{-\alpha y}$$

$$f(y) = k\, y^{\beta-1}.e^{-\alpha y}$$

Using the fact that,

$$\int_0^{\infty} f(y) = 1, \text{ then}$$

$$k = \frac{\alpha^{\beta}}{\Gamma\beta} \text{ and}$$

$$f(x) = \frac{1}{\Gamma\beta}\alpha(\alpha x)^{\beta-1}e^{-\alpha x}$$

which is Gamma distribution.

Appendix

Proof of Lemma 1.

$$\begin{aligned} F(y) - F(x) &= F(y) - F(x) + F(y)F(x) - F(y)F(x) \\ &= F(y)[1 - F(x)] - F(x)[1 - F(y)] \\ &= F(y)\bar{F}(x) - F(x)\bar{F}(y) \end{aligned}$$

Proof of Lemma 2.

$$\begin{aligned} A_1 f(x) &= A_1 f(x) + A_1 f(x)F(x) - A_1 f(x)F(x) \\ &= A_1 f(x)\bar{F}(x) + A_1 f(x)F(x) \end{aligned}$$

Proof of Lemma 4. By using Lemma 2:

$$-A_1 f(y) + A_2 f(x) = -A_1 f(y)F(x) - A_1 f(y)\bar{F}(x) + A_2 f(x)F(y) + A_2 f(x)$$

By dividing both sides on $\bar{F}(x)\bar{F}(y)$, then

$$\frac{A_1 F(y) + A_2 f(x)}{\bar{F}(x)\bar{F}(y)} = \frac{-A_1 f(y)F(x) - A_1 f(y)\bar{F}(x) + A_2 f(x)F(y) + A_2 f(x)}{\bar{F}(x)\bar{F}(y)}$$

Proof of Lemma 5. From Lemma 1:

$$F(y) - F(x) = F(y)\bar{F}(x) - F(x)\bar{F}(y)$$

Dividing both sides on $\bar{F}(x)\bar{F}(y)$, then

$$\frac{F(y) - F(x)}{\bar{F}(x)\bar{F}(y)} = \frac{F(y)\bar{F}(x) - F(x)\bar{F}(y)}{\bar{F}(x)\bar{F}(y)}$$

Using Definition 1, then

$$\frac{F(y) - F(x)}{\bar{F}(x)\bar{F}(y)} = \frac{h(y)}{r(y)} - \frac{h(x)}{r(x)} = \frac{h(y)r(x) - r(y)h(x)}{r(x)r(y)}$$

Proof of Lemma 6.

$$\begin{aligned} xf(x) - yf(y) &= xf(x) - yf(y) + xf(x)F(y) - xf(x)F(y) + yf(y)F(x) \\ &\quad - yf(y)F(x) \\ &= xf(x)\bar{F}(y) - yf(y)\bar{F}(x) + xf(x)F(y) - yf(y)F(x) \end{aligned}$$

Dividing both side on $\bar{F}(x)\bar{F}(y)$, then

$$\frac{\frac{xf(x)-yf(y)}{\bar{F}(y)\bar{F}(x)}}{} = \frac{xh(x)r(x)r(y)+xh(x)h(y)r(x)-yh(y)r(x)r(y)-yh(x)h(x)r(y)}{r(x)r(y)}$$

References

Ahmed, A.N. (1991).Characterization of Beta, Binomial and Poisson distribution. *IEEE Trans. Reliability* 40 (3), 290-295.

Arnold, B. (1980) .Two characterizations of the geometric distribution. *J. Applied Probability*, vol. 17, pp 570-573.

Dallas, A. (1987).Characterizing the geometric distribution using expectations of order statistics", *J. Applied Probability*, vol. 24, pp 534-539.

El-Arishi, S (2005). A conditional variance characterization of some discrete probability distributions, *Statist. Papers*, 46, 31- 45.

Galambos, J. and KOtz, S. (1978). *Characterization of Probability Distributions*; Springer-Verlage.

Gupta, R.C. (1975).On characterization of distributions by conditional expectations. *Comm. Statist.* 4 (1), 99-103.

Gupta, R.C (1987). On the monotonic properties of residual variance and their applications in reliability, *J. Statist. Plann. Inf.*, 16, 329-335 .

Gupta, R.C. and Kirmani, S.N.U.A (2000). Residual coefficient of variation and some characterization results, *J. Statist. Plann. Inf.*, 91, 23-31 .

Gupta, R.C. and Kirmani, S.N.U.A (2004). Some characterization of distributions by functions of failure rates and mean residual life, *Comm. Statist. Theory Methods*, 33, 3115- 3131.

Kagan, A. M. Linnik, Y. V. Rao, C. R. (1973) *Characterization Problems in Mathematical Statistics*; John Wiley & Sons.

Kotlarski, I. I. (1972). On a characterization of some probability distributions by conditional expectations. *Sankhya Ser. A.* 34:461–466.

Kotz, S. (1973). Characterization of statistical distribution: A supplement to recent surveys", *Rev. Znt. Statist.*, vol. 42X, pp 39-65.

Hitha, N. and Nair, N.U (1989). Characterization of some models by properties of residual life, *Calcutta Statist. Assoc. Bull.*, 38, 219-225.

Hossain, S.A. and Ahsanullah, M. (2010).Mean residual lives of some discrete distribution. *Pak. J. Statist.* Vol. 26(4), 565-568.

Muth, E. J. (1980). Memory as a property of probability distributions. *IEEE Trans. Reliability*, vol. R-29, pp160-166.

Nanda, A.K.(2010).Characterization of distributions through failure rate and mean residual life functions. *Statistics and Probability Letters*,80, 752-755.

Nassar, M. M. Mahmoud, M. R. (1985). On characterization of a mixture of exponential distributions", *IEEE Trans. on Reliability*, vol. R-34, 1985, pp 484-488.

Nair, N. U and Sudheesh. K. K (2008), Some Results on Lower Variance Bound Useful in Reliability and Estimation, *Ann. Inst. Stat. Math.*, 60, 591- 603.

Osaki, S. Li, X. (1988).Characterization of gamma and negative binomial distributions. *IEEE Trans. Reliability*, vol. R-37, pp 379-382.

Rao, C. and Rubin, H.(1964).On a characterization of the Poisson distribution", *Sankhya A*, vol. 26, pp 295-298.

Ruiz, J.M. and Navarro, J. (1994). Characterization of distributions by relationships between failure rate and mean residual life. *IEEE Trans. Reliability* 43 (4), 640-644.

- - -- (1995). Characterization of discrete distributions using expected values. *Statist. Papers*. 36 (3), 237-252.

---- (1996). Characterizations based on conditional expectations of the doubled truncated distribution. *Ann. Inst. Statist. Math.* 48 (3), 563-572.

Shanbhag, D. (1972).Some characterizations based on the charya matrix. *J. Applied Probability*, vol. 9, pp 580-587.

Shanbhag, D. N. and Bhaskara Rao, M. (1975). A note on characterizations of probability distributions based on conditional expected values. *Sankhya Ser. A*. 37: 297–300.

Shanbhag,D. and Clark, R.(1972).Some characterizations for the Poisson distribution starting with a power-series distribution", *Proc. Camb. Phil. Soc.*, pp 517-522.

Sheetrari, M. (1983).A characterization of the geometric distribution", *J. Applied Probability*, vol. 20, pp 209-212.

Srivastava, R. (1981).On some characterizations of the geometric distribution. in *Statistical Distributions in Scientific Work*, vol4, 981, pp 349-355; Taillie et al, Reidel, Dordrecht.

Srivastava, R. and Srivastava, A.(1974).On a characterization of a Poisson distribution", *J. Applied Probability*, vol. 7, pp 414-416.

Zoroa, P., Ruiz, J.M. and Marin, J. (1990). A characterization based on conditional expectations. *Comm. Statist. Theory Methods* 19 (8), 3127-3135.

Zijlstra, M. (1983).Characterizations of the geometric distribution by distributional properties", *J. Applied Probability*, vol. 20, pp 843-850.

In: Research in Applied Statistical Science
Editor: Mohammad Ahsanullah, pp. 183-197
ISBN: 978-1-63321-818-5

Chapter 16

ESTIMATION OF THE SEASONAL ACF OF PAR(1) MODEL IN THE PRESENCE OF ADDITIVE OUTLIERS

A. A. Samadi and A. M. Al-Quraam
Department of Statistics, Yarmouk University, Irbid, Jordan

Abstract

In this article the works of Berkoun et al. (2003) for the estimation of the first lag autocorrelation (of the autoregressive model of order one (AR(1)) as well as that of Smadi et al. (2009) for the estimation of the seasonal autocorrelation for the periodic autoregressive model of order one (PAR(1)) are considered. In the later work, the authors have generalized the estimators considered by the former towards the periodic case for the first lag seasonal autocorrelation. Here, we will again consider the PAR(1) model and the same estimators considered by Smadi et al. (2009), namely the moment estimator and two robust estimators. We will, however, generalize those estimators for the PAR(1) model but for all time lags. Some useful results about the theoretical seasonal autocorrelation function (SACF) of the PAR(1) model are derived. Then, a Monte-Carlo study is carried out to study the behaviors of all estimators in the presence of some additive outliers. The comparison of various estimators was done through the bias and MSE criterion. It is observed that the moment estimator is highly affected by the outliers whereas the other two estimators were less affected. A real application to river-flow time series is also presented.

Keywords: Periodic autoregression, Robust estimation, Seasonal auto-correlations, Additive outlier, Monte-Carlo simulation

1. Introduction

It is known that one of the most important models for time series is the class of autoregressive integrated moving average (ARIMA) models which are also known as Box-Jenkins models. Those models are extensively studied from a theoretical point of view and are widely accepted in practice. For instance, they are implemented in various statistical packages. The pure AR, the pure MA and mixed ARMA models are special cases of this model. Besides, several extensions of the ARIMA model are developed as for example the multiplicative seasonal ARIMA models and periodic ARMA (PARMA) models which are both developed for modeling seasonal time series with. A detailed account of the seasonal ARIMA models are found in many books of time series analysis as for example Box et al. (1994). Also, for more details on PARMA models, which are less popular than ARIMA models, see Hiple and McLeod (1994) and Franses and Paap (2004).

One of the most important special cases of the ARIMA model is the (zero-mean) stationary AR(1) model defined as:

$$X_t = \phi X_{t-1} + a_t, \tag{1.1}$$

where ϕ is the AR parameter and $\{a_t\}$ is the white noise (WN) process which is iid with mean zero and constant variance σ_a^2. A sufficient condition for the stationarity of the AR(1) process is that $|\phi| < 1$. Under the assumption of stationarity, the autocorrelation function (ACF) of $\{X_t\}$ becomes:

$$\rho_k = \frac{\gamma_k}{\gamma_0} = \phi^k \quad , k=0,1,\ldots$$

where γ_k is the autocovariance function (ACVF) of $\{X_t\}$ and k is the time lag. Figure (1) depicts the distinguished shapes of the ACF of AR(1) models for $0 < \phi < 1$ and $-1 < \phi < 0$.

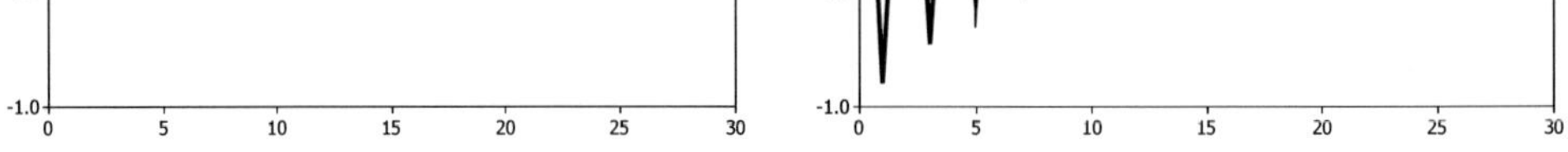

Figure 1. ACF of AR(1) model with $0 < \phi < 1$ (left) and $-1 < \phi < 0$ (right).

Berkoun et al. (2003) investigated robust estimation of the first lag autocorrelation of the AR(1) model, that is $\rho_1 = \phi$, when a single additive outlier is present. That is, if $\{X_t\}$ follows AR(1) as (1.1) and $Z_t = X_t + \Delta I_t(t_0)$, where Δ denotes the magnitude of the outlier and $I_t(t_0)$ is the indicator function at $t = t_0$, then $\{Z_t\}$ is the contaminated process that is affected by the additive outlier Δ at time t_0. Based on an observed realization $Z_1, Z_2, \ldots, Z_n$ of $\{Z_t\}$, they studied the following estimates of ρ_1:

$$\hat{\rho}_1 = \frac{\sum_{t=2}^{n} Z_t Z_{t-1}}{\sum_{t=2}^{n} Z_{t-1}^2} \tag{1.2}$$

$$\tilde{\rho}_1 = \text{Med}\left\{\frac{Z_2}{Z_1}, \frac{Z_3}{Z_2}, \ldots, \frac{Z_n}{Z_{n-1}}\right\} \tag{1.3}$$

$$\breve{\rho}_1 = \frac{\text{Med}\{Z_1 Z_2, Z_2 Z_3, \ldots, Z_{n-1} Z_n\}}{\text{Med}\{Z_1^2, Z_2^2, \ldots, Z_{n-1}^2\}} \tag{1.4}$$

where $\hat{\rho}_1$ is the ordinary moment estimator and the other two estimators are two robust estimators, originally proposed by Hurwicz (1950) and Haddad (2000), respectively. Berkoun et al. (2003) showed that the inference based on the moment estimator $\hat{\rho}_1$ is highly sensitive to a single additive outlier. Therefore, they suggest replacing $\hat{\rho}_1$ with the robust alternatives $\tilde{\rho}_1$ and $\breve{\rho}_1$ for inference purposes when some outliers (or suspect data) are present.

In this article our objective is to generalize the work of Berkoun et al. (2003) to periodic autoregressive (PAR) models. We will mainly focus on the $\text{PAR}_\omega(1)$ model which is written as:

$$X_t = \phi_1(t) X_{t-1} + a_t \tag{1.5}$$

where $\phi_1(t)$ is the periodic AR parameters of period ω and $\{a_t\}$ is a periodic WN process with mean zero and periodic variances $\sigma_a^2(t)$. Equation (1.5) is better written as:

$$X_{j\omega+v} = \phi_1(v) X_{j\omega+v-1} + a_{j\omega+v} \tag{1.6}$$

where v stands for season and j for the year. Notice that equation (1.6) summarizes ω–equations each of the AR(1) style corresponding to the different ω seasons with separate AR parameter and WN variance. The $\text{PAR}_\omega(1)$ model is a special case of the zero–mean $\text{PARMA}_\omega(p(v),q(v))$ model written as:

$$X_{j\omega+v} = \phi_1(v) X_{j\omega+v-1} + \ldots + \phi_{p(v)}(v) X_{j\omega+v-p(v)} + a_{j\omega+v} - \theta_1(v) a_{j\omega+v-1} - \ldots - \theta_{q(v)}(v) a_{j\omega+v-q(v)}$$

where $\phi_1(v), \ldots, \phi_{p(v)}(v)$ and $\theta_1(v), \ldots, \theta_{q(v)}(v)$ are the AR and MA parameters of season v, respectively. Pure PAR models are obtained when q(v) = 0 for all v. Due to the large number of parameters involved in PARMA models and other issues, the PAR(1) model is of particular importance in practice.

In general, there are several motivations for using PARMA models instead of the seasonal ARIMA models for modeling seasonal time series. In particular, it is more natural as it gives separate equations for different seasons. Besides, in many applications it is found capable of modeling "hidden" periodicity in the time series that could not be cached by seasonal ARIMA models (for example see McLeod 1993).

PARMA models are not stationary in the ordinary weak sense. They are rather examined for a weaker type of stationarity named as periodic stationarity. This means that the mean and the variance of the time series are constants for each season and periodic with period ω and the autocovariance function depends on the time lag and season only. For the $PAR_{\omega}(1)$ model the periodic stationarity condition reduces to $\left|\prod_{v=1}^{\omega}\phi_1(v)\right| < 1$ (Obeysekera and Salas,1986).

Now, for periodic stationary PARMA models, the seasonal autocorrelation function (SACF) is defined as:

$$\rho_k(v) = \mathrm{Corr}(X_{j\omega+v}, X_{j\omega+v-k}) = \frac{\gamma_k(v)}{\sqrt{\gamma_0(v)\gamma_0(v-k)}}$$

where $\gamma_k(v)$ denotes the seasonal autocovariance function and $\gamma_0(v)$ denotes the variance of the process for season v and time lag k = 0,1,... The moment estimators of $\gamma_k(v)$ and $\rho_k(v)$ are, respectively,

$$C_k(v) = \frac{\sum_{j=0}^{n-1}(X_{j\omega+v} - \overline{X}_v)(X_{j\omega+v-k} - \overline{X}_{v-k})}{n-1}$$

and

$$r_k(v) = \frac{C_k(v)}{\sqrt{C_0(v)C_0(v-k)}},$$

where $\overline{X}_v$ is the sample mean of data in season v and n is the number of years of data (see, Mcleod,1994). If $\{X_{j\omega+v}\}$ is Gaussian, i.e., if the white noise terms are independent and normally distributed, Pagano (1978) showed that, for all v and k, $r_k(v)$ is a consistent estimator for $\rho_k(v)$ while Ula and Smadi (2003) derived asymptotic mean and variance of $r_k(v)$.

Now, assuming that $\{X_{j\omega+v}\}$ follows the $PAR_{\omega}(1)$ model (1.6) and that $\{Z_{j\omega+v}\}$ is the same as $\{X_{j\omega+v}\}$ but contaminated with an additive outlier Δ at time $t = t_0$ (which can be written as $j_0\omega + v_0$), Smadi et al. (2009) generalized the estimators (1.2)–(1.4) above to the first lag seasonal autocorrelation ($\rho_1(v)$) of the periodic autoregressive model of order one. Based on an observed realization of length n years $\{Z_1, Z_2, \ldots, Z_{n\omega}\}$ they are given by:

$$\hat{\rho}_1(v) = \frac{\sum_{j=1}^{n}(Z_{(j-1)\omega+v} - \overline{Z}_v)(Z_{(j-1)\omega+v-1} - \overline{Z}_{v-1})}{\sqrt{\sum_{j-1}^{n}(Z_{(j-1)\omega+v} - \overline{Z}_v)^2 \sum_{j-1}^{n}(Z_{(j-1)\omega+v-1} - \overline{Z}_{v-1})^2}} \quad (1.7)$$

$$\tilde{\rho}_1(v) = \underset{j}{\mathrm{Med}}\left(\frac{Z^*_{(j-1)\omega+v}}{Z^*_{(j-1)\omega+v-1}}\right)\sqrt{\frac{\underset{j}{\mathrm{Med}}(Z^*_{(j-1)\omega+v-1})^2}{\underset{j}{\mathrm{Med}}(Z^*_{(j-1)\omega+v})^2}} \quad (1.8)$$

$$\breve{\rho}_1(v) = \frac{\underset{j}{\text{Med}}(Z^*_{(j-1)\omega+v} Z^*_{(j-1)\omega+v-1})}{\sqrt{\underset{j}{\text{Med}}(Z^*_{(j-1)\omega+v})^2 \, \underset{j}{\text{Med}}(Z^*_{(j-1)\omega+v-1})^2}} \tag{1.9}$$

where $Z^*_{(j-1)\omega+v} = Z_{(j-1)\omega+v} - \underset{j}{\text{Med}}(Z_{(j-1)\omega+v})$ is the seasonally median subtracted time series and $\underset{j}{\text{Med}}(Z_{(j-1)\omega+v})$ is the median of the data in season v, $v = 1, \ldots, \omega$. The estimator $\hat{\rho}_1(v)$ is the moment estimator of $\rho_1(v)$ which corresponds to $\hat{\rho}_1$ for the AR(1) model in (1.2) whereas $\widetilde{\rho}_1(v)$ and $\breve{\rho}_1(v)$ are two robust estimators of $\rho_1(v)$ corresponding to (1.3) and (1.4) for the AR(1) model. Using Monte-Carlo simulation, Smadi et al. (2009) showed, in view of bias and MSE, that the ordinary moment estimator $\hat{\rho}_1(v)$ is affected by the additive outlier while the other two estimators were apparently robust to such outliers. They reported that in the presence of outliers, $\widetilde{\rho}_1(v)$ was better than the other two estimators. Therefore, our main objective is to generalize the estimators (1.7)–(1.9) of $\rho_1(v)$ to $\rho_k(v)$ for all time lags k, that is the SACF for the $PAR_\omega(1)$ model. In the next section we study the theoretical SACF of $PAR_\omega(1)$ model. Later on (1.7)–(1.9) are generalized to all time lags k. Then the proposed estimators are investigated and compared via Monte–Carlo simulation in the presence of some additive outliers. Also, a real application of our results is given on river–flow time series.

Finally, we conclude this section with a brief review of literature on related issues. The impact of outliers on the parameter estimation of ARIMA models has been studied by several researchers including but not limited to Denby and Martin (1979), Chang and Tiao (1983), Barnett and Lewis (1994) and Bianco et al. (1996). Bustos and Yohai (1986) took the autocovariance structure of time series into consideration when robustifying the estimators. Robust estimation of the parameters of PAR models is recently studied by Shao (2007) and Sarnaglia eta al. (2010). We notice here that our approach is directed to the estimation of the ACF prior to the estimation of parameters. This, later on, may be helpful for the estimation of the parameters of AR and PAR models by methods as Yule-Walker equations.

2. The SACF Of PAR(1) Model

Assume that $\{X_{j\omega+v}\}$ is a periodic stationary $PAR_\omega(1)$ process. The following proposition is useful for the computation of the SACF of $\{X_{j\omega+v}\}$.

Proposition 1. For periodic stationary $PAR_\omega(1)$ models:

$$\rho_k(v) = \rho_1(v)\rho_1(v-1)\rho_1(v-2)\ldots\rho_1(v-k+1), k \geq 1. \tag{2.1}$$

Proof. It is easy to show that for $\{X_{j\omega+v}\}$:

$$\gamma_1(v) = \text{Cov}(X_{j\omega+v}, X_{j\omega+v-1}) = \text{Cov}(\phi_1(v)X_{j\omega+v-1} + a_{j\omega+v}, X_{j\omega+v-1}) = \phi_1(v)\gamma_0(v-1)$$

so that

$$\rho_1(\nu) = \frac{\gamma_1(\nu)}{\sqrt{\gamma_0(\nu)\gamma_0(\nu-1)}} = \phi_1(\nu)\sqrt{\frac{\gamma_0(\nu-1)}{\gamma_0(\nu)}}. \tag{2.2}$$

Similarly, $\gamma_2(\nu) = \phi_1(\nu)\gamma_1(\nu-1)$ so that

$$\rho_2(\nu) = \frac{\gamma_2(\nu)}{\sqrt{\gamma_0(\nu)\gamma_0(\nu-2)}} = \phi_1(\nu)\rho_1(\nu-1)\sqrt{\frac{\gamma_0(\nu-1)}{\gamma_0(\nu)}} = \rho_1(\nu)\rho_1(\nu-1).$$

Iterating up to lag k, equation (2.1) is obtained.♦

The result in Proposition (1) says that the SACF of the PAR(1) is a function of the first lag seasonal autocorrelations for all seasons. One implication of this point is that second order auto-correlations (or higher order) at a specific season is strongly affected by autocorrelations of lower orders of the other seasons. In other words, the seasonal autocorrelations for season 1, for instance, is not a function of the terms in season 1 only.

In Figure (1) we have seen the special exponentially decaying ACF of the AR(1) process. The following corollary, which is a simple conclusion of Proposition (1), generalizes this fact to the PAR(1) process.

Corollary 1. For periodic stationary $PAR_\omega(1)$ models and $k = m\omega$,

$$\rho_k(\nu) = (\prod_{\nu=1}^{\omega}\rho_1(\nu))^m = (\prod_{\nu=1}^{\omega}\phi_1(\nu))^m, m \geq 0.$$

Proof. The first equality is a direct result of (2.1) for $k = m\omega$. For the second equality, using (2.2) we have:

$$\prod_{\nu=1}^{\omega}\rho_1(\nu) = \rho_1(1)\rho_1(2)\cdots\rho_1(\omega) = \phi_1(1)\sqrt{\frac{\gamma_0(\omega)}{\gamma_0(1)}}\phi_1(2)\sqrt{\frac{\gamma_0(1)}{\gamma_0(2)}}\cdots\phi_1(\omega)\sqrt{\frac{\gamma_0(\omega-1)}{\gamma_0(\omega)}} = \prod_{\nu=1}^{\omega}\phi_1(\nu).$$ ♦

Corollary (1) says that the seasonal autocorrelations at each seasonal lag (i.e. multiple of ω) are the same for all seasons. Besides, Those autocorrelations show similar pattern as that of the AR(1) model in Figure (1). That is, a damped exponential decay appears when $\prod_{\nu=1}^{\omega}\phi_1(\nu) > 0$, else an alternating exponential decay appears. We illustrate Corollary (1) and the usage of Proposition (1) through the following example.

Example 1. Consider the zero-mean periodic stationary $PAR_4(1)$ model with AR parameters

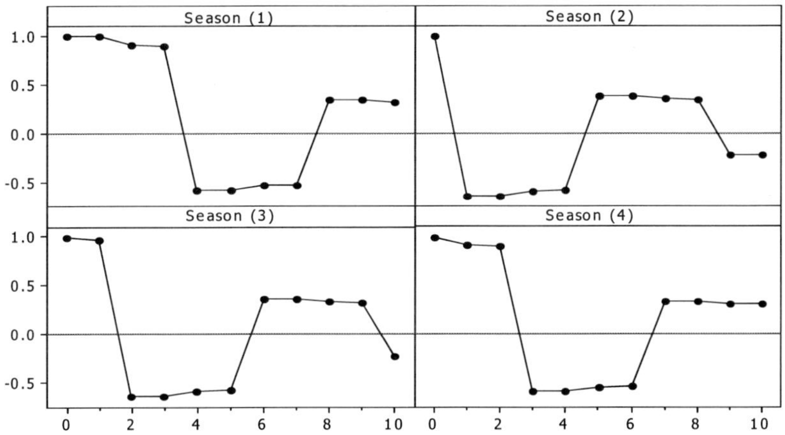

Figure 2. The SACF for Model (1).

$\phi_1(1)= 1.1$, $\phi_1(2)= -0.8$, $\phi_1(3)= 0.95, \phi_1(4)= 0.7$ and seasonal WN variances $\sigma_a^2(1) = 1$, $\sigma_a^2(2) = 64$, $\sigma_a^2(3) = 4$ and $\sigma_a^2(4) = 9$. This model will be referred to as Model (1). Now, using the fact that for the PAR$_\omega$(1) model $\gamma_0(\nu) = (\phi_1(\nu))^2\gamma_0(\nu-1)+\sigma_a^2(\nu)$, $\nu = 1, \ldots, \omega$. Thus for Model (1), $\gamma_0(1)= 73.8$, $\gamma_0(2)= 111.2$, $\gamma_0(3)= 104.4$ and $\gamma_0(4)= 60.1$. Using (2.2) we get $\rho_1(1)= 0.9932$, $\rho_1(2)= -0.6516$, $\rho_1(3)= 0.9807$ and $\rho_1(4)= 0.9221$. Now, using (2.1) the SACF is easily computed for all time lags. Figure (2) below shows the SACF of Model (1) for k = 0, …, 10. It can be seen that for each season the ACF has an exponential sinusoidal decay with periodicity 4. Also, notice that for all seasons, the autocorrelations at lags 4 and 8 are equal.

In the next section, Proposition (1) will be utilized to generalize estimates of $\rho_1(\nu)$ in (1.7)–(1.9) for all time lags. Also, it will later on be used for the computation of exact SACF for some specific PAR(1) models which is then used to compute biases of various estimates of SACF.

3. Robust Estimation of the SACF of PAR(1) Model

Assume that $\{X_{j\omega+\nu}\}$ is a periodic stationary PAR$_\omega$(1) process. Also, let $\{Z_1, Z_2, \ldots, Z_{n\omega}\}$ be an observed realization of length n periods (years, say) from $\{X_{j\omega+\nu}\}$ but contaminated with some additive outliers. Then the estimates of $\rho_1(\nu)$ given in (1.7)–(1.9) generalize to $\rho_k(\nu)$ as follows:

$$\hat{\rho}_k(\nu) = \frac{\sum_{j=1}^{n}(Z_{(j-1)\omega+\nu} - \overline{Z}_\nu)(Z_{(j-1)\omega+\nu-k} - \overline{Z}_{\nu-k})}{\sqrt{\sum_{j=1}^{n}(Z_{(j-1)\omega+\nu} - \overline{Z}_\nu)^2 \sum_{j=1}^{n}(Z_{(j-1)\omega+\nu-k} - \overline{Z}_{\nu-k})^2}} \tag{3.1}$$

$$\tilde{\rho}_k(\nu)=\tilde{\rho}_1(\nu)\tilde{\rho}_1(\nu-1)\cdots\tilde{\rho}_1(\nu-k+1) \tag{3.2}$$

$$\breve{\rho}_k(\nu)=\breve{\rho}_1(\nu)\breve{\rho}_1(\nu-1)\cdots\breve{\rho}_1(\nu-k+1). \tag{3.3}$$

The estimator in (3.1) is the moment estimator of $\rho_k(\nu)$ whereas those in (3.2) and (3.3) generalize the robust estimators of $\rho_1(\nu)$ in (1.8) and (1.9), respectively, for all time lags by making use of Proposition (1). For Gaussian periodic stationary processes $\hat{\rho}_k(\nu)$ is a consistent estimator of $\rho_k(\nu)$. However, as for ordinary ACF, it is expected to be affected by the presence of outliers.

Table 1. The absolute bias and MSE (in brackets) of the three estimators of the SACF of Model (1), n = 100 with no outliers

		Lag							
Season	Est.	1	2	3	4	5	6	7	8
	$\hat{\rho}_k(1)$	0.0104 (0.0002)	0.0112 (0.0007)	0.0110 (0.0008)	0.0121 (0.0063)	0.0186 (0.0068)	0.0188 (0.0079)	0.0192 (0.0082)	0.0222 (0.0122)
1	$\tilde{\rho}_k(1)$	0.0020 (0.0033)	0.0092 (0.0079)	0.0092 (0.0089)	0.0141 (0.0141)	0.0150 (0.0150)	0.0154 (0.0169)	0.0145 (0.0175)	0.0025 (0.0176)
	$\breve{\rho}_k(1)$	0.0290 (0.0037)	0.0724 (0.0179)	0.0801 (0.0234)	0.1588 (0.0473)	0.1685 (0.0507)	0.1674 (0.0531)	0.1651 (0.0546)	0.1386 (0.0388)
	$\hat{\rho}_k(2)$	0.0063 (0.0050)	0.0131 (0.0053)	0.0134 (0.0064)	0.0135 (0.0066)	0.0181 (0.0116)	0.0223 (0.0120)	0.0225 (0.0126)	0.0235 (0.0128)
2	$\tilde{\rho}_k(2)$	0.0065 (0.0170)	0.0087 (0.0167)	0.0134 (0.0152)	0.0141 (0.0141)	0.0004 (0.0205)	0.0009 (0.0201)	0.0024 (0.0182)	0.0025 (0.0176)
	$\breve{\rho}_k(2)$	0.1351 (0.0399)	0.1488 (0.0433)	0.1582 (0.0461)	0.1588 (0.0473)	0.1416 (0.0407)	0.1466 (0.0413)	0.1412 (0.0389)	0.1386 (0.0388)
	$\hat{\rho}_k(3)$	0.0004 (0.0000)	0.0061 (0.0053)	0.0130 (0.0056)	0.0129 (0.0067)	0.0132 (0.0069)	0.0178 (0.0117)	0.0222 (0.0121)	0.0226 (0.0126)
3	$\tilde{\rho}_k(3)$	0.0013 (0.0045)	0.0065 (0.0169)	0.0087 (0.0166)	0.0141 (0.0141)	0.0123 (0.0159)	0.0008 (0.0204)	0.0005 (0.0199)	0.0025 (0.0176)
	$\breve{\rho}_k(3)$	0.0121 (0.0049)	0.1374 (0.0417)	0.1506 (0.0449)	0.1588 (0.0473)	0.1573 (0.0497)	0.1393 (0.0408)	0.1442 (0.0413)	0.1386 (0.0388)
	$\hat{\rho}_k(4)$	0.0019 (0.0004)	0.0020 (0.0006)	0.0060 (0.0061)	0.0126 (0.0065)	0.0132 (0.0075)	0.0138 (0.0078)	0.0185 (0.0121)	0.0226 (0.0124)
4	$\tilde{\rho}_k(4)$	0.0061 (0.0078)	0.0063 (0.0086)	0.0114 (0.0153)	0.0141 (0.0141)	0.0139 (0.0171)	0.0130 (0.0177)	0.0008 (0.0185)	0.0025 (0.0176)
	$\breve{\rho}_k(4)$	0.0477 (0.0129)	0.0564 (0.0184)	0.1475 (0.0446)	0.1588 (0.0473)	0.1591 (0.0506)	0.1570 (0.0524)	0.1343 (0.0385)	0.1386 (0.0388)

To investigate the bias and MSE of the estimators (3.1)–(3.3) above but in the absence of outliers, we have generated 1000 realizations each of length n = 100 years from the $PAR_4(1)$ model (Model (1)) in Example (1) above. The simulations and computations in this article were carried out by an R-code (version 2.10.1) written by the authors. The results are summarized in Table (1). It can be seen that in view of both bias and MSE, $\breve{\rho}_k(\nu)$ was the worst. Also, the moment estimator $\hat{\rho}_k(\nu)$ slightly dominates $\tilde{\rho}_k(\nu)$ in view of MSE while both gave very close absolute bias.

4. Methodology and Simulation Results

In this section we use Monte-Carlo simulation to study the estimation of SACF of the $PAR_\omega(1)$ model using the proposed estimators in (3.1)–(3.3) for the presence of additive outliers. The investigation will be carried out through bias and MSE.

As far as the simulation-work is concerned; again, 1000 realizations each of length n = 30 years are simulated from the $PAR_\omega(1)$ model considered in Example (1) assuming that the white noise process is normal with mean zero and periodic variances $\sigma_a^2(\nu)$ as defined therein. Regarding the additive outliers we consider the following cases:

Case 1. A single additive outlier (Δ=100) is added at t =13 (season one).

Case 2. Four additive outliers (Δ_1=100, Δ_2=80, Δ_3=120, Δ_4=90) are added at t =13, 14, 15 and 16.

For Case (1), the simulation results are presented in Table (2) and Figures (3)–(4). For Case (2) we summarize the results in Figures (5)–(6).

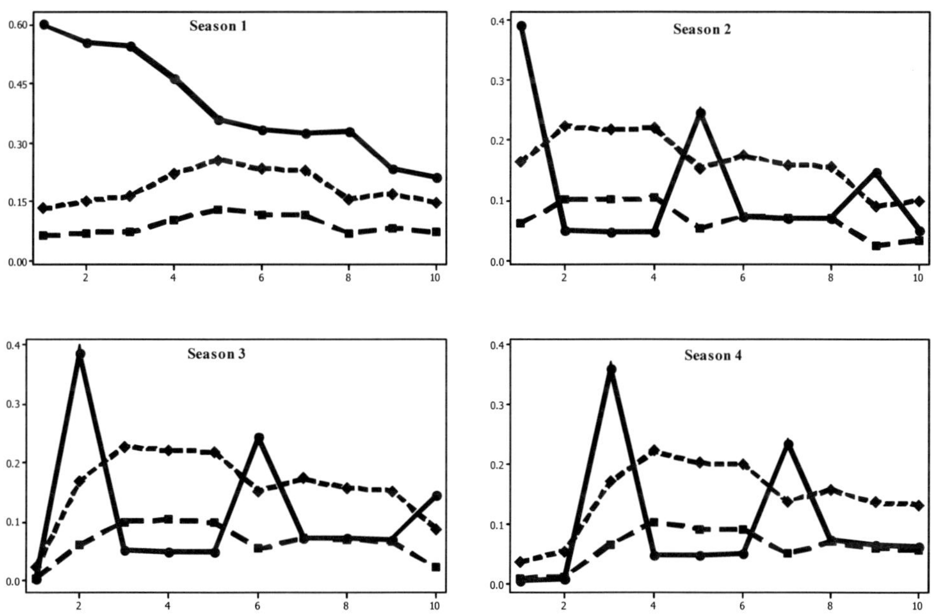

Figure 3. The absolute bias for the three estimators of the SACF for Model (1) with a single additive outlier at season one and n=30 (— : $\hat{\rho}_k(\nu)$, — : $\tilde{\rho}_k(\nu)$, - - - : $\breve{\rho}_k(\nu)$) with time-lag k on the X-axis.

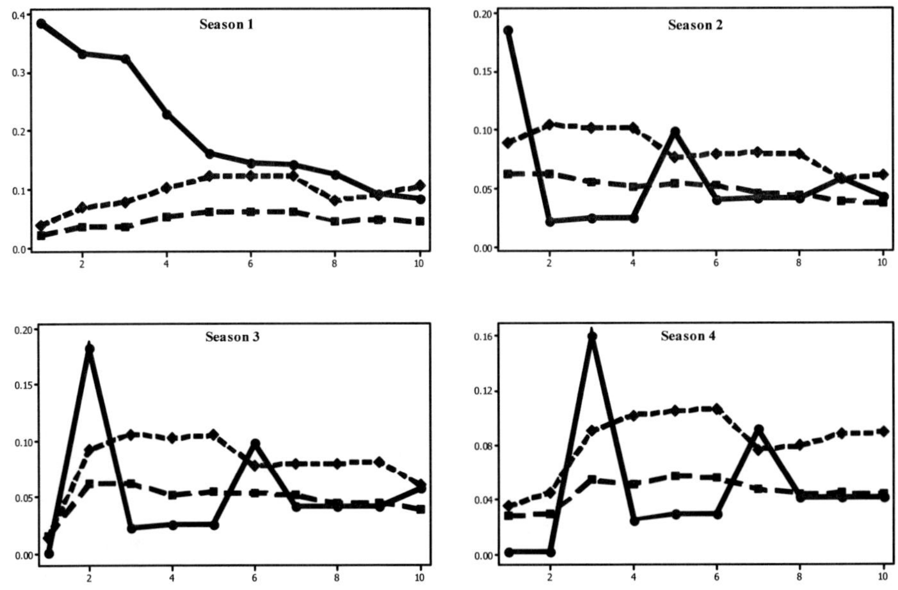

Figure 4. The MSE for the three estimators of the SACF for Model (1) with a single additive outlier at season one and n=30 (**—** : $\hat{\rho}_k(\nu)$, — : $\tilde{\rho}_k(\nu)$, - - - : $\breve{\rho}_k(\nu)$) with time-lag k on the X-axis.

5. Discussion of Results

We can see from Table (2) as well as Figures (3)–(6) that the moment estimator of the SACF, $\hat{\rho}_k(\nu)$ was highly affected by the presence of outliers in view of bias and MSE whereas the other two estimators were less affected. The differences in bias and MSE between $\hat{\rho}_k(\nu)$ and other estimators are clearly seen in the first few lags but decrease for larger time lags. More specifically, in Case (1) which has a single additive outlier (added at season 1) the bias and MSE of $\hat{\rho}_k(\nu)$ were the largest for all time lags in season (1) as seen in Figures (3) and (4). The situation for the SACF in other seasons was different. The additive outlier has left a clear affect on the bias and MSE at some specific lags with periodicity $\omega = 4$. For instance, in Figures (3) and (4) the bias and MSE of $\hat{\rho}_k(\nu)$ in season (2) were the largest in lags 1, 5 and 9. This is due to the inter-relations among periodic autocorrelations of all seasons as given by (2.1).

For the results of Case (2), in which four outliers are added in all seasons, Figures (5) and (6) show that $\hat{\rho}_k(\nu)$ was dominated by the other two estimators in view of bias and MSE for all time lags.

Finally, the estimator $\tilde{\rho}_k(\nu)$ defined by (1.8) and (3.2) was better than $\breve{\rho}_k(\nu)$ in all cases and for all time lags. A similar conclusion was obtained by Smadi et al. (2009) for the estimation of the first lag seasonal autocorrelation, $\rho_1(\nu)$. This result could be attributed to the fact that $\tilde{\rho}_k(\nu)$ is a model-based estimator, i.e. it assumes that the true model of the time series is PAR(1) whereas the estimators $\hat{\rho}_k(\nu)$ and $\breve{\rho}_k(\nu)$ are model-free estimators.

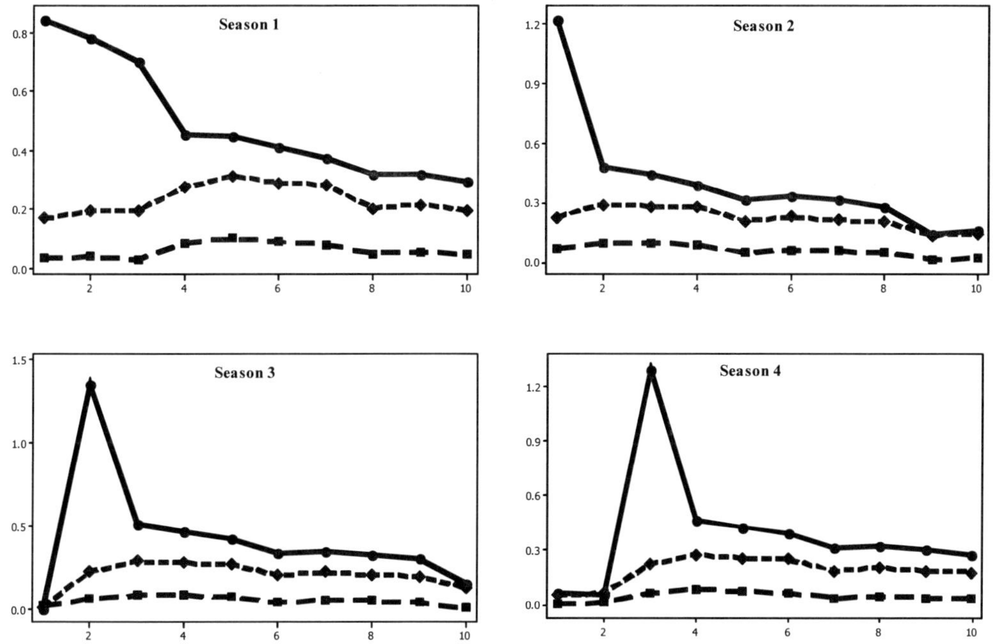

Figure 5. The absolute bias for the three estimators of the SACF for Model (1) with four additive outliers (Case (2)) and n=30 (— : $\hat{\rho}_k(\nu)$, — : $\widetilde{\rho}_k(\nu)$, - - - : $\breve{\rho}_k(\nu)$) with time-lag k on the X-axis.

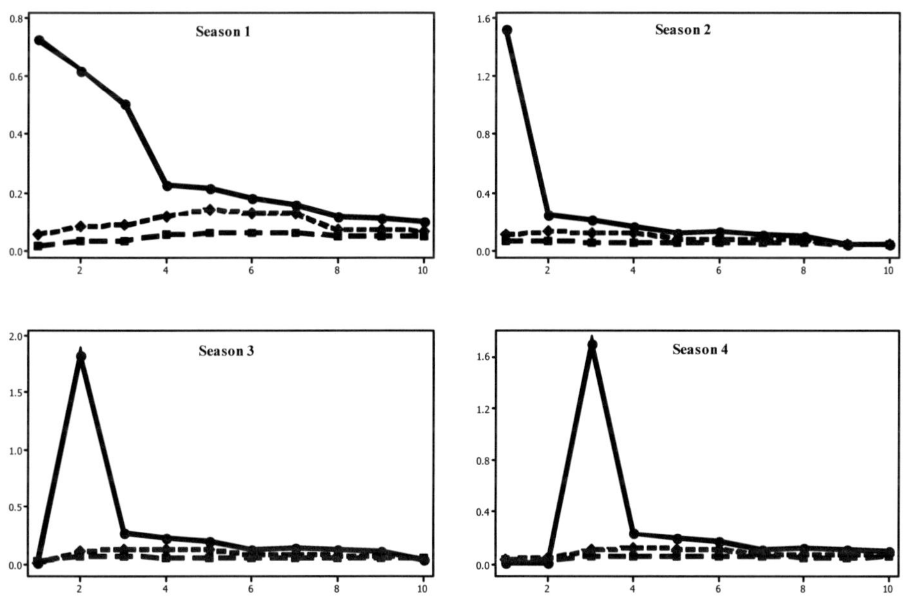

Figure 6. The MSE for the three estimators of the SACF for Model (1) with four additive outliers (Case (2)) and n=30 (— : $\hat{\rho}_k(\nu)$, — : $\widetilde{\rho}_k(\nu)$, - - - : $\breve{\rho}_k(\nu)$) with time-lag k on the X-axis.

Table 2. The absolute bias and MSE (in brackets) of the three estimators of the SACF of Model (1) with a single additive outlier at season one, n = 30

		Lag							
Season	**Est.**	1	2	3	4	5	6	7	8
1	$\hat{\rho}_k(1)$	0.6023	0.5544	0.5461	0.4641	0.3592	0.3328	0.3270	0.3307
		(0.3862)	(0.3332)	(0.3245)	(0.2285)	(0.1616)	(0.1454)	(0.1422)	(0.1235)
	$\tilde{\rho}_k(1)$	0.0617	0.0697	0.0712	0.1032	0.1293	0.1178	0.1154	0.0693
		(0.0204)	(0.0355)	(0.0359)	(0.0516)	(0.0600)	(0.0605)	(0.0594)	(0.0440)
	$\breve{\rho}_k(1)$	0.1311	0.1509	0.1634	0.2206	0.2560	0.2323	0.2296	0.1561
		(0.0375)	(0.0675)	(0.0764)	(0.1021)	(0.1207)	(0.1218)	(0.1212)	(0.0795)
2	$\hat{\rho}_k(2)$	0.3935	0.0508	0.0473	0.0474	0.2470	0.0721	0.0708	0.0712
		(0.1867)	(0.0222)	(0.0248)	(0.0250)	(0.0988)	(0.0406)	(0.0417)	(0.0416)
	$\tilde{\rho}_k(2)$	0.0607	0.1004	0.1015	0.1032	0.0539	0.0739	0.0691	0.0693
		(0.0634)	(0.0630)	(0.0560)	(0.0516)	(0.0542)	(0.0523)	(0.0465)	(0.0440)
	$\breve{\rho}_k(2)$	0.1633	0.2239	0.2182	0.2206	0.1537	0.1751	0.1583	0.1561
		(0.0898)	(0.1048)	(0.1022)	(0.1021)	(0.0766)	(0.0801)	(0.0805)	(0.0795)
3	$\hat{\rho}_k(3)$	0.0016	0.3870	0.0503	0.0477	0.0485	0.2430	0.0722	0.0713
		(0.0001)	(0.1814)	(0.0225	(0.0255)	(0.0258)	(0.0972)	(0.0408)	(0.0418)
	$\tilde{\rho}_k(3)$	0.0030	0.0606	0.0994	0.1032	0.0983	0.0528	0.0723	0.0693
		(0.0149)	(0.0617)	(0.0614)	(0.0516)	(0.0537)	(0.0529)	(0.0510)	(0.0440)
	$\breve{\rho}_k(3)$	0.0212	0.1690	0.2275	0.2206	0.2183	0.1516	0.1726	0.1561
		(0.0136)	(0.0918)	(0.1058)	(0.1021)	(0.1045)	(0.0771)	(0.0771)	(0.0795)
4	$\hat{\rho}_k(4)$	0.0060	0.0074	0.357	0.0490	0.0489	0.0499	0.2333	0.0728
		(0.0015)	(0.0022)	(0.1608	(0.0251)	(0.0289)	(0.0295)	(0.0916)	(0.0416)
	$\tilde{\rho}_k(4)$	0.0089	0.0117	0.0656	0.103	0.0914	0.0896	0..0499	0.0693
		(0.0284)	(0.0290)	(0.0543)	(0.0516)	(0.0571)	(0.0558)	(0.0473)	(0.0440)
	$\breve{\rho}_k(4)$	0.0372	0.0540	0.1698	0.2206	0.2004	0.1982	0.1371	0.1561
		(0.0349)	(0.0450)	(0.0904)	(0.1021)	(0.1052)	(0.1067)	(0..0768)	(0.0795)

6. An Application on Real Data

In this section we will apply the various estimators of SACF given by (3.1)–(3.3) on the time series of river flows of Fraser river. The data expand for 78 years from January 1914 to December 1991 (Hiple and McLeod, 1994). For simplicity, we transformed (aggregated) this time series into quarterly time series. We have selected this time series because of the nature of hydrological time series that are usually well fitted by PAR models (Hiple and McLeod, 1994). The parallel box-plot of this time series is shown in Figure (7), from which we conclude that some outliers are detected. This approach for detecting outliers in such time series was done for example by Shao (2007). Also note form Figure (7) that in this time series the medians as well the variability are different across various seasons.

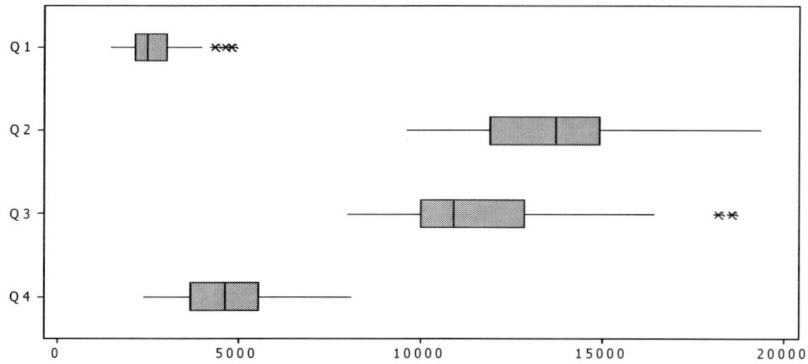

Figure 7. The box plot of quarterly (marginal) river-flow data.

Assuming that this quarterly time series follows the $PAR_4(1)$ model (for a justification, see Thomas and Fiering (1962) and McLeod (1994)), we will estimate the SACF of this time series using (3.1)–(3.3). To measure the accuracy of various estimates of the SACF we used bootstrap method by selecting 20 random samples each of length 10 (consecutive) years. Then, the average estimates of SACF and its standard error are obtained based on the 20 samples for all estimators of SACF. The standard errors of various estimates are shown in Figure (8).

Again, Figure (8) shows that the moment estimator of the SACF, $\hat{\rho}_k(\nu)$ had the largest standard errors except for the first few lags. The other two estimators had nearly similar standard errors which also decrease to zero for large time lags.

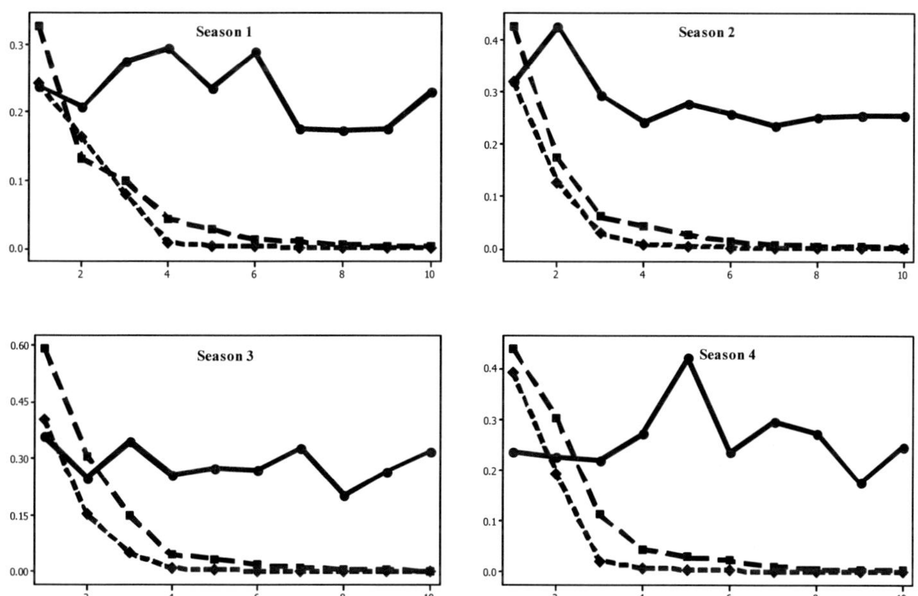

Figure 8. The SE for the three estimators of the SACF of the real data for $PAR_4(1)$ model. (— : $\hat{\rho}_k(\nu)$, — : $\tilde{\rho}_k(\nu)$, - - - : $\breve{\rho}_k(\nu)$) with time-lag k on the X-axis.

7. Conclusion

This article focused on the robust estimation of the SACF of PAR(1) models. When the time series contained some additive outliers, the moment estimator of the SACF was badly affected in view of bias and MSE. This fact was more apparent in the sub-SACF for seasons in which the outliers were present. The other two proposed estimators were more robust for the additive outliers. Therefore, in model identification and fitting of PAR models, one should screen the data for any plausible outliers prior to computing the sample (moment) SACF as it will badly estimate the theoretical SACF once some outliers are involved in the time series. Future work should focus on other aspects of robust estimation of PAR as well as PARMA models toward other types of outliers and the shape of the distribution of the white noise process, which is usually assumed normal.

References

Barnett, V. and Lewis, T. (1994). *Outliers in statistical Data*. 3rd Edition. Wiley, Chichester.

Berkoun, Y., Fellag, H. and Zielinski, R. (2003). Robust Testing Serial Correlation in AR(1) Processes in the Presence of a Single Additive Outlier. *Communications in Statistics Theory and Method*. 32, 1527-1540.

Bianco, A., Garcia, B., Martinez, E. and Yohai, V. (1996). Robust procedures for regression models with ARIMA errors. In *COMPSTAT 96, Proceedings in Computational Statistics*, Part A (ed.). Physica-Verlag: Berlin, 27-38.

Box, G., Jenkins, G. and Reinsel, G. (1994). *Time Series Analysis, Forecasting and Control.* 3rd Edition. Prentice-Hall, New Jersey.

Bustos, O. and Yohai, V. (1986). *Robust estimates of ARMA models*. JASA. 81, 69-155.

Chang I. and Tiao, G. (1983). Effect of exogenous intervention on the estimation of time series parameters. *Proceeding of the American Statistical Association*, Business and Economics Statistics Section. 532-537.

Denby, L. and Martin, R. (1979). Robust estimation of the first-order autoregressive parameter, *JASA*. 74, 140-146.

Franses P. and Paap, R. (2004). *Periodic Time Series Models*, Oxford Univ. Press.

Haddad, J. (2000). On robust estimation in the first-order autoregressive processes. *Comm. in Statistics, Theory and Methods*. 29, 45-54.

Hiple, W. and McLeod, A. (1994). *Time Series Modeling of Water Resources and Environmental Systems*. Elsevier Publisher.

Hurwicz, L. (1950). Least-Squares Bias in Time Series. In: Koopmans, T. C., ed. *Statistical Inference in Dynamic Economic Models*. Wiley.

McLeod, A. (1993). Parsimony, Model Adequacy and Periodic Correlation in Time Series Forecasting. *International Statistical Review*. 61, 387-393.

McLeod, A. (1994). Diagnostic Checking Periodic Autoregression Models with Application. *The Journal of Time Series Analysis*. 15, 221-233.

Obeysekera, J. and Salas, J. (1986). Modeling of Aggregated Hydrologic Time Series. *J. Hydrol*. 86, 197-219.

Pagano, M. (1978). On Periodic and Multiple Autoregressions. *Annals of Stat.* 6, 1310-1317.

Sarnaglia, A., Reisen, V. and Lévy-Leduc, C. (2010). Robust estimation of periodic autoregressive processes in the presence of additive outliers, *J. Mult. Analysis*, 101, 2168-2183.

Shao, Q. (2007). Robust estimation for periodic autoregressive time series. *Journal of Time Series Analysis*. 29, 251-263.

Smadi, A., Abu-Afouna, N. and Al-Quraan, A. (2009). Robust Estimation of the Seasonal Autocorrelation of The PAR(1) Model. *Jordan Journal of Mathematics and Statistics*. 2, 105-118.

Thomas, H. and Fiering, M. (1962). Mathematical synthesis of stream flow sequence for the analysis of river basins by simulation. In Maass, A., Hufshmidt, M., Dorfman, R., Thomas, Jr., Marglin, S., Fair, M., Editors, *Design of Water Resources System*.

Ula, T. and Smadi, A. (2003). Identification of Periodic Moving-Average Models. Communications In *Statistics, Theory and Methods*. 32, 2465–2475.

In: Research in Applied Statistical Science
Editor: Mohammad Ahsanullah, pp. 199-208
ISBN: 978-1-63321-818-5

Chapter 17

Regression Models for Poisson Data Based on Accelerated Lifetime Testing

***Francisco Louzada-Neto*[1] *and Vera Lucia Damasceno Tomazella*[2,*]**
[1]Department of Applied Mathematics and Statistics,
Universidade de São Paulo, Brazil
[2]Department of Statistics, Universidade Federal de São Carlos,
São Carlos, Brazil

Abstract

The accelerated failure time model (AFT) is an important alternative to the Cox proportional hazards model (PH) in survival analysis. In this paper we present a natural extension of the conventional accelerated failure time model to accommodate recurrent events. This extension parallels the Lawless (1987) extensions of the classical Cox proportional hazard model. Like Lawless we formulate the effects of covariate on the intensity function of Poisson process for recurrent events. The proposed model retains the direct physical interpretation of the original accelerated failure time model in that the role of the covariates is to accelerate or decelerate the time to each recurrence. These include parametric approaches to model fitting, we consider estimate the vector of regression parameters under this model and the baseline intensity are considered specified up to a parameter θ. This methodology is illustrated with a known real data set.

Keywords: Accelerated failure time model, Survival analysis, Intensity function, recurrent events

1. Introduction

In long-term follow-up studies, individual subjects may experience recurrent or repeated events. Lifetime data where more than one event is observed on each subject arise in areas such as, biomedical studies, criminology, demography, manufacturing and industrial reliability. For example, in medical research include the sequence of asthmatic attacks,

*E-mail address: vera@ufscar.br (Corresponding author)

epileptic seizures, several tumors may be observed for an individual and recurrent pneumonia episodes arise in patients with human immunodeficiency syndrome while in industry recurrent events a piece of equipment may experience repeated failures or warranty claims. Processes involving recurrent events are common in survival analysis and many other areas. Several articles have approached the problem of recurrent events suggesting different methodologies Lawless & Nadeans (1995) apply the Poisson process to develop models that focus on the expected number of events occurred in determined time interval and Lin et al. (1998) presented the (AFT) to formulate the effects of covariate on the mean function of the counting process for recurrent events. In this article the problem is treated under the focus of punctual counting process where we formulate the effects of covariate on the intensity function of Poisson process for recurrent events. The development of statistical models based on counting process to analysis by recurrent events data were originally introduced by Aalen (1978). There is an extensive literature about point process models, (e.g., Cox & Isham 1980), this approach offers us tools powerful enough able to generalize several situation. Poisson process has been well studied many recent discussion about lifetime and stochastic process transition data have focused on modelling.

One of the most significant advances in classical survival data, was the introduction of the Cox proportional hazard model (Cox, 1972). Anderson & Gill (1982) introduced of the counting process model with the Cox-type intensity function for recurrent events. An alternative to the Cox proportional hazards model (PH) (Cox, 1972) is the accelerated failure time (AFT) model (Kalbfleisch & Prentice, 1980, pp. 32-4; Cox & Oakes, 1984, pp. 64-5). However, as argued by Wei (1992) and Cox (1997), the accelerated failure time (AFT) model has the advantage of being more intuitively interpretable than the PH model. The accelerated failure time model, relates the logarithm of the failure time linearly to the covariates (Kalbfleisch & Prentice, 1980, pp. 32-4; Cox & Oakes, 1984, pp. 64-5). So the accelerated failure model is described by the modification in the failure time due to the application of stress or a treatment or an exposure.

This article is directed toward situations where individuals can experience repeated events, and data on the i-th individual consists of the total number, m_i of the events (lifetimes) observed over the time period $(0, T]$ and the ordered epoch of the m_i recurrent at times $0 \leq t_{i1} < t_{i2} < \cdots < t_{im} \leq T_i$. Additionally, we may have covariate information on each subject defined by a vector of censoring indicators. In such studies, we provide a natural extension of the aforementioned accelerated failure time model to accommodate reccurents events. This extension parallels the Lawless (Lawless, 1987) extensions of the classical Cox proportional hazards model. Like Lawless, we formulate the intensity functions of Poisson process. The proposed model retains the direct physical interpretation of the original accelerated failure time model in that the role of the covariates is to accelerate or decelerate the time to each recurrence. Methods of regression analysis are presented, these include parametric approaches to model fitting. Section 2 reviews parametric methods, where we describe associated inference procedures, Section 3 presents an example and Section 4 concludes remarks.

2. Parametric Poisson Process Methods

The Cox proportional hazards model (PH) (Cox, 1972) is probably the most widely used model in survival regression analysis. The great limitation of the Cox ′s model is the proportionality assumption enters the risk functions in different levels of the covariates. As pointed out by D. R. Cox (Reid, 1994, p. 450), accelerated life model are in many way more appealing than the proportional hazards model because of their quite direct physical interpretation.

The methods in the article are based on the so-called accelerated time intensity Poisson process model, defined as follows. An individual with $k \times 1$ covariate vector $\mathbf{x}$ has repeated events occur according to a nonhomogeneous Poisson process with intensity function

$$\lambda\left(t|\mathbf{x}\right) = \lambda_0\left(e^{\mathbf{x}'\beta}t\right)e^{\mathbf{x}'\beta}, \tag{1}$$

where $\lambda_0\left(t\right)$ is a baseline intensity function and $e^{\mathbf{x}'\beta}$ is a positive-valued function of the $\mathbf{x}$ and a vector of parameters β. This conditional hazard function scales the baseline hazard function in time depending on the covariates. So the effect of the covariates is to accelerate or decelerate the aging process, thus influencing the time of failure, depending on the relevant characteristics of the individual.

Define $m_i(t)$ as the number of events that have occurred on the ith subject by time t . The mean function of the counting process $m_i(t)$ associated with a p-vector of covariates $\mathbf{x}_i$ it is $\lambda_0\left(e^{\mathbf{x}'\beta}t\right)$, that is, the expected number of events by time t under $\mathbf{x}_i$ equals the expected number of events by $e^{\mathbf{x}'\beta}$ under $\mathbf{x}_i = 0$. In other words, the covariates $\mathbf{x}$ affects the frequency of recurrences over time by expanding or contracting the time scale on which the events occur by a multiplicative factor of $e^{\mathbf{x}'\beta}$ relative to that of a zero-valued covariate vector.

The corresponding cumulative or integrated intensity function is

$$\Lambda_x\left(t\right) = \int_0^t \lambda_x\left(u\right)du = \Lambda_0\left(e^{\mathbf{x}'\beta}t\right)e^{\mathbf{x}'\beta}, \tag{2}$$

where $\Lambda_0\left(t\right) = \int_0^t \lambda_0\left(e^{\mathbf{x}'\beta}u\right)du$.

Supposed that there are independent observations on n individual; individual i is observed over the time interval (S_i, T_i) and m_i events are observed to occur, at time $t_{i1} < \cdots < t_{im_i}$. For simplicity of exposition we assume that all of the $S_i's$ equal 0. If individual i has covariate vector $\mathbf{x}_i$, then considering the Poisson process model (1) with (2) the likelihood function in the form count process is given by

$$L\left(\theta,\beta\right) = \prod_{i=1}^{n}\left\{\prod_{j=1}^{m_i}\lambda_{x_i}\left(e^{\mathbf{x}_i'\beta}t_{ij}\right)\right\}\exp\left\{-\Lambda_{x_i}\left(e^{\mathbf{x}_i'\beta}T_i\right)\right\}, \tag{3}$$

which can be decomposed as

$$L(\theta,\beta) = \left\{ \prod_{i=1}^{n} \prod_{j=1}^{m_i} \frac{\lambda_0\left(e^{\mathbf{x}_i'\beta} t_{ij}\,;\,\theta\right)}{\Lambda_0\left(e^{\mathbf{x}_i'\beta} T_i\,;\,\theta\right)} \right\} \tag{4}$$

$$\times \prod_{i=1}^{n} \exp\left[\Lambda_0\left(e^{\mathbf{x}_i'\beta} T_i\,;\,\theta\right) e^{x_i'\beta}\right] \left[\Lambda_0\left(e^{\mathbf{x}_i'\beta} T_i\,;\,\theta\right) e^{x_i'\beta}\right]^{m_i}$$

$$= L_1(\theta,\beta)\, L_2(\theta,\beta). \tag{5}$$

The likelihood kernel $L_2(\theta,\beta)$ arises form the Poisson distribution of the counts $m_1, \cdots,$
m_n and the kernel $L_1(\theta,\beta)$ arises from the conditional distribution of the event times, given the counts.

2.1. Weibull Accelerated Failure Time Model

The Weibull distribution has been extensively used to model the baseline hazard function due to its simplicity and flexibility (e.g., Crow, 1974, Lee and Lee, 1978). where its risk function is defined by

$$\lambda_0(t) = \delta t^{\delta-1}. \tag{6}$$

Considering the model (1) we have

$$\lambda_0\left(e^{\mathbf{x}'\beta} t\right) = \delta\left(e^{\mathbf{x}'\beta} t\right)^{\delta-1}, \;\; \Lambda_0\left(e^{\mathbf{x}'\beta} t\right) = \left(e^{\mathbf{x}'\beta}\right)^{\delta-1} t^{\delta}. \tag{7}$$

From (4) the likelihood function for δ and β is given by

$$L(\delta,\beta) = \left\{ \prod_{i=1}^{n} \prod_{j=1}^{m_i} \frac{\delta t_{ij}^{\delta-1}}{T_i^{\delta}} \right\} \prod_{i=1}^{n} \exp\left[\left(e^{\mathbf{x}_i'\beta} T_i\right)^{\delta}\right] \left[\left(e^{\mathbf{x}_i'\beta} T_i\right)^{\delta}\right]^{m_i}, \tag{8}$$

where we have $L_1(\delta)$ and $L_2(\delta,\beta)$ are the same as in (4).

The maximization of (8) with respect to δ and β is generally straightforward. At the same time, however, we note the possibility of estimating δ from $L_1(\delta)$ without having to consider β, and also observe that, given δ, the likelihood kernel $L_2(\delta,\beta)$ is simply that for standard Poisson log-linear model (e.g., see McCullagh & Nelder(1993, chap.6).

The log-likelihood function for δ and β is given by

$$l(\delta,\beta) = n\log\delta + (\delta-1)\sum_{i=1}^{n}\sum_{j=1}^{m_i}\log t_{ij} - \sum_{i=1}^{n}\left(e^{\mathbf{x}_i'\beta} T_i\right)^{\delta} + \delta\sum_{i=1}^{n} m_i x_i'\beta, \tag{9}$$

where $n = \sum_{i=1}^{n} m_i$ in these expressions:

The first and second derivatives of (9) are given by $\frac{\partial l}{\partial\delta} = \frac{n}{\delta} + \sum_{i=1}^{n}\sum_{i=1}^{m_i}\log t_{ij} -$
$\sum_{i=1}^{n}\left(e^{\mathbf{x}_i'\beta} T_i\right)^{\delta}\log\left(e^{\mathbf{x}_i'\beta} T_i\right) + \sum_{i=1}^{n} m_i \mathbf{x}_i'\beta;$

$$\frac{\partial l}{\partial \beta_r} = -\delta \sum_{i=1}^{n} \left(e^{\mathbf{x}_i'\beta} T_i\right)^{\delta} x_{ir} + \sum_{i=1}^{n} m_i x_{ir}\,, \qquad r = 0, 1, \ldots, k;$$

$$\frac{\partial l}{\partial \delta^2} = -\frac{n}{\delta^2} - \sum_{i=1}^{n} \left(e^{\mathbf{x}_i'\beta} T_i\right)^{\delta} \log\left(e^{\mathbf{x}_i'\beta} T_i\right)^2;$$

$$\frac{\partial l}{\partial \beta_r \beta_s} = -\delta^2 \sum_{i=1}^{n} \left(e^{\mathbf{x}_i'\beta} T_i\right)^{\delta} x_{ir} x_{is}, \qquad r, s = 0, 1, \ldots, k;$$

$$\frac{\partial l}{\partial \delta \partial \beta_r} = -\sum_{i=1}^{n} \left(e^{\mathbf{x}_i'\beta} T_i\right)^{\delta} x_{ir} \left[\delta \log\left(e^{\mathbf{x}_i'\beta} T_i\right) + 1\right] + \sum_{i=1}^{n} m_i x_{ir}, \qquad r = 0, 1, \ldots, k.$$

So considering $L_1(\delta)$ the maximum likelihood estimates (mle) of δ is given by

$$\overset{\wedge}{\delta} = \frac{-m}{\sum\limits_{i=1}^{n} \sum\limits_{j=1}^{m_i} \log(\frac{t_{ij}}{T})}. \tag{10}$$

The mle $\overset{\wedge}{\beta}$ can then be found by solving the equations $\frac{\partial l}{\partial \beta_r} = 0$ $(r = 0, 1, 2, ...k)$ iteratively, with δ set equal to $\overset{\wedge}{\delta}$. Theses are simply the maximum likelihood equations for the Poisson log-linear model.

Furthermore, Fisher information matrix entries are

$$I_{\delta\delta} = E\left\{-\frac{\partial^2 l}{\partial \delta^2}\right\} = \frac{n}{\delta^2} + \sum_{i=1}^{n} \left(e^{\mathbf{x}'\boldsymbol{\beta}} T_i\right)^{\delta} \log\left(e^{\mathbf{x}'\boldsymbol{\beta}} T_i\right)^2;$$

$$I_{\beta_r \beta_s} = E\left\{-\frac{\partial^2 l}{\partial \beta_r \beta_s}\right\} = \delta^2 \sum_{i=1}^{n} \left(e^{\mathbf{x}'\boldsymbol{\beta}} T_i\right)^{\delta} x_{ir} x_{is}, \quad r, s = 0, 1, \ldots, k;$$

$$I_{\beta_r \beta_s} = E\left\{-\frac{\partial^2 l}{\partial \delta \partial \beta_r}\right\} = \sum_{i=1}^{n} \left(e^{\mathbf{x}'\boldsymbol{\beta}} T_i\right)^{\delta} x_{ir} \left[\delta \log\left(e^{\mathbf{x}'\boldsymbol{\beta}} T_i\right)\right], \quad r = 0, 1, \ldots, k.$$

Interval estimates and hypothesis tests for the parameters can be performed, in principle, by considering the asymptotic normal distribution of the mle's and the asymptotic chi-squared distribution of the likelihood ratio statistics, respectively (Lawles,1982).

In the accelerated failure-time model with baseline hazard Weibull have exactly the expression that we have for the proportional hazard rate model, hence, the equivalence of a proportional hazard rate model and an accelerated failure-time model is a characterization of the Weibull distribution (that is, only the Weibull has this property), the accelerated failure time (AFT) approach is an alternative strategy for the analysis of time-to-event data and can be suitable even when hazards are not proportional.

2.2. Log-Logistic Accelerated Failure Time Model

A random variable is loglogistically distributed if the logarithm of the random variable is logistically distributed (see eg.Meeker & Escobar, 1998). Because of this, there are many mathematical similarities between the two distributions. Such model appears to be very interesting, particularly for its hazard rate:

$$\lambda_0(t) = \frac{\lambda\alpha(\lambda t)^{\alpha-1}}{1 + (\lambda t)^{\alpha}}, \tag{11}$$

wehre α is a shape parameter, and λ is a scale parameter, which is always decreasing with time if $\alpha \leq 1$; first increasing, then decreasing with time if $\alpha > 1$. This and other motivations, make the Log-logistic failure time distribution a good candidate used to the analysis of time-to-event data.

To examine and illustrate the parametric approach we consider the log-logistic model (11) with parameter of scale $\lambda = 1$. Then considering the model (1) we have

$$\lambda_0\left(e^{\mathbf{x}'\beta}t\right) = \frac{\alpha(e^{\mathbf{x}'\beta}t)^{\alpha-1}}{1+(e^{\mathbf{x}'\beta}t)^{\alpha}}, \tag{12}$$

where $\mathbf{x}$ and a vector of parameters β are defined in (1). The corresponding cumulative function is

$$\Lambda_0\left(e^{\mathbf{x}'\beta}t\right) = \left(e^{\mathbf{x}'\beta}\right)^{-1}\log[1+(e^{\mathbf{x}'\beta}t)^{\alpha}]. \tag{13}$$

From (4) the likelihood function fo α and β is given by

$$\begin{aligned} L(\alpha,\beta) &= \left\{\prod_{i=1}^{n}\prod_{j=1}^{m_i}\frac{\alpha(e^{\mathbf{x}'\beta})^{\alpha}(t_{ij})^{\alpha-1}}{(1+(e^{\mathbf{x}'\beta}t_{ij})^{\alpha})\log[1+(e^{\mathbf{x}'\beta}T_i)^{\alpha}]}\right\}\times \\ &\prod_{i=1}^{n}\exp\left[-\log[1+(e^{\mathbf{x}'\beta}T_i)^{\alpha}]\right]\left[\log[1+(e^{\mathbf{x}'\beta}T_i)^{\alpha}]\right]^{m_i}. \end{aligned} \tag{14}$$

The log-likelihood function for α and β is given by

$$\begin{aligned} l(\alpha,\beta) &= n\log\alpha+(\alpha-1)\sum_{i=1}^{n}\sum_{j=1}^{m_i}\log t_{ij}+\alpha\sum_{i=1}^{n}m_i x_i'\beta - \\ &\sum_{i=1}^{n}\sum_{j=1}^{m_i}\log\left(1+(e^{\mathbf{x}'\beta}t_{ij})^{\alpha}\right)-\sum_{i=1}^{n}\log\left(1+(e^{\mathbf{x}'\beta}T_i)^{\alpha}\right). \end{aligned} \tag{15}$$

The maximization of (14) with respect to α and β is generally straightforward. So interval estimates and hypothesis tests for the parameters can be performed, by considering the asymptotic normal distribution of the mle's or the asymptotic chi-squared distribution of the likelihood ratio statistics, (Lawles,1982).

3. Example

In this work the methodology is illustrated using the animals carcinogenicity data describe by Gail et al. (1980) (Table 1). The experiment used 48 female rats to the development of tumor mammary. The animals in question there were 23 rats in Group 1 (Treatment) and 25 in Group 2 (Control) and the data are the days on which new tumors occurred for each animal; a given animal may experience any number of tumors. The main objective of analysis was to assess the difference between Groups 1 and 2 with regard to the development of tumors. The rats in study were all induced to remain tumor-free for 60 days and were observed over the period 60 to 182 days.

Considering the Weibull accelerated failure time model with intensity function (7), suppose that the ith individual rat has tumors occur according to a Poisson process with intensity function (7), where the $\lambda_0(t)$, have form Weibull and $\mathbf{x}_i$ is a dummy covariate indicating whether an individual is Group 1 $(x_i = 1)$ or Group 2 $(x_i = 0)$. Time t is defined to be the number of days from the start of the study minus 60, so that the observation intervals $(0, T_i)$ are $(0, 122)$ for all animals. The observation interval is the same for all individuals here.

Table 1. Passed times in days until the occurrence of new tumors for 48 rats

n° of tumors	Treatment (23 rats)	n° of tumors	Control (25 rats)
1	182	7	63,102,119,161,161,172,179,182†
0	182†	11	88,91,95,105,112,119,119,137,145,167,172,182†
2	63,68,182†	9	91,98,108,112,134,137,161,179,182†
1	152,182†	2	71,174,182†
4	130,134,145,152,182†	9	95,105,134,134,137,140,145,150,150,182†
3	98,152,182	4	68,68,130,137,182†
6	98,95,105,130,137,167,182†	6	77,95,112,137,161,174,182†
1	152,182†	7	81,84,126,134,161,161,174,182†
1	181,182†	6	68,77,98,102,102,102,182†
5	71,84,126,134,152,182†	1	112,182†
2	116,130,182†	13	88,88,91,98,112,134,134,137,137,140,140,
1	91,182†	2	152,152,182†
5	63,68,84,95,152,182†	2	77,179,182†
2	105,152,182†	1	112,182†
3	63,102,152,182†	10	71,71,74,77,112,116,116,140,140,140,167,182†
4	63,77,112,140,182†	4	77,95,126,150,182†
5	77,119,152,161,167,182†	5	88,126,130,130,134,182†
5	105,112,145,161,182	11	63,74,84,84,88,91,95,108,134,137,179,182†
1	152,182†	11	81,88,105,116,123,140,145,152,161,161,179,182†
2	81,95,182†	9	88,95,112,119,126,126,150,157,179,182†
6	84,91,102,108,130,130,134,182†	12	68,68,84,84,102,105,119,123,123,137,161,179,182
0	182†	1	140,182†
1	91,182†	3	152,182,182
		1	81,182†
		3	63,88,134,182†
		3	84,134,182

† indicates that in this time no new tumor was found

The mle's and estimated standard deviations (DP) and 95% confidence interval (IC) of the parameters for the model proposed are given in the Table 2. In this simple two-sample problem, it is easily seen that the significance of the treatment.

We have estimate that hazard taxe for each time was dependending on whether $x = 0$ or 1 (Group control or Group treatment)

$$\lambda\left(t|\mathbf{x}\right) = 1.107 time^{0.107}\left[\exp\left(-3.179 - 0.743\, x\right)\right]^{1.107}$$

That is, for fixed value of covariate, the hazad taxe for time to each recurrence is decelerated. So we note that $\hat{\delta} = 1.107$, with standard error 0.076, suggesting a time -nonhomogeneous Poisson process for individuals.

Similarily the Weibull accelerated failure time model considering the Log-Logistic accelerated failure time model with intensty function (12), the mle's and estimated standard deviations (DP) and 95% confidence interval (IC) of the parameters for the model proposed are given in the Table 3. In that case we observed that the effect of the treatment is not significant

Table 2. MLE , D.P and I.C (95%) to parameter of the model (7).

Parameter	mle	D.P.	IC (95%)
δ	1.107	0.076	[0.964 ; 1.263]
β	−0.743	0.134	[−1.051 ; −0.472]
β_0	−3.179	0.146	[−3.421 ; −2.894]

Also have that the estimate Log-Logistic hazard taxe for each time is

$$\lambda\left(t|\mathbf{x}\right) = \frac{1.341\ time^{0.341}[\exp\left(-4.419 + 0.019\ x\right)]^{1.341}}{1 + [time \exp\left(-4.419 + 0.019\ x\right)]^{1.341}}$$

Table 3. mle , D.P and I.C (95%) to parameter of the model(12).

Parameter	mle	D.P.	IC (95%)
α	1.341	0.084	[1.184; 1.514]
β	0.019	0.171	[−0.313; 0.361]
β_0	−4.419	0.092	[−4.594 ; −4.231]

4. Concluding Remarks

Although there is always uncertainty about the most appropriate form for the regression function for time to event data, Cox proportional hazard models are more typically used. As a different but equally simple formulation, one may consider Accelerated Failure Time. In situations where covariates are thought to affect the rate at which events occur, the combined accelerated failure time intensity/ Poisson process assumptions of this article often provide a simple and convenient initial framework for analysis. Another appealing feature of our approach is it the feasible interpretation of the AFT model.

Acknowledgment

The author thanks the referees and the Editor. The work was partially supported by CNPq-Brazil.

References

Aalen, O.O. (1978). Nonparametric inference for a family of counting processes, *The Annals of Statistics*, 6, 701–726.

Andersen,P.K.; Gill, R. D.(1982). Cox 's regression model for counting processes: A large sample study, *The Annals of Statistics* 10: 1100-1120.

Bickel, P.J.;Doksum, K.A.(1981). In Analysis of Transformations Revisited, *Journal of the American Statistical Association*, 76, 296-311.

Cox D.R. (1972a). Regression models and life tables (with discussion), *Journal of the Royal Statistical Society B* 34: 187-220.

Cox D.R. (1972b).The statistical analysis of dependencies in point processes. In *Stochastic point Processes*, Ed. P.A.W. Lewis, pp.55-66. New York: Wiley.

Cox D.R. (1997). Some remarks on the analysis of survival data. In *Proceedings of the the Frist Seattle Symposium in Biostatistics*; Survival Analysis , Springer. New York.

Cox, D.R.;Isham, V. (1980), *Point Process*, London: Chapman & Hall.

Cox,D.R.,Hinkley, D.V.(1974). *Theoretical Statistics*. London: Chapaman and Hall,511p.

Cox, D.R.; Oakes, D. (1984). *Analysis of Survival Data*. London: Chapman & Hall

Crow,L. H. (1974). Reliability Analysis for Complex, Reparable Systems, in *Reliability and Biometry*, eds. F. Proschan and R. J. Serfling, Philadelphia: SIAM,pp 379-410.

Gail, M. H. Santner , T. J. and Brown, C.C. (1980). An Analysis of Comparative Carcinogenesis Experiments Based on Multiple Times to Tumor, *Biometrics*, 36, 255-266.

Kalbfleisch, J. D. & Prentice, R. L. (1980). *The Statistical Analysis of Survival of Failure Time Data*. new york; Wiley.

Lawless, J.F. (1982). *Statistical Models and Methods for Lifetime Data*. New York: Wiley.

Lawless, J.F.(1987a).Regression Methods for Poisson process model, *Journal of the American Statistical Association* vol. 82 n°399. pp.808-815.

Lawless, J.F.(1987b).Negative Binomial and Mixed Poisson Regression.*Canadian Journal of Statistics*,15, 209-225.

Lawless, J.F. (1995). The analysis of recurrent events for multiple subjects. *Applied Statistics* 44, 487-98.

Lawless, J.F. and Nadeu, C. (1995). Some simple robust methods for the analysis of recurrent events. *Technometrics* 37, 158-68.

Lee, L. and S. K. Lee (1978). Some result on inference for the Weibull process. *Technometrics*, 20, 41-45.

Lin, Y. D. ; Wei, J. L. and Ying, Z. (1998). Accelerated failure time models for counting processes. *Biometrika*, 8, 3, 605-618.

Meeker, W.Q., ; Escobar, L.A.(1998.). *Statistical Methods for Reliability Data*, John Wiley Sons, Inc., New York, 1998.

Nielsen, G. G., Gill,R.D., Andersen, P.K. & Sorensen,T.I.A. (1992). A counting process approach to maximum likelihood estimation in frailty models, *Scandinavian Journal of Statistics* 19:25-43.

Reid, N. (1994). A conversation with Sir David Cox. *Statist. Sci.* 9, 439-55.

Wei, L.J. (1992). The accelerated failure time model: A useful alternative to the Cox regression model in survival analysis (with discussion). *Statistics in Medicine* 11, 1871-1879.

In: Research in Applied Statistical Science
Editor: Mohammad Ahsanullah, pp. 209-218
ISBN: 978-1-63321-818-5

Chapter 18

ON METHODS OF CONSTRUCTION OF PARTIALLY BALANCED INCOMPLETE BLOCK DESIGNS WITH THREE-ASSOCIATE-CLASS ASSOCIATION SCHEME

***Swati Raj*[3], *Jagdish Prasad*[1], *D. K. Ghosh*[2] *and Sarla Pareek*[3]**
[1]Department of Statistics, University of Rajasthan, Jaipur, India
[2]Department of Mathematics and Statistics, Saurashtra University, Rajkot, India
[3]Centre for Mathematical Sciences, Banasthali University, Rajasthan, India

Abstract

Rectangular association scheme is three-associate-class association scheme. In this paper we have constructed rectangular association scheme using difference set for a balanced incomplete block design and symmetrical balanced incomplete block design. Here we have also derived some parametric relations of rectangular association scheme. Further based on these results, a table is also provided in the range of $4 \leq r, k \leq 20$.

Keywords: Symmetrical BIB design, Association scheme, Rectangular association scheme, Galois field

1. Introduction

Bose and Nair (1939) introduced a class of binary, equireplicate and proper designs, which are called Partially Balanced Incomplete Block (PBIB) Designs. In these designs the variance of every estimated elementary contrast among treatment effect is not the same. For defining a PBIB designs, the concept of an association scheme on v symbols is needed. Nair and Rao (1942) modified the original definition of PBIB designs. For m= 2, Bose and Shimamoto (1952) classified the known PBIB designs into (1) Group Divisible (GD), (2) Simple (S.I), (3) Triangular (T), (4) Latin Square Type (L_i) and (5) Cyclic Designs. Vartak (1955) introduced three-associate-class association scheme which is known as a rectangular association scheme. Let there be $v=mn$ symbols arranged in a rectangle of m rows and n columns. With respect to each symbol, the first associates are the other (n-1) symbols of the same row, the second associates are the other (m-1) symbols of the same column, and the remaining (m-1)(n-1) symbols are third associates. For the association scheme we have

$$n_1 = n-1,\ n_2 = m-1,\ n_3 = (m-1)(n-1),$$

$$P_1 = \begin{bmatrix} n-2 & 0 & 0 \\ 0 & 0 & m-1 \\ 0 & m-1 & (m-1)(n-2) \end{bmatrix}$$

$$P_2 = \begin{bmatrix} 0 & 0 & n-1 \\ 0 & m-2 & 0 \\ n-1 & 0 & (m-2)(n-1) \end{bmatrix}$$

$$P_3 = \begin{bmatrix} 0 & 1 & n-2 \\ 1 & 0 & m-2 \\ n-2 & m-2 & (n-2)(m-2) \end{bmatrix}$$

The characteristic roots of *NN'* are $\theta_0 = rk$, $\theta_1 = r-\lambda_1 + (m-1)(\lambda_2-\lambda_3)$, $\theta_2 = r- \lambda_2 + (n-1)(\lambda_1-\lambda_3)$ and $\theta3 = r- \lambda_1 - \lambda_2 + \lambda_3$, with the respective multiplicities $\alpha_0 = 1$, $\alpha_1 = n-1$, $\alpha_2 = m-1$ and $\alpha_3 = (n-1)(m-1)$. For further properties of the rectangular association scheme we refer to Shah (1964) and Vartak (1955, 1959) and Raghavarao (1970).

The parametric relations of rectangular association schemes are also derived in this investigation, which are as follows

(a) $vr=bk$

(b) $n_1 + n_2 + n_3 = v-1$

(c) $n_1\lambda_1 + n_2\lambda_2 + n_3\lambda_3 = r(k-1)$

(d) $\sum_{k=1}^{3} P^{i}_{jk} = n_j-1$; if $i=j$

$= n_j$; if $i\neq j$

(e) $n_1P^{1}_{23} = n_2P^{2}_{13} = n_3P^{3}_{12}$

$n_1P^{1}_{33} = n_3P^{1}_{11}$

$n_1P^{1}_{33} = n_3P^{3}_{13}$

$n_2P^{2}_{33} = n_3P^{3}_{22}$

$n_2P^{2}_{33} = n_3P^{3}_{23}$

In this paper, some methods to construct three associate rectangular PBIB designs are given. Based on these results a table has been provided at the end of the main results, in which only those cases are considered where r, k ≤ 20.

2. Methods of Construction

Theorem 2.1. If (4t-1) is a prime or prime power when t=1, 2, 3…, then there always exists a rectangular association scheme with parameters v= 8t, b=2(4t+1), r = 4t+1, k=4t, $\lambda_1 = 0$, $\lambda_2 = 2t$, $\lambda_3=2t+1$, m = 4t, n = 2, which is obtained from the incidence matrix N defined as,

$$N = \begin{bmatrix} N_1^{*}{}' & N_2^{*}{}' \\ N_2^{*}{}' & N_1^{*}{}' \end{bmatrix} \tag{2.1}$$

where, N_1^* and N_2^* are incidence matrices of two incomplete block designs and $N_1^*{}'$ and $N_2^*{}'$ are the transpose of N_1^* and N_2^* respectively

Proof: Let $\alpha_0= 0$, $\alpha_1= x^0=1$, $\alpha_2= x$, $\alpha_3= x^2$, $\alpha_4= x^3$........ , $\alpha_{4t+1}= x^{4t}$ be the elements of GF(4t-1) where x is the primitive element of GF(4t-1). Consider the two initial sets $(0,x^0,x^2,x^4,x^6..... ,x^{4(t-1)})$ and $(\infty,x,x^3,x^5,x^7..... ,x^{4t-3})$ form a difference set for a BIB design with parameters $v' = 4t$, b'=2(4t-1), $r' = 4t-1$, $k'= 2t$ and $\lambda' = 2t-1$. Consider two initial sets as two initial blocks. With the help of these two initial blocks, two separate design d_0 and d_2 are obtained using reduced mod(4t-1). Let N_0 and N_2 denotes the incidence matrices of design d_0 and d_2 respectively. Next we consider two new incidence matrices N_1^* and N_2^* of dimension $(v'+1) \times (\frac{b'}{2}+1)$, defined as

$$N_1^* = \begin{bmatrix} 1 & E_{1\frac{b'}{2}} \\ E_{\frac{b'}{2}1} & 1 \end{bmatrix} \text{ and } N_2^* = \begin{bmatrix} 0 & 0_{1\frac{b'}{2}} \\ 0_{\frac{b'}{2}1} & 0 \end{bmatrix}$$

where, $E_{1b'/2}$ and $E_{v'1}$ are unitary matrices whose all elements are one while $0_{1b'/2}$ and $0_{v'1}$ are null matrices with $N_1 = \begin{bmatrix} N_0 \\ 0_{1\frac{b'}{2}} \end{bmatrix}$

Now, NN' can be expressed as

$$NN' = \begin{bmatrix} N_1^*{}'N_1^* + N_2^*{}'N_2^* & N_1^*{}'N_2^* + N_2^*{}'N_1^* \\ N_2^*{}'N_1^* + N_1^*{}'N_2^* & N_2^*{}'N_2^* + N_1^*{}'N_1^* \end{bmatrix} \tag{1.1}$$

Solving for each elements of the matrix (1.1) we have

$$N_1^*{}'N_1^* + N_2^*{}'N_2^* = \begin{bmatrix} 1+4t E_{11} & (2t+1)E_{1\frac{b'}{2}} \\ (2t+1)E_{\frac{b'}{2}1} & (2t+1)E_{\frac{b'}{2}} + 2tI_{\frac{b'}{2}} \end{bmatrix}$$

Similarly,

$$N_1^*{}'N_2^* + N_2^*{}'N_1^* = \begin{bmatrix} 0 & 2tE_{1\frac{b'}{2}} \\ 2tE_{\frac{b'}{2}1} & 2t(E_{\frac{b'}{2}} - I_{\frac{b'}{2}}) \end{bmatrix}$$

Substituting the values in the matrix (1.1)

$$NN' = \begin{bmatrix} \begin{pmatrix} (1+4t)E_{11} & (2t+1)E_{1\frac{b'}{2}} \\ (2t+1)E_{\frac{b'}{2}1} & (2t)I_{\frac{b'}{2}} + (2t+1)E_{\frac{b'}{2}} \end{pmatrix} & \begin{pmatrix} 0 & (2t)E_{1\frac{b'}{2}} \\ (2t)E_{\frac{b'}{2}1} & (2t)(E_{\frac{b'}{2}} - I_{\frac{b'}{2}}) \end{pmatrix} \\ \begin{pmatrix} 0 & (2t)E_{1\frac{b'}{2}} \\ (2t)E_{\frac{b'}{2}1} & (2t)(E_{\frac{b'}{2}} - I_{\frac{b'}{2}}) \end{pmatrix} & \begin{pmatrix} (1+4t)E_{11} & (2t+1)E_{1\frac{b'}{2}} \\ (2t+1)E_{\frac{b'}{2}1} & (2t)I_{\frac{b'}{2}} + (2t+1)E_{\frac{b'}{2}} \end{pmatrix} \end{bmatrix}_{8t\times 8t}$$

Here it is obvious that diagonal elements are (4t+1) and off diagonal elements are 0, 2t and (2t+1). Further we verified that all the primary and secondary parameters of PBIB designs are satisfied which shows the existing PBIB design with three associate classes has rectangular association scheme with parameters $v= 8t$, $b= 2(4t+1)$, $r = 4t+1$, $k=4t$, $\lambda_1 = 0$, $\lambda_2 = 2t$, $\lambda_3 = 2t+1$, $n_1= 1$, $n_2 = 4t-1$, $n_3 = 4t-1$, $m = 4t$, $n = 2$.

Example. If t=2, then we obtain a three associate rectangular partially balanced incomplete block design with parameters $v = 16$, $b= 18$, $r = 9$, $k= 8$, $\lambda_1 = 0$, $\lambda_2 = 5$, $\lambda_3 = 4$, m = 8, n = 2. Consider these two initial sets $(0, x^0, x^2, x^4)$ and (∞, x, x^3, x^5) form a difference set for a BIB design.

Since x = 3 is a primitive root of GF (7) hence the initial sets are [0 1 2 4] and [∞ 3 6 5]. Consider two initial sets as two initial blocks. Next by developing these two blocks we get two incomplete block design, say, d_0 and d_2 whose blocks are following

$$\begin{bmatrix} 0 & 1 & 2 & 3 & 4 & 5 & 6 \\ 1 & 2 & 3 & 4 & 5 & 6 & 0 \\ 2 & 3 & 4 & 5 & 6 & 0 & 1 \\ 4 & 5 & 6 & 0 & 1 & 2 & 3 \end{bmatrix} \quad \begin{bmatrix} \infty & \infty & \infty & \infty & \infty & \infty & \infty \\ 3 & 4 & 5 & 6 & 0 & 1 & 2 \\ 6 & 0 & 1 & 2 & 3 & 4 & 5 \\ 5 & 6 & 0 & 1 & 2 & 3 & 4 \end{bmatrix}$$

Here 0 and ∞ are considered as 7th and 8th treatments of the blocks respectively. Now, let N_0 and N_2 be the incidence matrices of d_0 and d_2 respectively. Here it is obvious that d_0 is a BIB design with parameters $v' = 7 = b'$, $r' = 4 = k', \lambda' = 2$ while d_2 is the IBD with unequal block size. Next using Theorem 2.1, we obtain the blocks of three associate class association scheme, which are as follows

Here the diagonal elements are 7 and off diagonal elements are 0, 5 and 4. Further we verified that all the primary and secondary parameters of PBIB design with three associate classes along with v=mn where, m= 8 (rows) and n=2 (columns), $n_1 = 1$, $n_2 = 7$ and $n_3 = 7$ are satisfied which shows the existing PBIB design with three associate classes has rectangular association scheme.

1	1	1	1	1	3	4	2	1	9	4	5	2	3	2	2	3	2
2	2	2	3	2	5	6	5	10	10	5	6	6	7	4	3	4	3
3	3	3	4	4	6	7	7	11	11	7	8	7	8	8	5	6	4
4	6	4	5	5	7	8	8	12	12	9	9	9	9	9	9	9	5
5	8	7	8	6	10	10	11	13	13	10	10	11	10	11	12	10	6
6	12	13	10	11	12	11	12	14	14	11	11	12	12	13	14	13	7
7	13	14	14	15	16	13	14	15	15	14	12	13	13	14	15	15	8
8	15	16	15	16	16	13	14	16	16	16	15	16	14	15	16	16	9

Theorem 2.2: If (4t-1) is a prime or prime power where t is any positive integer then there exist a rectangular association scheme with parameters v= 8t, b=2(4t-1), r = 4t-1, k = 4t, $\lambda1 = 0$, $\lambda2 = 2t-1$, $\lambda3=2t$, m = 4t, n = 2, which is obtained from the incidence matrix N defined as

$$N = \begin{bmatrix} N_1 & N_2 \\ N_2 & N_1 \end{bmatrix} \tag{2.2}$$

where, $N_1 = \begin{bmatrix} N_0 \\ 0_{1\frac{b'}{2}} \end{bmatrix}$, N_0 and N_2 are some other incidence matrices of two incomplete block designs.

Proof. Let $\alpha_0=0$, $\alpha_1=x^0=1$, $\alpha_2=x$, $\alpha_3=x^2$, $\alpha_4=x^3$........ , $\alpha_{4t+1}=x^{4t}$ be the elements of GF(4t-1), where x is the primitive element of GF(4t-1). Consider the initial sets $(0,x^0,x^2,x^4,x^6\text{.....},x^{4(t-1)})$ and $(\infty,x,x^3,x^5,x^7\text{.....},x^{4t-3})$ which form a difference set for a BIB design with parameters v' = 4t, b'= 2(4t-1), r' = 4t-1, k'= 2t and λ' = 2t-1. With the help of two initial blocks we have obtained two incidence matrices, say, N_0 and N_2. Next on the similar line of Theorem 2.1, using (2.2), we have

$$NN' = \begin{bmatrix} \begin{pmatrix} (2t-1)E_{\frac{b'}{2}}+2tI_{\frac{b'}{2}} & (2t-1)E_{1\frac{b'}{2}} \\ (2t-1)E_{\frac{b'}{2}1} & (4t-1)E_{11} \end{pmatrix} & \begin{pmatrix} (2t)(E_{\frac{b'}{2}}-I_{\frac{b'}{2}}) & (2t)E_{1\frac{b'}{2}} \\ (2t)E_{\frac{b'}{2}1} & 0 \end{pmatrix} \\ \begin{pmatrix} (2t)(E_{\frac{b'}{2}}-I_{\frac{b'}{2}}) & (2t)E_{1\frac{b'}{2}} \\ (2t)E_{\frac{b'}{2}1} & 0 \end{pmatrix} & \begin{pmatrix} (2t-1)E_{\frac{b'}{2}}+2tI_{\frac{b'}{2}} & (2t-1)E_{1\frac{b'}{2}} \\ (2t-1)E_{\frac{b'}{2}1} & (4t-1)E_{11} \end{pmatrix} \end{bmatrix}_{8t\times 8t}$$

Here the diagonal elements are (4t-1) and off diagonal elements are 0, 2t and (2t-1). Further it satisfies all the primary and secondary parameters of PBIB design with three associate classes hence the existing PBIB design has a rectangular association scheme with parameters $v = 8t$, $b=2(4t-1)$, $r = 4t-1$, $k=4t$, $\lambda_1 = 0$, $\lambda_2 = 2t-1$, $\lambda_3=2t$, $m = 4t$, $n = 2$.

Example. If t=2, then a three associate rectangular partially balanced incomplete block design with parameters $v = 16$, $b= 14$, $r = 7$, $k= 8$, $\lambda_1 = 0$, $\lambda_2 = 3$, $\lambda_3 =4$, m = 8, n = 2 is obtained by considering 2 initial sets $(0, x^0, x^2, x^4)$ and (∞, x, x^3, x^5) which form a difference set for a BIB design.

Since x = 3 is a primitive root of GF(7) hence the initial sets are [0 1 2 4] and [∞ 3 6 5]. Consider two initial sets as two initial blocks. Next by developing these two blocks we get two incomplete block design, say, d_0 and d_2 whose blocks are following

$$\begin{bmatrix} 0 & 1 & 2 & 3 & 4 & 5 & 6 \\ 1 & 2 & 3 & 4 & 5 & 6 & 0 \\ 2 & 3 & 4 & 5 & 6 & 0 & 1 \\ 4 & 5 & 6 & 0 & 1 & 2 & 3 \end{bmatrix} \quad \begin{bmatrix} \infty & \infty & \infty & \infty & \infty & \infty & \infty \\ 3 & 4 & 5 & 6 & 0 & 1 & 2 \\ 6 & 0 & 1 & 2 & 3 & 4 & 5 \\ 5 & 6 & 0 & 1 & 2 & 3 & 4 \end{bmatrix}$$

Here 0 and ∞ are considered to be the 7th and 8th treatments of the blocks respectively. Now, let N_0 and N_2 be the incidence matrices of d_0 and d_2 respectively. Here it is obvious that d_0 is a BIB design with parameters $v' = 7 = b'$, $r' = 4 = k', \lambda' = 2$ while d_2 is the IBD with unequal block size. Next using Theorem 2.2, we obtain the blocks of three associate class association scheme, which are as follows

1	1	2	3	1	2	1	3	4	1	1	2	1	2
2	2	3	4	4	5	3	5	6	5	2	3	3	4
4	3	4	5	5	6	6	6	7	7	6	7	4	5
7	5	6	7	6	7	7	8	8	8	8	8	8	8
11	12	9	9	10	9	10	9	9	9	11	9	10	9
13	14	13	10	11	11	12	10	10	10	12	12	13	11
14	15	15	14	15	12	13	12	11	11	13	13	14	14
16	16	16	16	16	16	16	15	13	13	15	14	15	15

Here the diagonal elements are 7 and off diagonal elements are 0, 3 and 4. Further we verified that all the primary and secondary parameters of PBIB design with three associate class along with v=mn where, m= 8 (rows) and n=2 (columns),), $n_1 = 1$, $n_2 = 7$ and $n_3 = 7$ are satisfied which shows the existing PBIB design with three associate class has a rectangular association scheme.

Theorem 2.3. *If (4t+3) is prime or prime power where t is any positive integer then there exists a rectangular association scheme with parameters* $v = 2(4t+3) = b$, $r = 2(2t+1) =k$, $\lambda_1 = 0$, $\lambda_2 = 2t$, $\lambda_3=2t+1$, $m = 4t+3$, $n = 2$, *using incidence matrix*

$$N = \begin{bmatrix} N_1 & N_2 \\ N_2 & N_1 \end{bmatrix} \tag{2.3}$$

where, N_1 and N_2 are some other incidence matrices of two incomplete block designs.

Proof. Let α0=0, α1=x0=1, α2=x, α3=x2, α4=x3…….. , α4t+2=x4t+1 be the elements of GF(4t+3) by considering the initial blocks (x0,x2,x4,x6….. ,x4t) and (x,x3,x5,x7….. ,x4t+1). From these two initial blocks we obtain two SBIB designs with parameters v' =4t+3 =b', r' = 2t+1 =k', λ' = t. With the help of two SBIB designs we have obtained two incidence matrices, say, N_1 and N_2 of order $v' \times b'$.

Now, using (2.3), we have

$$NN' = \begin{bmatrix} 2(t\,E_{v'} + (t+1)\,I_{v'}) & (2t+1)(E_{v'} - I_{v'}) \\ (2t+1)(E_{v'} - I_{v'}) & 2(t\,E_{v'} + (t+1)\,I_{v'}) \end{bmatrix}_{2(4t+1)\times 2(4t+1)}$$

Here the diagonal elements are 2(2t+1) and off diagonal elements are 0, 2t and (2t+1). Further it satisfies all the primary and secondary parameters of PBIB design hence the existing PBIB design is a rectangular association scheme with parameters $v = 2(4t+3) = b$, $r = 2(2t+1) = k$, $\lambda_1 = 0$, $\lambda_2 = 2t$, $\lambda_3 = 2t+1$, $m = 4t+3$, $n = 2$.

Example. If t=1, then a three associate rectangular partially balanced incomplete block design with parameters $v = 14 = b$, $r = 6 = k$, $\lambda_1 = 0$, $\lambda_2 = 2$, $\lambda_3 = 3$, m = 7, n = 2 is obtained by considering 2 initial sets (x^0, x^2, x^4) and (x, x^3, x^5).

Since x = 3 is a primitive root of GF (7) hence the initial sets are [1 2 4] and [3 6 5]. Consider two initial sets as two initial blocks. Next by developing these two blocks we get two symmetrical balanced incomplete block design, say, d_1 and d_2 whose blocks are following:

$$\begin{bmatrix} 1 & 2 & 3 & 4 & 5 & 6 & 0 \\ 2 & 3 & 4 & 5 & 6 & 0 & 1 \\ 4 & 5 & 6 & 0 & 1 & 2 & 3 \end{bmatrix} \begin{bmatrix} 3 & 4 & 5 & 6 & 0 & 1 & 2 \\ 6 & 0 & 1 & 2 & 3 & 4 & 5 \\ 5 & 6 & 0 & 1 & 2 & 3 & 4 \end{bmatrix}$$

Let N_1 and N_2 be the incidence matrices of d_1 and d_2 respectively. Using theorem 2.3 the blocks of three-associate-class association schemes are following

1	2	3	4	1	2	1	3	4	1	1	2	1	2
2	3	4	5	5	6	3	5	6	5	2	3	3	4
4	5	6	7	6	7	7	6	7	7	6	7	4	5
10	11	8	8	9	8	9	8	9	10	11	8	9	8
12	13	12	9	10	10	11	9	10	11	12	12	13	10
13	14	14	13	14	11	12	11	12	13	14	13	14	14

As v=mn where, m= 7 (rows) and n=2 (columns) so the existing design is a PBIB design with three associate classes. Here the diagonal elements are 6 and off diagonal elements are 0, 2 and 3 as well as it satisfies the primary and secondary parameters of PBIB design with three associate classes, hence the design has the rectangular association scheme.

Theorem 2.4, *If (4t+1) is prime or prime power where t is any positive integer then there exists a rectangular association schemes with parameters* $v= 2(4t+1) = b$, $r = 4t =k$, $\lambda_1 = 0$, $\lambda_2 = 2t-1$, $\lambda_3=2t$, $m = 4t+1$, $n = 2$, *using incidence matrix*

$$N = \begin{bmatrix} N_1 & N_2 \\ N_2 & N_1 \end{bmatrix} \tag{2.4}$$

where, N_1 and N_2 are some other incidence matrices of two incomplete block designs.

Proof. Let $\alpha_0=0$, $\alpha_1=x^0=1$, $\alpha_2=x$, $\alpha_3=x^2$, $\alpha_4=x^3$........ , $\alpha_{4t+1}=x^{4t}$ be the elements of GF(4t+1) where x is the primitive element of GF(4t+1). Considering the two initial blocks $(x^0,x^2,x^4,x^6 \ldots\ldots ,x^{4t})$ and $(x,x^3,x^5,x^7 \ldots\ldots ,x^{4t-1})$ which form a difference set for a BIB design with parameters v'=4t+1, b'= 2(4t+1), r'=4t, k'=2t, λ'=2t-1. Next with the help of two initial blocks we have obtained two incidence matrices, say, N_1 and N_2 .

Now, using (2.4), we have

$$NN' = \begin{bmatrix} (2t-1)E_{v'} + (2t+1)I_{v'} & 2t(E_{v'} - I_{v'}) \\ 2t(E_{v'} - I_{v'}) & (2t-1)E_{v'} + (2t+1)I_{v'} \end{bmatrix}_{2(4t+1)\times 2(4t+1)}$$

Here the diagonal elements are 4t and off diagonal elements are 0, (2t-1) and 2t. Further it satisfies all the primary and secondary parameters of PBIB design with three associate classes hence the existing PBIB design with three associate classes has a rectangular association scheme with parameters $v= 2(4t+1)=b$, $r =4t=k$, $\lambda_1 =0$, $\lambda_2=2t-1$, $\lambda_3=2t$, $m = 4t+1$, $n = 2$.

Example If t=1, then a three associate rectangular partially balanced incomplete block design with parameters $v = 10 = b$, $r = 4 =k$, $\lambda_1 = 0$, $\lambda_2 =1$, $\lambda_3 = 2$, m = 5, n − 2 is obtained by considering two initial sets (x^0, x^2, x^4) and (x, x^3, x^5) .

Since x = 2 is a primitive root of GF (5) hence the initial sets are [1 4] and [2 3]. Consider two initial sets as two initial blocks. Next by developing these two blocks we get two incomplete block design, say, d_1 and d_2 whose blocks are following

$$\begin{bmatrix} 1 & 2 & 3 & 4 & 0 \\ 4 & 0 & 1 & 2 & 3 \end{bmatrix} \begin{bmatrix} 2 & 3 & 4 & 0 & 1 \\ 3 & 4 & 0 & 1 & 2 \end{bmatrix}$$

Let N_1 and N_2 be the incidence matrices of d_1 and d_2 respectively. Using *theorem 2.4*, we get the following blocks of three-associate-class rectangular PBIB design.

1	2	1	2	3	2	3	4	1	1
4	5	3	4	5	3	4	5	5	2
7	8	9	6	6	6	7	6	7	8
8	9	10	10	7	9	10	8	9	10

Here the diagonal elements are 4 and off diagonal elements are 0, 1 and 2. Further we verified that all the primary and secondary parameters of PBIB design with associate classes along with v=mn where, m= 5 (rows) and n=2 (columns),), $n_1 = 1$, $n_2 = 4$ and $n_3 = 4$ are satisfied which shows the existing PBIB design with three associate classes has a rectangular association scheme.

Table 1. Rectangular Designs with *r, k* ≤ 20

S. No.	v	b	r	k	λ_1	λ_2	λ_3	n_1	n_2	n_3	m	n	Source
1	8	10	5	4	0	2	3	1	3	3	4	2	Theorem 2.1
2	8	6	3	4	0	1	2	1	3	3	4	2	Theorem 2.2
3	10	10	4	4	0	1	2	1	4	4	5	2	Theorem 2.4
4	14	14	6	6	0	2	3	1	7	7	7	2	Theorem 2.3
5	16	14	7	8	0	3	4	1	7	7	8	2	Theorem 2.2
6	16	18	9	8	0	4	5	1	7	7	8	2	Theorem 2.1
7	18	18	8	8	0	3	4	1	8	8	9	2	Theorem 2.4
8	22	22	10	10	0	4	5	1	10	10	11	2	Theorem 2.3
9	24	22	11	12	0	5	6	1	11	11	12	2	Theorem 2.2
10	24	26	13	12	0	6	7	1	11	11	12	2	Theorem 2.1
11	26	26	12	12	0	5	6	1	12	12	13	2	Theorem 2.4
12	34	34	16	16	0	7	8	1	16	16	17	2	Theorem 2.4
13	38	38	18	18	0	8	9	1	18	18	19	2	Theorem 2.3
14	40	38	19	20	0	9	10	1	19	19	20	2	Theorem 2.2

References

Bose, R.C. and Nair, K.R. (1939). Partially Balanced Incomplete Block Designs. *Sankhya*. 4, 307-372.

Bose, R.C. and Shimamoto, T. (1952). Classification and analysis of partially balanced incomplete block designs with two associate classes. *J. Amer. Statist. Assoc.*, 47, 151-184.

Das, M.N. and Giri, N.C.(1986). *Design and Analysis of Experiments.* Wiley Eastern Limited, New Delhi.

Nair, K. R. and Rao, C. R. (1942). A note on partially balanced incomplete block designs. *Science and Culture,* 7, 568-569.

Raghavarao, D. (1960). A generalization of group divisible designs. *Ann. Math. Statist.*, 31, 756-771.

Raghavarao, D. (1970). Constructions and Combinatorial Problems in *Design of Experiments*. Wiley Eastern Limited, New Delhi.

Shah, S. M. (1964). An upper bound for the number of disjoint blocks in certain PBIB designs. *Ann. Math. Stat.,* 35, 398-407.

Vartak, M. N. (1955). On an application of Kronecker product of matrices to statistical designs. *Ann. Math. Stat.,* 26, 420-438.

Vartak, M. N. (1959). The non-existence of certain PBIB designs. *Ann. Math. Stat.*, 30, 1051-1062.

In: Research in Applied Statistical Science
Editor: Mohammad Ahsanullah, pp. 219-235
ISBN: 978-1-63321-818-5

Chapter 19

LOWER GENERALIZED ORDER STATISTICS OF GENERALIZED INVERSE WEIBULL DISTRIBUTION

R. U. Khan*, Benazir Zia† and Haseeb Athar‡
Department of Statistics and Operations Research,
Aligarh Muslim University, Aligarh, India

Abstract

In this paper, some recurrence relations satisfied by single and product moments of lower generalized order statistics from generalized inverse Weibull distribution are derived and characterization of this distribution based on the conditional expectation of function of lower generalized order statistics and a recurrence relation for single moments are given. Further, some important deductions and computational works are carried out.

Keywords: Lower generalized order statistics, order statistics, lower record values, single moments, product moments, recurrence relations, generalized inverse Weibull distribution, characterization

AMS Subject Classifications: 62G30, 62E10

1. Introduction

Kamps (1995) introduced the concept of generalized order statistics (gos). It is known that ordinary order statistics, sequential order statistics, Stigler's order statistics and upper record values are special cases of gos. In this article we will consider the lower generalized order statistics $(l\,gos)$ defined as follows:

Let $n \in N$, $k \geq 1$, $m \in \Re$, be the parameters such that

* E-mail address: aruke@rediffmail.com (Corresponding author)
† E-mail address: benazir.stats@gmail.com
‡ E-mail address: haseebathar@hotmail.com

$$\gamma_r = k + (n-r)(m+1) > 0 \text{, for all } 1 \le r \le n .$$

Then $X^*(1, n, m, k), \ldots, X^*(n, n, m, k)$ are called $l\,gos$ from an absolutely continuous distribution function (df) $F()$ with the probability distribution function (pdf) $f()$ if their joint pdf has the form

$$k\left(\prod_{j=1}^{n-1} \gamma_j\right)\left(\prod_{i=1}^{n-1} [F(x_i)]^m f(x_i)\right)[F(x_n)]^{k-1} f(x_n) \tag{1.1}$$

for $F^{-1}(1) > x_1 \ge x_2 \ge \ldots \ge x_n > F^{-1}(0)$.

The marginal pdf of $r-$th $l\,gos$, $X^*(r, n, m, k)$ is

$$f_{X^*(r,n,m,k)}(x) = \frac{C_{r-1}}{(r-1)!}[F(x)]^{\gamma_r - 1} f(x)\, g_m^{r-1}(F(x)). \tag{1.2}$$

The joint pdf of $X^*(r, n, m, k)$ and $X^*(s, n, m, k)$, $1 \le r < s \le n$, is expressed from (1.1) as

$$f_{X^*(r,n,m,k), X^*(s,n,m,k)}(x, y) = \frac{C_{s-1}}{(r-1)!(s-r-1)!}[F(x)]^m f(x)\, g_m^{r-1}(F(x))$$
$$\times [h_m(F(y)) - h_m(F(x))]^{s-r-1} [F(y)]^{\gamma_s - 1} f(y), \; x > y, \tag{1.3}$$

where

$$C_{r-1} = \prod_{i=1}^{r} \gamma_i \,, \; h_m(x) = \begin{cases} -\dfrac{1}{m+1} x^{m+1}, & m \ne -1 \\ -\ln x, & m = -1 \end{cases}$$

and

$$g_m(x) = h_m(x) - h_m(1), \; x \in [0,1).$$

We shall also take $X^*(0, n, m, k) = 0$. If $m = 0$, $k = 1$, then $X^*(r, n, m, k)$ reduces to the $(n-r+1)-$th order statistic, $X_{n-r+1:n}$ from the sample $X_1, X_2, \ldots, X_n$ and when $m = -1$, then $X^*(r, n, m, k)$ reduces to the $r-$th k record value (Pawlas and Szynal, 2001). The work of Burkschat *et al.* (2003) may also refer for lower generalized order statistics.

Recurrence relations for single and product moments of lower generalized order statistics from the inverse Weibull distribution are derived by Pawlas and Szynal (2001). Khan *et al.*

(2008) and Khan and Kumar (2010) have established recurrence relations for moments of lower generalized order statistics from exponentiated Weibull and Pareto distributions. Ahsanullah (2004) and Mbah and Ahsanullah (2007) characterized the uniform and power function distributions based on distributional properties of lower generalized order statistics, respectively.

Kamps (1998) investigated the importance of recurrence relations of order statistics in characterization.

In this paper, we establish some recurrence relations for single and product moments of $l\,gos$ from generalized inverse Weibull distribution. Results for order statistics and lower record values are deduced as special cases and two theorems for characterizing this distribution have been stated and proved on using conditional expectation of function of $l\,gos$ and a recurrence relation for single moments. Further, we use the established explicit expressions to calculate the first four moments and variances of order statistics and lower records.

A random variable X is said to have generalized inverse Weibull distribution (Gusmão *et al.*, 2011) if its *pdf* is of the form

$$f(x) = \theta\beta\alpha^{\beta} x^{-(\beta+1)} e^{-\theta(\alpha/x)^{\beta}}, x>0,\ \alpha>0,\ \beta>0,\ \theta>0 \tag{1.4}$$

and the corresponding *df* is

$$F(x) = e^{-\theta(\alpha/x)^{\beta}}, x>0, \alpha>0, \beta>0, \theta>0. \tag{1.5}$$

It is easy to see that

$$x^{\beta+1} f(x) = \theta\beta\alpha^{\beta} F(x). \tag{1.6}$$

The inverse Weibull and inverse exponential distributions are considered as special cases of the generalized inverse Weibull distribution when $\theta = 1$ and $\beta = 1$, $\theta = 1$, respectively.

More details on inverse distributions and their applications can be found in Klugman *et al.* (1998).

2. Relations for Single Moments

We shall first establish some results which may be helpful in proving the main result.

Lemma 2.1. For generalized inverse Weibull distribution as given in (1.5) and any non-negative and finite integers a and b with $m \neq -1$,

$$I_j(a,0) = \frac{\theta^{j/\beta}\alpha^{j}}{(a+1)^{1-j/\beta}}\Gamma(1-j/\beta),\ \beta > j \text{ and } j = 0,1,\ldots, \tag{2.1}$$

where

$$I_j(a,b)=\int_0^\infty x^j [F(x)]^a f(x)\, g_m^b(F(x))dx \cdot \tag{2.2}$$

Proof. From (2.2), we have

$$I_j(a,0)=\int_0^\infty x^j [F(x)]^a f(x)dx. \tag{2.3}$$

By setting $t=-\ln F(x)$ in (2.3), we obtain

$$I_j(a,0)=\theta^{j/\beta}\alpha^j \int_0^\infty t^{-j/\beta} e^{-(a+1)t} dt$$

and hence the result given in (2.1).

Lemma 2.2. For the generalized inverse Weibull distribution as given in (1.5) and any non-negative and finite integers a and b,

$$\begin{aligned} I_j(a,b) &= \frac{1}{(m+1)^b}\sum_{u=0}^{b}(-1)^u \binom{b}{u} I_j(a+u(m+1),0) \\ &= \frac{\theta^{j/\beta}\alpha^j \Gamma(1-j/\beta)}{(m+1)^b}\sum_{u=0}^{b}(-1)^u \binom{b}{u}\frac{1}{[a+u(m+1)+1]^{1-j/\beta}}, \\ & \qquad m \neq -1 \end{aligned} \tag{2.4}$$

$$=\frac{\theta^{j/\beta}\alpha^j}{(a+1)^{b+1-j/\beta}}\Gamma(b+1-j/\beta),\ m=-1, \tag{2.5}$$

where $I_j(a,b)$ is as given in (2.2).

Proof. On expanding $g_m^b(F(x))=\left[\frac{1}{m+1}\{1-[F(x)]^{m+1}\}\right]^b$ binomially in (2.2), we get when $m \neq -1$:

$$\begin{aligned} I_j(a,b) &= \frac{1}{(m+1)^b}\sum_{u=0}^{b}(-1)^u \binom{b}{u}\int_0^\infty x^j [F(x)]^{a+u(m+1)} f(x)dx \\ &= \frac{1}{(m+1)^b}\sum_{u=0}^{b}(-1)^u \binom{b}{u} I_j(a+u(m+1),0). \end{aligned}$$

Making use of Lemma 2.1, we establish the result given in (2.4).

When $m=-1$, $I_j(a,b)$ is undefined as $\sum_{u=0}^{b}(-1)^u\binom{b}{u}=0$, therefore, we have

$$I_j(a,b)=\theta^{j/\beta}\alpha^j\Gamma(1-j/\beta)\sum_{u=0}^{b}(-1)^u\binom{b}{u}\frac{[a+u(m+1)+1]^{-(1-j/\beta)}}{(m+1)^b}. \quad (2.6)$$

Differentiating numerator and denominator of (2.6) b times with respect to m, we get

$$I_j(a,b)=\frac{\theta^{j/\beta}\alpha^j\Gamma(b+1-j/\beta)}{b!}\sum_{u=0}^{b}(-1)^{u+b}\binom{b}{u}\frac{u^b}{[a+u(m+1)+1]^{b+1-j/\beta}},\ b>0.$$

On applying the L' Hospital rule, we have

$$\lim_{m\to-1}I_j(a,b)=\frac{\theta^{j/\beta}\alpha^j\Gamma(b+1-j/\beta)}{b!(a+1)^{b+1-j/\beta}}\sum_{u=0}^{b}(-1)^{u+b}\binom{b}{u}u^b. \quad (2.7)$$

But for all integers $n\geq 0$ and for all real numbers x, we have Ruiz (1996)

$$\sum_{i=0}^{n}(-1)^i\binom{n}{i}(x-i)^n=n!. \quad (2.8)$$

Therefore,

$$\sum_{u=0}^{b}(-1)^{u+b}\binom{b}{u}u^b=b!. \quad (2.9)$$

Now on substituting (2.9) in (2.7), we have the result given in (2.5).

Theorem 2.1. For the generalized inverse Weibull distribution as given in (1.5) and $1\leq r\leq n$, $k=1,2,\ldots$ and $m\neq -1$,

$$E[X^{*j}(r,n,m,k)]=\frac{C_{r-1}}{(r-1)!}I_j(\gamma_r-1,r-1) \quad (2.10)$$

$$=\frac{\theta^{j/\beta}\alpha^j C_{r-1}\Gamma(1-j/\beta)}{(r-1)!(m+1)^{r-1}}\sum_{u=0}^{r-1}(-1)^u\binom{r-1}{u}\frac{1}{(\gamma_{r-u})^{1-j/\beta}},$$

$$\beta>j \text{ and } j=0,1,\ldots, \quad (2.11)$$

where $I_j(\gamma_r-1,r-1)$ is as defined in (2.2).

Proof. From (1.2) and (2.2), we have

$$E[X^{*j}(r,n,m,k)] = \frac{C_{r-1}}{(r-1)!}\int_0^\infty x^j [F(x)]^{\gamma_r - 1} f(x) g_m^{r-1}(F(x))\,dx$$

$$= \frac{C_{r-1}}{(r-1)!} I_j(\gamma_r - 1, r-1).$$

Making use of the Lemma 2.2, we establish the relation given in (2.11).

Special cases:

i) Putting $m=0$, $k=1$ in (2.11), the explicit formula for the single moments of order statistics of the generalized inverse Weibull distribution can be obtained as

$$E[X_{n-r+1:n}^j] = \theta^{j/\beta}\alpha^j \Gamma(1-j/\beta) C_{r:n} \sum_{u=0}^{r-1} (-1)^u \binom{r-1}{u} \frac{1}{(n-r+1+u)^{1-j/\beta}},$$

where

$$C_{r:n} = \frac{n!}{(r-1)!(n-r)!}.$$

That is

$$E[X_{r:n}^j] = \theta^{j/\beta}\alpha^j \Gamma(1-j/\beta) C_{r:n} \sum_{u=0}^{n-r} (-1)^u \binom{n-r}{u} \frac{1}{(r+u)^{1-j/\beta}}$$

as obtained by Gusmão et al. (2011).

ii) Putting $m=-1$ in (2.11), we deduce the explicit expression for the single moments of lower k record values for the generalized inverse Weibull distribution in view of (2.10) and (2.5) in the form

$$E[X^{*j}(r,n,-1,k)] = E[(Z_r^{(k)})^j] = \frac{(k\theta)^{j/\beta}\alpha^j}{(r-1)!}\Gamma(r-j/\beta)$$

and hence for lower records

$$E[(Z_r^{(1)})^j] = E[X_{L(r)}^j] = \frac{\theta^{j/\beta}\alpha^j}{(r-1)!}\Gamma(r-j/\beta)$$

which is the result obtained by Khan and Zia (2011).

Recurrence relations for single moments of $l\ gos$ from (1.5) can be obtained in the following theorem.

Theorem 2.2. For the distribution given in (1.5) and for $2 \le r \le n$, $n \ge 2$ and $k = 1,2,\ldots$,

$$E[X^{*j+\beta+1}(r,n,m,k)] = \frac{\theta\beta\alpha^{\beta}\gamma_r}{j+1} \times \{E[X^{*j+1}(r-1,n,m,k)] - E[X^{*j+1}(r,n,m,k)]\} \cdot \tag{2.12}$$

Proof. From (1.2) and (1.6), we have

$$E[X^{*j+\beta+1}(r,n,m,k)] = \frac{\theta\beta\alpha^{\beta}C_{r-1}}{(r-1)!}\int_0^{\infty} x^j [F(x)]^{\gamma_r} g_m^{r-1}(F(x))dx .$$

Integrating by parts treating x^j for integration and the rest of the integrand for differentiation, we get

$$E[X^{*j+\beta+1}(r,n,m,k)] = \frac{\theta\beta\alpha^{\beta}\gamma_r}{j+1}$$

$$\times \left\{ \frac{C_{r-2}}{(r-2)!}\int_0^{\infty} x^{j+1}[F(x)]^{\gamma_{r-1}-1} f(x) g_m^{r-2}(F(x))dx - \frac{C_{r-1}}{(r-1)!}\int_0^{\infty} x^{j+1}[F(x)]^{\gamma_r-1} f(x) g_m^{r-1}(F(x))dx \right\}$$

and hence the result.

Remark 2.1. Setting $m=0$, $k=1$ in (2.12), we obtain a recurrence relation for single moments of order statistics of the generalized inverse Weibull distribution in the form

$$E[X_{n-r+1:n}^{j+\beta+1}] = \frac{\theta\beta\alpha^{\beta}(n-r+1)}{j+1}\{E[X_{n-r+2:n}^{j+1}] - E[X_{n-r+1:n}^{j+1}]\}.$$

Replacing $(n-r+1)$ by $(r-1)$, we have

$$E[X_{r:n}^{j+1}] = E[X_{r-1:n}^{j+1}] + \frac{j+1}{\theta\beta\alpha^{\beta}(r-1)} E[X_{r-1:n}^{j+\beta+1}].$$

Remark 2.2. Putting $m=-1$, in Theorem 2.2, we get a recurrence relation for single moments of lower k record values from generalized inverse Weibull distribution in the form

$$E[(Z_r^{(k)})^{j+\beta+1}] = \frac{\theta\beta\alpha^{\beta}k}{j+1}\{E[(Z_{r-1}^{(k)})^{j+1}] - E[(Z_r^{(k)})^{j+1}]\}.$$

Remark 2.3. Putting $\theta = 1$, in (2.12), we deduce the recurrence relation for single moments of $l\ gos$ from inverse Weibull distribution, established by Pawlas and Szynal (2001).

Remark 2.4. For $\theta = 1$ and $\beta = 1$, the relation in (2.12) gives the recurrence relation for single moments of $l\ gos$ from inverse exponential distribution.

Table 2.1. First four moments of order statistics

n	r	$j=1$, $\beta=2$				$j=1$, $\beta=3$			
		$\alpha=1$		$\alpha=2$		$\alpha=1$		$\alpha=2$	
		$\theta=1$	$\theta=2$	$\theta=1$	$\theta=2$	$\theta=1$	$\theta=2$	$\theta=1$	$\theta=2$
1	1	1.77245	2.50663	3.54491	5.01326	1.35412	1.70608	2.70824	3.41216
2	1	2.50663	3.54491	5.01326	7.08982	1.70608	2.14953	3.41216	4.29906
	2	1.03828	1.46835	2.07656	2.93670	1.00215	1.26264	2.00431	2.52527
3	1	3.06998	4.34161	6.13996	8.68322	1.95298	2.46060	3.90595	4.92119
	2	1.37992	1.95151	2.75985	3.90302	1.21229	1.52739	2.42459	3.05479
	3	0.86746	1.22677	1.73491	2.45354	0.89708	1.13026	1.79417	2.26051
4	1	3.54491	5.01326	7.08982	10.02651	2.14953	2.70824	4.29906	5.41647
	2	1.64520	2.32666	3.29039	4.65332	1.36332	1.71767	2.72664	3.43535
	3	1.11465	1.57636	2.22930	3.15271	1.06127	1.33711	2.12253	2.67423
	4	0.78506	1.11024	1.57012	2.22048	0.84236	1.06130	1.68471	2.12261
5	1	3.96333	5.60499	7.92666	11.20998	2.31551	2.91736	4.63102	5.83472
	2	1.87123	2.64632	3.74246	5.29264	1.48561	1.87174	2.97121	3.74349
	3	1.30615	1.84717	2.61230	3.69435	1.17989	1.48657	2.35978	2.97314
	4	0.98699	1.39581	1.97397	2.79162	1.35412	1.70608	2.70824	3.41216
	5	0.73458	1.03885	1.46916	2.07770	1.70608	2.14953	3.41216	4.29906

n	r	$j=1$, $\beta=4$				$j=2$, $\beta=3$			
		$\alpha=1$		$\alpha=2$		$\alpha=1$		$\alpha=2$	
		$\theta=1$	$\theta=2$	$\theta=1$	$\theta=2$	$\theta=1$	$\theta=2$	$\theta=1$	$\theta=2$
1	1	1.22542	1.45727	2.45083	2.91455	2.67894	4.25255	10.71576	17.01020
2	1	1.45727	1.73300	2.91455	3.46600	4.25255	6.75050	17.01020	27.00201
	2	0.99356	1.18155	1.98712	2.36310	1.10533	1.75460	4.42131	7.01839
3	1	1.61274	1.91788	3.22548	3.83576	5.57242	8.84566	22.28967	35.38265
	2	1.14634	1.36324	2.29269	2.72648	1.61282	2.56019	6.45127	10.24075
	3	0.91717	1.09070	1.83433	2.18140	0.85158	1.35180	3.40633	5.40722
4	1	1.73300	2.06090	3.46600	4.12180	6.75050	10.71576	27.00201	42.86303
	2	1.25195	1.48883	2.50391	2.97767	2.03816	3.23538	8.15265	12.94152
	3	1.04074	1.23765	2.08147	2.47530	1.18747	1.88500	4.74989	7.53998
	4	0.87598	1.04172	1.75195	2.08344	0.73962	1.17407	2.95848	4.69629
5	1	1.83243	2.17913	3.66485	4.35827	7.83327	12.43453	31.33306	49.86303
	2	1.33530	1.58795	2.67061	3.17590	2.41946	3.84065	9.67783	15.36259
	3	1.12693	1.34015	2.25386	2.68031	1.46622	2.32748	5.86489	9.30993
	4	0.98327	1.16932	1.96655	2.33863	1.00164	1.59001	4.00657	6.36003
	5	0.84915	1.00982	1.69830	2.01964	0.67411	1.07009	2.69646	4.28036

n	r	$j=2$, $\beta=4$				$j=3$, $\beta=4$			
		$\alpha=1$		$\alpha=2$		$\alpha=1$		$\alpha=2$	
		$\theta=1$	$\theta=2$	$\theta=1$	$\theta=2$	$\theta=1$	$\theta=2$	$\theta=1$	$\theta=2$
1	1	1.77245	2.50663	7.08982	10.02651	3.62561	6.09753	29.00488	48.78020
2	1	2.50663	3.54491	10.02651	14.17963	6.09753	10.25477	48.78020	82.03819
	2	1.03828	1.46835	4.15312	5.87340	1.15370	1.94028	9.22956	15.52221
3	1	3.06998	4.34161	12.27992	17.36643	8.26460	13.89935	66.11683	111.19480
	2	1.37992	1.95151	5.51970	7.80603	1.76337	2.96562	14.10694	23.72496
	3	0.86746	1.22677	3.46983	4.90708	0.84886	1.42760	6.79087	11.42083
4	1	3.54491	5.01326	14.17963	20.05303	10.25477	17.24640	82.03819	137.97120
	2	1.64520	2.32666	6.58079	9.30664	2.29409	3.85819	18.35274	30.86551
	3	1.11465	1.57636	4.45861	6.30542	1.23264	2.07305	9.86114	16.58439
	4	0.78506	1.11024	3.14023	4.44096	0.72093	1.21246	5.76745	9.69966
5	1	3.96333	5.60499	15.85331	22.41997	12.12296	20.38831	96.98367	163.10640
	2	1.87123	2.64632	7.48492	10.58527	2.78204	4.67881	22.25629	37.43038
	3	1.30615	1.84717	5.22460	7.38869	1.56218	2.62726	12.49741	21.01806
	4	0.98699	1.39581	3.94794	5.58323	1.01295	1.70357	8.10361	13.62859
	5	0.73458	1.03885	2.93831	4.15540	0.64792	1.08967	5.18340	8.71740

n	r	$j=3$, $\beta=5$				$j=4$, $\beta=5$			
		$\alpha=1$		$\alpha=2$		$\alpha=1$		$\alpha=2$	
		$\theta=1$	$\theta=2$	$\theta=1$	$\theta=2$	$\theta=1$	$\theta=2$	$\theta=1$	$\theta=2$
1	1	2.21816	3.36210	17.74528	26.89681	4.59084	7.99312	73.45351	127.89000
2	1	3.36210	5.09599	26.89681	40.76795	7.99312	13.91684	127.89000	222.66940
	2	1.07422	1.62821	8.59375	13.02568	1.18856	2.06941	19.01703	33.11057
3	1	4.28811	6.49956	34.30486	51.99644	11.05578	19.24924	176.89250	307.98780
	2	1.51009	2.28887	12.08073	18.31096	1.86780	3.25204	29.88486	52.03256
	3	0.85628	1.29788	6.85025	10.38304	0.84894	1.47810	13.58311	23.64957
4	1	5.09599	7.72408	40.76794	61.79265	13.91684	24.23062	222.66940	387.68990
	2	1.86445	2.82598	14.91559	22.60781	2.47263	4.30509	39.56203	68.88151
	3	1.15573	1.75177	9.24588	14.01413	1.26298	2.19898	20.20771	35.18367
	4	0.75646	1.14659	6.05171	9.17268	0.71093	1.23781	11.37494	19.80492
5	1	5.82606	8.83065	46.60847	70.64523	16.63675	28.96627	266.18800	463.46030
	2	2.17573	3.29779	17.40585	26.38234	3.03718	5.28803	48.59480	84.60847
	3	1.39753	2.11825	11.18020	16.94601	1.62581	2.83069	26.01289	45.29107
	4	0.99454	1.50745	7.95635	12.05957	1.02110	1.77785	16.33768	28.44555
	5	0.69694	1.05637	5.57556	8.45096	0.63339	1.10280	10.13426	17.64477

Table 2.2. Variance of order statistics

n	r	$\beta=3$				$\beta=4$			
		$\alpha=1$		$\alpha=2$		$\alpha=1$		$\alpha=2$	
		$\theta=1$	$\theta=2$	$\theta=1$	$\theta=2$	$\theta=1$	$\theta=2$	$\theta=1$	$\theta=2$
1	1	0.84529	1.34184	3.38119	5.36736	0.27079	0.38299	1.08325	1.53191
2	1	1.34184	2.13002	5.36736	8.52009	0.38299	0.54162	1.53191	2.16647
	2	0.10103	0.16034	0.40405	0.64140	0.05112	0.07229	0.20447	0.28916
3	1	1.75829	2.79111	7.03322	11.16454	0.46905	0.66335	1.87619	2.65338
	2	0.14317	0.22727	0.57263	0.90901	0.06582	0.09309	0.26327	0.37234
	3	0.04683	0.07431	0.18758	0.29731	0.02626	0.10126	0.10506	0.14857
4	1	2.14953	3.38119	8.52009	13.52488	0.54162	0.76595	2.16647	3.06379
	2	0.17952	0.28499	0.71808	1.13989	0.07782	0.11005	0.31122	0.44012
	3	0.06118	0.09714	0.24476	0.38847	0.03151	0.04458	0.12609	0.17831
	4	0.03005	0.04771	0.12023	0.19082	0.01772	0.02506	0.07090	0.10024
5	1	2.47168	3.92354	9.88671	15.81907	0.60553	0.85638	2.42218	3.42545
	2	2.12423	0.33724	0.84974	1.34887	0.08820	0.12473	0.35276	0.49893
	3	0.07408	0.11759	0.29633	0.47037	0.03681	0.05117	0.14472	0.20463
	4	0.03696	0.05868	0.14782	0.23465	0.02017	0.02850	0.80062	0.11404
	5	0.02222	0.03527	0.08888	0.14109	0.01352	0.01911	0.05409	0.07645

Table 2.3. First four moments of lower records

r	$j=1,\ \beta=2$				$j=1,\ \beta=3$			
	$\alpha=1$		$\alpha=2$		$\alpha=1$		$\alpha=2$	
	$\theta=1$	$\theta=2$	$\theta=1$	$\theta=2$	$\theta=1$	$\theta=2$	$\theta=1$	$\theta=2$
1	1.77245	2.50663	3.54491	5.01326	1.35412	1.70608	2.70824	3.41216
2	0.88623	1.25331	1.77245	2.50663	0.90275	1.13739	1.80549	2.27478
3	0.66467	0.93999	1.32934	1.87997	0.75229	0.94782	1.50458	1.89565
4	0.55389	0.78332	1.10778	1.56664	0.66870	0.84251	1.33740	1.68502
5	0.48466	0.68541	0.96931	1.37081	0.61298	0.77230	1.22595	1.54460
r	$j=1,\ \beta=4$				$j=2,\ \beta=3$			
	$\alpha=1$		$\alpha=2$		$\alpha=1$		$\alpha=2$	
	$\theta=1$	$\theta=2$	$\theta=1$	$\theta=2$	$\theta=1$	$\theta=2$	$\theta=1$	$\theta=2$
1	1.22542	1.45727	2.45083	2.91455	2.67894	4.25255	10.71576	17.01020
2	0.91906	1.09296	1.83812	2.18591	0.89298	1.41752	3.57192	5.67007
3	0.80418	0.95634	1.60836	1.91267	0.59532	0.94501	2.38128	3.78004
4	0.73716	0.87664	1.47433	1.75328	0.46303	0.73501	1.85211	2.94003
5	0.69109	0.82185	1.38218	1.64370	0.38586	0.61251	1.54342	2.45003
r	$j=2,\ \beta=4$				$j=3,\ \beta=4$			
	$\alpha=1$		$\alpha=2$		$\alpha=1$		$\alpha=2$	
	$\theta=1$	$\theta=2$	$\theta=1$	$\theta=2$	$\theta=1$	$\theta=2$	$\theta=1$	$\theta=2$
1	1.77245	2.50663	7.08982	10.02651	3.62561	6.09753	29.00488	48.78020
2	0.88623	1.25331	3.54491	5.01326	0.90640	1.52438	7.25122	12.19505
3	0.66467	0.93999	2.65868	3.75994	0.56650	0.95274	5.32012	7.62191
4	0.55389	0.78332	2.21557	3.13329	0.42488	0.71455	3.39901	5.71643
5	0.48466	0.68541	1.93862	2.74163	0.34521	0.58057	2.76169	4.64460
r	$j=3,\ \beta=5$				$j=4,\ \beta=5$			
	$\alpha=1$		$\alpha=2$		$\alpha=1$		$\alpha=2$	
	$\theta=1$	$\theta=2$	$\theta=1$	$\theta=2$	$\theta=1$	$\theta=2$	$\theta=1$	$\theta=2$
1	2.21816	3.36210	17.74528	26.89681	4.59084	7.99312	73.45351	127.89000
2	0.88726	1.34484	7.09811	10.75872	0.91817	1.59862	14.69070	25.57799
3	0.62108	0.94139	4.96868	7.53110	0.55090	0.95917	8.81442	15.34679
4	0.49687	0.75311	3.97494	6.02488	0.40399	0.70339	6.46391	11.25432
5	0.42234	0.64014	3.37870	5.12115	0.32320	0.56272	5.17113	9.00345

Table 2.4. Variance of lower records

r	$\beta=3$				$\beta=4$			
	$\alpha=1$		$\alpha=2$		$\alpha=1$		$\alpha=2$	
	$\theta=1$	$\theta=2$	$\theta=1$	$\theta=2$	$\theta=1$	$\theta=2$	$\theta=1$	$\theta=2$
1	0.84529	1.34184	3.38119	5.36736	0.27079	0.38299	1.08325	1.53191
2	0.07802	0.12386	0.31213	0.49545	0.04156	0.05875	0.16622	0.23506
3	0.02938	0.04665	0.11752	0.18655	0.01796	0.02540	0.07186	0.10163
4	0.01587	0.02519	0.06347	0.10074	0.01049	0.01482	0.04192	0.05929
5	0.01012	0.01606	0.04047	0.06424	0.00705	0.00997	0.02819	0.03988

3. Relations for Product Moments

Making use of (1.3) and (1.6), we can derive recurrence relations for product moments of $l\,gos$ from (1.5).

Theorem 3.1. For the given generalized inverse Weibull distribution in (1.5) and $n \geq 2$, $m \in \Re$, $1 \leq r < r+1 \leq n$,

$$E[X^{*i}(r,n,m,k)X^{*j+\beta+1}(r+1,n,m,k)] = \frac{\theta\beta\alpha^{\beta}\gamma_{r+1}}{j+1}$$

$$\times \{E[X^{*i+j+1}(r,n,m,k)] - E[X^{*i}(r,n,m,k)X^{*j+1}(r+1,n,m,k)]\} \quad (3.1)$$

and for $1 \leq r < s \leq n$, $s-r \geq 2$ and $i, j \geq 0$

$$E[X^{*i}(r,n,m,k)X^{*j+\beta+1}(s,n,m,k)]$$

$$= \frac{\theta\beta\alpha^{\beta}\gamma_s}{j+1}\{E[X^{*i}(r,n,m,k)X^{*j+1}(s-1,n,m,k)]$$

$$- E[X^{*i}(r,n,m,k)X^{*j+1}(s,n,m,k)]\}. \quad (3.2)$$

Proof. From (1.3), we have

$$E[X^{*i}(r,n,m,k)X^{*j+\beta+1}(s,n,m,k)]$$

$$= \frac{C_{s-1}}{(r-1)!(s-r-1)!}\int_0^{\infty} x^i [F(x)]^m f(x) g_m^{r-1}(F(x)) I(x)dx, \quad (3.3)$$

where

$$I(x) = \int_0^x y^{j+\beta+1}[h_m(F(y)) - h_m(F(x))]^{s-r-1}[F(y)]^{\gamma_s - 1} f(y)dy$$

$$= \theta\beta\alpha^{\beta}\int_0^x y^j [h_m(F(y)) - h_m(F(x))]^{s-r-1}[F(y)]^{\gamma_s} dy$$

upon using the relation in (1.6). Integrating now by parts treating y^j for integration and the rest of the integrand for differentiation, we obtain when $s = r+1$ that

$$I(x) = \frac{\theta\beta\alpha^{\beta}\gamma_{r+1}}{j+1}\left\{\frac{x^{j+1}}{\gamma_{r+1}}[F(x)]^{\gamma_{r+1}} - \int_0^x y^{j+1}[F(y)]^{\gamma_{r+1}-1} f(y)dy\right\}$$

and when $s > r+1$ that

$$I(x) = \frac{\theta\beta\alpha^{\beta}\gamma_s}{j+1}$$

$$\times\left\{\frac{(s-r-1)}{\gamma_s}\int_0^x y^{j+1}[h_m(F(y)) - h_m(F(x))]^{s-r-2}[F(y)]^{\gamma_s+m} f(y)dy\right.$$

$$-\int_0^x y^{j+1}[h_m(F(y))-h_m(F(x))]^{s-r-1}[F(y)]^{\gamma_s-1}f(y)dy\Bigg\}.$$

Upon substituting the above expressions for $I(x)$ in (3.3), we have, after simplifications, the recurrence relations (3.1) and (3.2).

Remark 3.1. Setting $m=0,\ k=1$, in (3.2), we obtain recurrence relations for product moments of order statistics for the generalized inverse Weibull distribution in the form

$$E[X_{n-r+1:n}^{i}\,X_{n-s+1:n}^{j+\beta+1}]$$

$$=\frac{\theta\beta\alpha^{\beta}(n-s+1)}{j+1}\{E[X_{n-r+1:n}^{i}\,X_{n-s+2:n}^{j+1}]-E[X_{n-r+1:n}^{i}\,X_{n-s+1:n}^{j+1}]\}.$$

That is

$$E[X_{r:n}^{i+1}\,X_{s:n}^{j}]=E[X_{r-1:n}^{i+1}\,X_{s:n}^{j}]+\frac{i+1}{\theta\beta\alpha^{\beta}(r-1)}E[X_{r-1:n}^{i+\beta+1}\,X_{s:n}^{j}].$$

Remark 3.2. Setting $m=-1$ and $k\geq 1$, in Theorem 3.1, we get the recurrence relations for product moments of lower $k-$th record values from generalized inverse Weibull distribution in the form

$$E[(Z_r^{(k)})^i(Z_s^{(k)})^{j+\beta+1}]$$

$$=\frac{\theta\beta\alpha^{\beta}k}{j+1}\{E[(Z_r^{(k)})^i(Z_{s-1}^{(k)})^{j+1}]-E[(Z_r^{(k)})^i(Z_s^{(k)})^{j+1}]\}.$$

Remark 3.3. Putting $\theta=1$, in (3.1) and (3.2), we deduce the recurrence relations for product moments of $l\,gos$ from inverse Weibull distribution, established by Pawlas and Szynal (2001).

Remark 3.4. For $\theta=1$ and $\beta=1$, the relations in (3.1) and (3.2) give the recurrence relations for product moments of $l\,gos$ from inverse exponential distribution.

4. Characterization

Let $X^*(r,n,m,k)$, $r=1,2,\ldots,n$ be $l\,gos$ from a continuous population with df $F(x)$ and pdf $f(x)$, then the conditional pdf of $X^*(s,n,m,k)$ given $X^*(r,n,m,k)=x$, $1\leq r<s\leq n$, in view of (1.2) and (1.3), is

$$f_{X^*(s,n,m,k)|X^*(r,n,m,k)}(y|x) = \frac{C_{s-1}}{(s-r-1)!C_{r-1}}[F(x)]^{m-\gamma_r+1}$$
$$\times[h_m(F(y)) - h_m(F(x))]^{s-r-1}[F(y)]^{\gamma_s-1} f(y). \tag{4.1}$$

Theorem 4.1. Let X be a non-negative random variable having an absolutely continuous distribution function $F(x)$ with $F(0)=0$ and $0<F(x)<1$ for all $x>0$, then

$$E[\xi(X^*(s,n,m,k)) | X^*(l,n,m,k) = x] = e^{-\theta\alpha^\beta x^{-\beta}} \prod_{j=1}^{s-l}\left(\frac{\gamma_{l+j}}{\gamma_{l+j}+1}\right),$$
$$l = r, r+1 \tag{4.2}$$

if and only if

$$F(x) = e^{-\theta(\alpha/x)^\beta}, \ x>0, \ \alpha>0, \beta>0, \theta>0,$$

where

$$\xi(x) = e^{-\theta\alpha^\beta y^{-\beta}}.$$

Proof. From (4.1), we have

$$E[\xi(X^*(s,n,m,k)) | X^*(r,n,m,k) = x] = \frac{C_{s-1}}{(s-r-1)!C_{r-1}(m+1)^{s-r-1}}$$
$$\times \int_0^x e^{-\theta\alpha^\beta y^{-\beta}} \left[1-\left(\frac{F(y)}{F(x)}\right)^{m+1}\right]^{s-r-1} \left(\frac{F(y)}{F(x)}\right)^{\gamma_s-1} \frac{f(y)}{F(x)} dy. \tag{4.3}$$

By setting $u = \frac{F(y)}{F(x)} = \frac{e^{-\theta\alpha^\beta y^{-\beta}}}{e^{-\theta\alpha^\beta x^{-\beta}}}$ from (1.5) in (4.3), we obtain

$$E[\xi(X^*(s,n,m,k)) | X^*(r,n,m,k) = x] = \frac{C_{s-1}}{(s-r-1)!C_{r-1}(m+1)^{s-r-1}}$$
$$\times e^{-\theta\alpha^\beta x^{-\beta}} \int_0^1 u^{\gamma_s}(1-u^{m+1})^{s-r-1} du. \tag{4.4}$$

Again by setting $t = u^{m+1}$ in (4.4), we get

$$E[\xi(X^*(s,n,m,k)) | X^*(r,n,m,k) = x] = \frac{C_{s-1}}{(s-r-1)!C_{r-1}(m+1)^{s-r}}$$

$$\times e^{-\theta\alpha^{\beta}x^{-\beta}} \int_0^1 t^{\frac{k+1}{m+1}+n-s-1}(1-t)^{s-r-1}dt$$

$$= \frac{C_{s-1}e^{-\theta\alpha^{\beta}x^{-\beta}}}{C_{r-1}(m+1)^{s-r}} \frac{\Gamma\left(\frac{k+1}{m+1}+n-s\right)}{\Gamma\left(\frac{k+1}{m+1}+n-r\right)}$$

$$= e^{-\theta\alpha^{\beta}x^{-\beta}} \prod_{j=1}^{s-r}\left(\frac{\gamma_{r+j}}{\gamma_{r+j}+1}\right)$$

and hence the relation given in (4.2).

To prove sufficient part, we have from (4.1) and (4.2)

$$\frac{C_{s-1}}{(s-r-1)!C_{r-1}(m+1)^{s-r-1}} \int_0^x e^{-\theta\alpha^{\beta}y^{-\beta}} [(F(x))^{m+1}-(F(y))^{m+1}]^{s-r-1}$$

$$\times [F(y)]^{\gamma_s-1} f(y)dy = [F(x)]^{\gamma_{r+1}} H_r(x), \tag{4.5}$$

where

$$H_r(x) = e^{-\theta\alpha^{\beta}x^{-\beta}} \prod_{j=1}^{s-r}\left(\frac{\gamma_{r+j}}{\gamma_{r+j}+1}\right).$$

Differentiating (4.5) both the sides with respect to x, we get

$$\frac{C_{s-1}[F(x)]^m f(x)}{(s-r-2)!C_{r-1}(m+1)^{s-r-2}} \int_0^x e^{-\theta\alpha^{\beta}y^{-\beta}} [(F(x))^{m+1}-(F(y))^{m+1}]^{s-r-2}$$

$$\times [F(y)]^{\gamma_s-1} f(y)dy = H_r'(x)[F(x)]^{\gamma_{r+1}} + \gamma_{r+1}H_r(x)[F(x)]^{\gamma_{r+1}-1} f(x)$$

or

$$\gamma_{r+1}H_{r+1}(x)[F(x)]^{\gamma_{r+2}+m} f(x)$$

$$= H_r'(x)[F(x)]^{\gamma_{r+1}} + \gamma_{r+1}H_r(x)[F(x)]^{\gamma_{r+1}-1} f(x),$$

where

$$H_r'(x) = \theta\beta\alpha^{\beta} x^{-(\beta+1)} e^{-\theta\alpha^{\beta}x^{-\beta}} \prod_{j=1}^{s-r}\left(\frac{\gamma_{r+j}}{\gamma_{r+j}+1}\right),$$

$$H_{r+1}(x) - H_r(x) = \frac{e^{-\theta\alpha^{\beta}x^{-\beta}}}{\gamma_{r+1}} \prod_{j=1}^{s-r}\left(\frac{\gamma_{r+j}}{\gamma_{r+j}+1}\right).$$

Therefore,

$$\frac{f(x)}{F(x)} = \frac{H_r'(x)}{\gamma_{r+1}[H_{r+1}(x) - H_r(x)]}$$
$$= \theta\beta\alpha^{\beta} x^{-(\beta+1)}$$

which proves that

$$F(x) = e^{-\theta(\alpha/x)^{\beta}}, x > 0,\ \alpha > 0, \beta > 0, \theta > 0.$$

Theorem 4.2. Let X be a non-negative random variable having an absolutely continuous distribution function $F(x)$ with $F(0) = 0$ and $0 < F(x) < 1$ for all $x > 0$, then

$$E[X^{*j+\beta+1}(r,n,m,k)]$$
$$= \frac{\theta\beta\alpha^{\beta}\gamma_r}{j+1}\{E[X^{*j+1}(r-1,n,m,k)] - E[X^{*j+1}(r,n,m,k)]\} \tag{4.6}$$

if and only if

$$F(x) = e^{-\theta(\alpha/x)^{\beta}}, x > 0,\ \alpha > 0,\ \beta > 0, \theta > 0.$$

Proof. The necessary part follows immediately from equation (2.12). On the other hand if the recurrence relation in equation (4.6) is satisfied, then on using equation (1.2), we have

$$\frac{C_{r-1}}{(r-1)!}\int_0^{\infty} x^{j+\beta+1}[F(x)]^{\gamma_r - 1} f(x) g_m^{r-1}(F(x))dx$$
$$= \frac{\theta\beta\alpha^{\beta}(r-1)}{j+1}\frac{C_{r-1}}{(r-1)!}\int_0^{\infty} x^{j+1}[F(x)]^{\gamma_r + m} f(x) g_m^{r-2}(F(x))dx$$
$$- \frac{\theta\beta\alpha^{\beta}\gamma_r}{j+1}\frac{C_{r-1}}{(r-1)!}\int_0^{\infty} x^{j+1}[F(x)]^{\gamma_r - 1} f(x) g_m^{r-1}(F(x))dx$$
$$= \frac{\theta\beta\alpha^{\beta}}{j+1}\frac{C_{r-1}}{(r-1)!}\int_0^{\infty} x^{j+1}[F(x)]^{\gamma_r} f(x) g_m^{r-2}(F(x))$$
$$\times \left\{(r-1)[F(x)]^m - \frac{\gamma_r g_m(F(x))}{F(x)}\right\}dx. \tag{4.7}$$

Let

$$h(x) = -[F(x)]^{\gamma_r} g_m^{r-1}(F(x)) \tag{4.8}$$

and

$$h'(x) = [F(x)]^{\gamma_r} f(x) g_m^{r-2}(F(x)) \left\{ (r-1)[F(x)]^m - \frac{\gamma_r g_m(F(x))}{F(x)} \right\}.$$

Thus,

$$\frac{C_{r-1}}{(r-1)!} \int_0^\infty x^{j+\beta+1} [F(x)]^{\gamma_r - 1} f(x) g_m^{r-1}(F(x)) dx$$

$$= \frac{\theta\beta\alpha^\beta}{j+1} \frac{C_{r-1}}{(r-1)!} \int_0^\infty x^{j+1} h'(x) dx . \quad (4.9)$$

Now integrating *RHS* in (4.9) by parts and using the value of $h(x)$ from (4.8), we obtain

$$\frac{C_{r-1}}{(r-1)!} \int_0^\infty x^{j+\beta+1} [F(x)]^{\gamma_r - 1} f(x) g_m^{r-1}(F(x)) dx$$

$$= \theta\beta\alpha^\beta \frac{C_{r-1}}{(r-1)!} \int_0^\infty x^j [F(x)]^{\gamma_r} g_m^{r-1}(F(x)) dx$$

which reduces to

$$\frac{C_{r-1}}{(r-1)!} \int_0^\infty x^j [F(x)]^{\gamma_r - 1} g_m^{r-1}(F(x)) \{\theta\beta\alpha^\beta F(x) - x^{\beta+1} f(x)\} dx = 0 . \quad (4.10)$$

Now applying a generalization of the Müntz-Szász Theorem (Hwang and Lin, 1984) to equation (4.10), we get

$$\frac{f(x)}{F(x)} = \theta\beta\alpha^\beta x^{-(\beta+1)}$$

which proves that

$$F(x) = e^{-\theta(\alpha/x)^\beta}, x > 0, \ \alpha > 0, \ \beta > 0, \ \theta > 0.$$

Acknowledgments

The authors are grateful to Professor M. Ahsanullah, Department of Management Sciences, Rider University, Lawrenceville, NJ, USA for helpful suggestions which resulted in the improved presentation of the paper.

References

Ahsanullah, M. (2004): A characterization of the uniform distribution by dual generalized order statistics. *Comm. Statist. Theory Methods*, 33, 2921-2928.

Burkschat, M., Cramer, E. and Kamps, U. (2003): Dual generalized order statistics. *Metron*, LXI, 13-26.

Gusmão, F.R.S., Ortega, E.M.M. and Cordeiro, G.M. (2011): The generalized inverse Weibull distribution. *Statist. Papers*, 52, 591-619.

Hwang, J.S. and Lin, G.D. (1984): On a generalized moments problem II. *Proc. Amer. Math. Soc.*, 91, 577-580.

Kamps, U. (1995): *A Concept of Generalized Order Statistics*. B.G. Teubner Stuttgart, Germany.

Kamps, U. (1998): Characterizations of distributions by recurrence relations and identities for moments of order statistics. In: Balakrishnan, N. and Rao, C.R., *Handbook of Statistics* 16, *Order Statistics: Theory & Methods*, North-Holland, Amsterdam, 291-311.

Khan, R.U., Anwar, Z. and Athar, H. (2008): Recurrence relations for single and product moments of dual generalized order statistics from exponentiated Weibull distribution. *Aligarh J. Statist.*, 28, 37-45.

Khan, R.U. and Kumar, D. (2010): On moments of lower generalized order statistics from exponentiated Pareto distribution and its characterization. *Appl. Math. Sci.* (*Ruse*), 4, 2711-2722.

Khan, R.U. and Zia, B. (2011): Recurrence relations for the moments of generalized inverse Weibull distribution based on record statistics and a characterization. *Submitted for Publication.*

Klugman, S.A., Panjer, H.H. and Willmot, G.E. (1998): *Loss Models: From Data to Decisions*. John Wiley, New York.

Mbah, A.and Ahsanullah, M. (2007): Some characterizations of the power function distribution based on lower generalized order statistics. *Pakistan J. Statist.*, 23, 139-146.

Pawlas, P. and Szynal, D. (2001): Recurrence relations for single and product moments of lower generalized order statistics from the inverse Weibull distribution. *Demonstratio Math.*, XXXIV, 353-358.

Ruiz, S.M. (1996): An algebraic identity leading to Wilson's theorem. *The Mathematical Gazette*, 80, 579-582.

In: Research in Applied Statistical Science
Editor: Mohammad Ahsanullah, pp. 237-247
ISBN: 978-1-63321-818-5

Chapter 20

CHARACTERIZATION OF CONTINUOUS DISTRIBUTIONS CONDITIONED ON A PAIR OF NON-ADJACENT GENERALIZED ORDER STATISTICS USING MEIJER'S G-FUNCTION

***M. J. S. Khan*[1,*]*, A. H. Khan*[2] *and M. I. Khan*[2]**
[1]Department of Applied Mathematics and Statistics,
Babasaheb Bheemrao Ambedkar University, Lucknow, India
[2]Department of Statistics and Operations Research,
Aligarh Muslim University, Aligarh, India

Abstract

In this paper, a generalized family of continuous distributions have been characterized through conditional expectation of generalized order statistics, conditioned on a pair of non-adjacent generalized order statistics using Meijer's G-function and its various deductions are discussed.

Keywords: Continuous distributions, Meijer's G-function, generalized order statistics, records and conditional expectation

AMS Subject Classification: 62E15, 62E10, 62G30

1. Introduction

Let $n \in N, (n \geq 2), k \geq 1$, $\tilde{m} = (m_1, m_2, \ldots, m_{n-1}) \in R^{n-1}$, $M_r = \prod_{j=r}^{n-1} m_j$, $1 \leq r \leq n-1$, be the parameters such that $\gamma_r = k + n - r + M_r \geq 1$, for all $r \in \{1, \cdots, n-1\}$. Then $U(r, n, \tilde{m}, k)$ are

* E-mail address: jahangirskhan@gmail.com (Corresponding author)

said to be uniform generalized order statistics (*gos*), if their joint probability density function (*pdf*) is given by (Kamps, 1995)

$$f_{U(1,n,\tilde{m},k),\cdots,U(n,n,\tilde{m},k)}(u_1,\cdots,u_n) = k\left(\prod_{j=1}^{n-1}\gamma_j\right)\left(\prod_{i=1}^{n-1}(1-u_i)^{m_i}(1-u_n)^{k-1}\right), \quad (1.1)$$

on the cone

$$0 < u_1 \le \cdots \le u_n < 1.$$

Based on any arbitrary distribution function (*df*) $F(x)$, generalized order statistic $X(r,n,\tilde{m},k)$ can be defined by quantile transformation $X(r,n,\tilde{m},k) = F^{-1}(U(r,n,\tilde{m},k)), 1 \le r \le n$, where F^{-1} denotes the quantile function of F defined by

$$F^{-1}(u) = \sup\{x \in (\alpha,\beta) : F(x) \le u\}, \ \forall \ u \in (0,1),$$

where $\alpha = \inf\{x \in R : F(x) > 0\}$ and $\beta = \sup\{x \in R : F(x) < 1\}$ are left and right end point of X.

An alternative definition of *gos* is given by Cramer and Kamps (2003) as follows:

Let $r \in N$ and let $\gamma_1, \ldots \gamma_r > 0$, be the parameters, then generalized order statistic $X(r,n,\tilde{m},k)$ based on a continuous distribution function

$$X(r,n,\tilde{m},k) \underline{\underline{d}} F^{-1}\left(1-\prod_{i=1}^{r} B_i\right),$$

where $B_1, \ldots B_r$ are independent random variables with respective distributions Beta $(\gamma_j, 1), 1 \le j \le r$.

Choosing the parameters appropriately, models such as ordinary order statistics ($m = 0$, $k = 1$ i.e. $\gamma_i = n - i + 1$), k^{th} record value ($m = -1$, $k \in N$ i.e. $\gamma_i = k$), sequential order statistics ($\gamma_i = (n-i+1)\beta_i; \beta_1, \beta_2, \cdots, \beta_n > 0$), order statistics with non integral sample size ($\gamma_i = \beta - i + 1; \beta > 0$), Pfeifer record values ($\gamma_i = \beta_i; \beta_1, \beta_2, \cdots, \beta_n > 0$) and progressive type II censored order statistics ($m \in N$, $k \in N$) are obtained.

The problem of characterization of probability distributions through regression, conditioned on adjacent and non-adjacent *gos*, have been discussed by Keseling (1999), Bieniek and Szynal (2003), Cramer *et al.* (2004 a), Khan and Alzaid (2004), Raqab and Abu-Lawi (2004), Ahsanullah and Raqab (2004), Khan *et al.* (2006) and Bieniek (2007) among others. Ahsanullah and Beg (2008), Ahsanullah *et al.* (2009) have characterized continuous distributions through conditional expectation of *gos*, conditioned on two *gos*.

Recently Bieniek (2009), Khan and Khan (2011) have characterized continuous distribution functions conditioned on non-adjacent *gos* using Meijer's G-function. We, in this paper have extended the result of Bieniek (2009), Khan and Khan (2011) when the regression is based on two non-adjacent *gos* . Also, the result is deduced for known results on *gos*. Further extension in this direction is not possible because of the Markovian property of the *gos* (Kamps, 1995).

Let P_F stand for the probability measure on R determined by $F(x)$, then the *pdf* of $X(r,n,\tilde{m},k)$ with respect to a measure P_F is given by (Cramer and Kamps, 2003)

$$f_r(x) = c_{r-1}\, G_r\,(\overline{F}(x)\,|\,\gamma_1,\ldots,\gamma_r)\; I_{(\alpha,\beta)}(x), \tag{1.2}$$

where I_A denotes the indicator function and $G_r(x) = G_{r,r}^{r,0}(x\,|\,\gamma_1,\ldots,\gamma_r)$ $= G_{r,r}^{r,0}\left(x\middle|\begin{matrix}\gamma_1,\ldots,\gamma_r\\ \gamma_1-1,\ldots,\gamma_r-1\end{matrix}\right)$ is the particular Meijer's G-function defined by

$$G_{r,r}^{r,0}\left(s\middle|\begin{matrix}\gamma_1,\ldots,\gamma_r\\ \gamma_1-1,\ldots,\gamma_r-1\end{matrix}\right) = \frac{1}{2\pi i}\int_L \frac{s^z}{\prod_{j=1}^{r}(\gamma_j-1-z)}\,dz, \tag{1.3}$$

and L is an appropriate chosen contour of integration. See Mathai (1993, Chapter 3) for the definition of G-function and its numerous properties and applications.

The joint $P_F \otimes P_F$ density of $X(r,\, n,\, \tilde{m},\, k)$ and $X(s,\, n,\, \tilde{m},\, k)$, $1 \le r < s \le n$, is given by

$$f_{r,s}(x,y) = c_{s-1}\, G_{s-r}\left(\frac{\overline{F}(y)}{\overline{F}(x)}\middle|\,\gamma_{r+1},\ldots,\gamma_s\right)\frac{G_r\,(\overline{F}(x)\,|\,\gamma_1,\ldots,\gamma_r)}{\overline{F}(x)},\quad \alpha < x < y < \beta \tag{1.4}$$

and the joint $\bigotimes_{j=1}^{3} P_F$ density function of $X(r,\, n,\, \tilde{m},\, k)$, $X(j,\, n,\, \tilde{m},\, k)$ and $X(s,\, n,\, \tilde{m},\, k)$, $1 \le r < j < s \le n$, may similarly be given as

$$f_{r,\,j,\,s}(x,t,y) = c_{s-1}\,\frac{1}{\overline{F}(x)}\frac{1}{\overline{F}(t)}\,G_{s-j}\left(\frac{\overline{F}(y)}{\overline{F}(t)}\middle|\,\gamma_{j+1},\ldots,\gamma_s\right)$$

$$\times\, G_{j-r}\left(\frac{\overline{F}(t)}{\overline{F}(x)}\middle|\,\gamma_{r+1},\ldots,\gamma_j\right) G_r\,(\overline{F}(x)\,|\,\gamma_1,\ldots,\gamma_r),\quad \alpha < x < t < y < \beta \tag{1.5}$$

Hence the conditional P_F density function of $X(j,n,\tilde{m},k)$ given $X(r,n,\tilde{m},k) = x$ and $X(s,n,\tilde{m},k) = y$, $1 \le r < j < s \le n$, is given by

$$f_{j|r,s}(t\,|\,x,y) = \frac{1}{\overline{F}(t)}\frac{G_{s-j}\left(\frac{\overline{F}(y)}{\overline{F}(t)}\middle|\,\gamma_{j+1},\ldots,\gamma_s\right) G_{j-r}\left(\frac{\overline{F}(t)}{\overline{F}(x)}\middle|\,\gamma_{r+1},\ldots,\gamma_j\right)}{G_{s-r}\left(\frac{\overline{F}(y)}{\overline{F}(x)}\middle|\,\gamma_{r+1},\ldots,\gamma_s\right)} I_{(x,y)}(t), \tag{1.6}$$

where $\overline{F}(x)=1-F(x)$, and $c_{r-1}=\prod_{i=1}^{r}\gamma_i$.

Auxiliary Results

Various results which are used in the subsequent sections are reproduced here:

(i) $G_1(x\,|\,\gamma_1)=x^{\gamma_1-1}$

(ii) $(\gamma_r-\gamma_1)G_r(x\,|\,\gamma_1,\ldots,\gamma_r)=G_{r-1}(x\,|\,\gamma_1,\ldots,\gamma_{r-1})-G_{r-1}(x\,|\,\gamma_2,\ldots,\gamma_r)$

(iii) $x^a\,G_r(x\,|\,\gamma_1,\ldots,\gamma_r)=G_r(x\,|\,\gamma_1+a,\ldots,\gamma_r+a),\qquad a\in R$

(iv) $\lim_{x\to 1-} G_r(x\,|\,\gamma_1,\ldots,\gamma_r)=\begin{cases}1, & r=1\\ 0, & r\ge 2\end{cases}$

and

$$\lim_{x\to 0+} G_r(x\,|\,\gamma_1,\ldots,\gamma_r)=\begin{cases}0, & \text{if } \gamma_{1:r}>1\\ \prod_{\substack{j=1\\ j\ne l}}^{r}\dfrac{1}{\gamma_j-\gamma_l}, & \text{if } \gamma_{1:r}=1<\gamma_{2:r}\\ \infty, & \text{if } \gamma_{1:r}=\gamma_{2:r}=1 \text{ or } \gamma_{1:r}<1\end{cases}$$

where $\gamma_{1:r}=\min(\gamma_1,\ldots,\gamma_r)$ and $l=\max(1\le j\le r:\gamma_j=\gamma_{1:r})$.

(v) $\dfrac{d}{dx}G_r(x\,|\,\gamma_1,\ldots,\gamma_r)=\dfrac{1}{x}\left[(\gamma_r-1)\,G_r(x\,|\,\gamma_1,\ldots,\gamma_r)-G_{r-1}(x\,|\,\gamma_1,\ldots,\gamma_{r-1})\right]$

(vi) $\dfrac{d}{dx}G_r(x\,|\,\gamma_1,\ldots,\gamma_r)=\dfrac{1}{x}\left[(\gamma_1-1)\,G_r(x\,|\,\gamma_1,\ldots,\gamma_r)-G_{r-1}(x\,|\,\gamma_2,\ldots,\gamma_r)\right]$

Proof. For the property (i), see Mathai (1993, p. 130), for property (ii), see Cramer and Kamps (2003), and for the property (iii), see Mathai (1993, p. 69). Property (iv) can easily be deduced from Lemma 2.2 of Cramer *et al.* (2004 b), whereas (vi) and (vii) can be established from (1.3).

2. Characterization of Distributions

Theorem 2.1. Let $X(i,\,n,\,\tilde{m},\,k)$, $i=1,\ldots,n$ be the ith *gos* from a continuous population with the *df* $F(x)$ and the *pdf* $f(x)$ over the support (α,β), and $h(t)$ be a monotonic and differentiable function of t. If for two consecutive values r and $r+1$, $1<r+1<j<s\le n$,

$$g_{j|l,s}(x,y) = E[h\{X(j,n,\tilde{m},k)\} \big| X(l,n,\tilde{m},k) = x, X(s,n,\tilde{m},k) = y],\ \ l = r, r+1,$$

exist, then

$$(\gamma_{r+1}-1)\frac{f(x)}{\overline{F}(x)} - \frac{\frac{\partial}{\partial x} G_{s-r}\left(\frac{\overline{F}(y)}{\overline{F}(x)} \Big| \gamma_{r+1},\ldots,\gamma_s\right)}{G_{s-r}\left(\frac{\overline{F}(y)}{\overline{F}(x)} \Big| \gamma_{r+1},\ldots,\gamma_s\right)} = \frac{\frac{\partial}{\partial x} g_{j|r,s}(x,y)}{[g_{j|r,s}(x,y) - g_{j|r+1,s}(x,y)]},$$

and

$$\frac{G_{s-r}\left(\frac{\overline{F}(y)}{\overline{F}(x)} \Big| \gamma_{r+1}-\gamma_{r+1}+1,\ldots,\gamma_s-\gamma_{r+1}+1\right)}{G_{s-r}(\overline{F}(y) \big| \gamma_{r+1}-\gamma_{r+1}+1,\ldots,\gamma_s-\gamma_{r+1}+1)} = \exp\left[-\int_{\alpha}^{x} D_1(t,y)\,dt\right], \tag{2.1}$$

where $g(\ )$ is a finite and differentiable function of x, and

$$D_1(x,y) = \frac{\frac{\partial}{\partial x} g_{j|r,s}(x,y)}{[g_{j|r,s}(x,y) - g_{j|r+1,s}(x,y)]}. \tag{2.2}$$

Proof. We have,

$$g_{j|r,s}(x,y)\, G_{s-r}\left(\frac{\overline{F}(y)}{\overline{F}(x)} \Big| \gamma_{r+1},\ldots,\gamma_s\right)$$

$$= \int_x^y \frac{h(t)}{\overline{F}(t)} G_{s-j}\left(\frac{\overline{F}(y)}{\overline{F}(t)} \Big| \gamma_{j+1},\ldots,\gamma_s\right) G_{j-r}\left(\frac{\overline{F}(t)}{\overline{F}(x)} \Big| \gamma_{r+1},\ldots,\gamma_j\right) f(t)\,dt \tag{2.3}$$

Differentiating both the sides of (2.3) *w.r.t.* x and using the property (vi), we have

$$\frac{\partial}{\partial x} g_{j|r,s}(x,y) G_{s-r}\left(\frac{\overline{F}(y)}{\overline{F}(x)} \Big| \gamma_{r+1},\ldots,\gamma_s\right)$$

$$+ g_{j|r,s}(x,y)\frac{f(x)}{\overline{F}(x)}\left[(\gamma_{r+1}-1)G_{s-r}\left(\frac{\overline{F}(y)}{\overline{F}(x)} \Big| \gamma_{r+1},\ldots,\gamma_s\right) - G_{s-r-1}\left(\frac{\overline{F}(y)}{\overline{F}(x)} \Big| \gamma_{r+2},\ldots,\gamma_s\right)\right]$$

$$= \frac{f(x)}{\overline{F}(x)}\left[g_{j|r,s}(x,y)(\gamma_{r+1}-1)G_{s-r}\left(\frac{\overline{F}(y)}{\overline{F}(x)} \Big| \gamma_{r+1},\ldots,\gamma_s\right) - g_{j|r+1,s}(x,y)G_{s-r-1}\left(\frac{\overline{F}(y)}{\overline{F}(x)} \Big| \gamma_{r+2},\ldots,\gamma_s\right)\right]$$

implying that

$$\frac{f(x)}{\overline{F}(x)} \frac{G_{s-r-1}\left(\frac{\overline{F}(y)}{\overline{F}(x)} \Big| \gamma_{r+2},\ldots,\gamma_s\right)}{G_{s-r}\left(\frac{\overline{F}(y)}{\overline{F}(x)} \Big| \gamma_{r+1},\ldots,\gamma_s\right)} = \frac{\frac{\partial}{\partial x} g_{j|r,s}(x,y)}{[g_{j|r,s}(x,y) - g_{j|r+1,s}(x,y)]}$$

Therefore,

$$(\gamma_{r+1}-1)\,\frac{f(x)}{\overline{F}(x)}-\frac{\frac{\partial}{\partial x}G_{s-r}\left(\frac{\overline{F}(y)}{\overline{F}(x)}\Big|\,\gamma_{r+1},\ldots,\gamma_s\right)}{G_{s-r}\left(\frac{\overline{F}(y)}{\overline{F}(x)}\Big|\,\gamma_{r+1},\ldots,\gamma_s\right)}=\frac{\frac{\partial}{\partial x}g_{j|r,s}(x,y)}{[g_{j|r,s}(x,y)-g_{j|r+1,s}(x,y)]}$$

and hence the result.

Corollary 2.1. Using residue theorem, it can be proved that

$$G_{s-r}\left(\frac{\overline{F}(y)}{\overline{F}(x)}\Big|\,\gamma_{r+1},\ldots,\gamma_s\right)=\left(\sum_{i=r+1}^{s}a_i^{(r)}(s)\left[\frac{\overline{F}(y)}{\overline{F}(x)}\right]^{\gamma_i-1}\right), \tag{2.4}$$

where $a_i^{(r)}(s)=\prod_{\substack{j=r+1\\ j\neq i}}^{s}\frac{1}{(\gamma_j-\gamma_i)}$, $\gamma_j\neq\gamma_i$, $r+1\le i\le s\le n$,

Therefore (2.1) reduces to

$$\frac{[\overline{F}(x)]^{\gamma_{r+1}}\,B_r^{\,s}(x,y)}{B_r^{\,s}(\alpha,y)}=\exp\left[-\int_\alpha^x D_1(t,y)\,dt\right], \tag{2.5}$$

where

$$B_r^{\,s}(x,y)=\sum_{i=r+1}^{s}a_i^{(r)}(s)\left(\frac{\overline{F}(y)}{\overline{F}(x)}\right)^{\gamma_i},$$

as obtained by Ahsanullah *et al.* (2009).

Also since for $m_1=\ldots=m_{n-1}=m\neq-1$,

$$a_i^{(r)}(s)=\frac{1}{\prod_{\substack{j=r+1\\ i\neq j}}^{s}(\gamma_j-\gamma_i)}=\frac{(-1)^{s-i}}{(m+1)^{s-r-1}(s-r-1)!}\binom{s-r-1}{s-i},$$

thus for $m_1=\ldots=m_{n-1}=m\neq-1$, (2.1) reduces to

$$\frac{1-\{\overline{F}(x)\}^{m+1}}{1-\{\overline{F}(y)\}^{m+1}}=1-\exp\left[-\frac{1}{(s-r-1)}\int_\alpha^x D_1(t,y)\,dt\right],\qquad m\neq-1, \tag{2.6}$$

and

$$\frac{\log \overline{F}(x)}{\log \overline{F}(y)} = 1 - \exp\left[-\frac{1}{(s-r-1)} \int_{\alpha}^{x} D_1(t, y)\, dt \right], \qquad m = -1 \tag{2.7}$$

as obtained by Ahsanullah *et al.* (2009).

Remark 2.1. At $\gamma_s = 0$ i.e. $s = k + n + M_s$, by convention $X(s, n, \tilde{m}, k) = y = \beta$, and hence $\overline{F}(y) = 0$. Therefore,

$$g_{j|r}(x) = E[h(X(j, n, \tilde{m}, k)) \mid X(r, n, \tilde{m}, k) = x]$$

and

$$\overline{F}(x) = \exp\left(-\frac{1}{\gamma_{r+1}} \int_{\alpha}^{x} \frac{g'_{j|r}(t)}{[g_{j|r}(t) - g_{j|r+1}(t)]} dt \right), \tag{2.8}$$

as given by Khan *et al.* (2006), Bieniek (2009).

Theorem 2.2. Let $X(i, n, \tilde{m}, k)$, $i = 1, \ldots, n$ be the ith *gos* from a continuous population with the *df* $F(x)$ and the *pdf* $f(x)$ over the support (α, β), and $h(t)$ be a monotonic and differentiable function of t. If for two consecutive values $s-1$ and s, $1 \le r < j < s-1 < n$,

$$g_{j|r,l}(x, y) = E[h\{(X(j,n,\tilde{m},k)\} \mid X(r,n,\tilde{m},k) = x,\ X(l,n,\tilde{m},k) = y],\ l = s-1, s \text{ exist,}$$

then

$$(\gamma_s - 1)\frac{f(y)}{\overline{F}(y)} + \frac{\frac{\partial}{\partial y} G_{s-r}\left(\frac{\overline{F}(y)}{\overline{F}(x)} \Big| \gamma_{r+1}, \ldots, \gamma_s \right)}{G_{s-r}\left(\frac{\overline{F}(y)}{\overline{F}(x)} \Big| \gamma_{r+1}, \ldots, \gamma_s \right)} = \frac{\frac{\partial}{\partial y} g_{j|r,s}(x, y)}{[g_{j|r,s-1}(x, y) - g_{j|r,s}(x, y)]}, \tag{2.9}$$

and

$$G_{s-r}\left(\frac{\overline{F}(y)}{\overline{F}(x)} \Big| \gamma_{r+1} - \gamma_s + 1, \ldots, \gamma_s - \gamma_s + 1 \right) = a_s^{(r)}(s) \exp\left[-\int_{y}^{\beta} D_2(x,t)\, dt \right], \forall\ \gamma_i > \gamma_s, i = r+1, \ldots, s-1 \tag{2.10}$$

and for $\gamma_{r+1} = \ldots = \gamma_s$,

$$\frac{1 + \log \overline{F}(y)}{1 + \log \overline{F}(x)} = 1 - \exp\left(-\frac{1}{(s-r-1)} \int_{y}^{q} D_2(x,t)\, dt \right), \tag{2.11}$$

where

$$q \in (\alpha, \beta): -\log \overline{F}(q) = 1, \tag{2.12}$$

$g(\ )$ is a finite and differentiable function of y, and

$$D_2(x, y) = \frac{\frac{\partial}{\partial y} g_{j|r,s}(x, y)}{[g_{j|r,s-1}(x, y) - g_{j|r,s}(x, y)]}. \tag{2.13}$$

Proof. Differentiating both the sides of (2.3) *w.r.t.* y and proceeding as in the Theorem 2.1, we get

$$\frac{f(y)}{\overline{F}(y)} \frac{G_{s-r-1}\left(\frac{\overline{F}(y)}{\overline{F}(x)} \Big| \gamma_{r+1}, \ldots, \gamma_{s-1}\right)}{G_{s-r}\left(\frac{\overline{F}(y)}{\overline{F}(x)} \Big| \gamma_{r+1}, \ldots, \gamma_{s}\right)} = \frac{\frac{\partial}{\partial y} g_{j|r,s}(x, y)}{[g_{j|r,s-1}(x, y) - g_{j|r,s}(x, y)]}$$

Thus,

$$(\gamma_s - 1)\frac{f(y)}{\overline{F}(y)} + \frac{\frac{\partial}{\partial y} G_{s-r}\left(\frac{\overline{F}(y)}{\overline{F}(x)} \Big| \gamma_{r+1}, \ldots, \gamma_{s}\right)}{G_{s-r}\left(\frac{\overline{F}(y)}{\overline{F}(x)} \Big| \gamma_{r+1}, \ldots, \gamma_{s}\right)} = \frac{\frac{\partial}{\partial y} g_{j|r,s}(x, y)}{[g_{j|r,s-1}(x, y) - g_{j|r,s}(x, y)]},$$

implying that

$$\frac{G_{s-r}\left(\frac{\overline{F}(y)}{\overline{F}(x)} \Big| \gamma_{r+1} - \gamma_s + 1, \ldots, \gamma_s - \gamma_s + 1\right)}{G_{s-r}\left(\frac{\overline{F}(\beta)}{\overline{F}(x)} \Big| \gamma_{r+1} - \gamma_s + 1, \ldots, \gamma_s - \gamma_s + 1\right)} = \exp\left[-\int_y^\beta D_2(x,t)\,dt\right], \tag{2.14}$$

It can be seen that for $\gamma_i > \gamma_s$, $i = r+1, \ldots, s-1$, $G_{s-r}(x | \gamma_{r+1} - \gamma_s + 1, \ldots, \gamma_s - \gamma_s + 1) \to a_s^{(r)}(s)$ as $x \to 0$, and therefore

$$G_{s-r}\left(\frac{\overline{F}(y)}{\overline{F}(x)} \Big| \gamma_{r+1} - \gamma_s + 1, \ldots, \gamma_s - \gamma_s + 1\right) = a_s^{(r)}(s) \exp\left[-\int_y^\beta D_2(x,t)\,dt\right], \forall\ \gamma_i > \gamma_s, i = r+1, \ldots, s-1$$

Also, for $\gamma_{r+1} = \ldots = \gamma_s$, (Cramer, 2002, p. 35)

$$G_{s-r}\left(\frac{\overline{F}(y)}{\overline{F}(x)}\Big|\,\gamma_{r+1},\ldots,\gamma_s\right)=\frac{1}{(s-r-1)!}\left(-\log\overline{F}(y)+\log\overline{F}(x)\right)^{s-r-1}\left[\frac{\overline{F}(y)}{\overline{F}(x)}\right]^{\gamma_{r+1}-1}$$

Thus in the case of record statistic,

$$\frac{1+\log\overline{F}(y)}{1+\log\overline{F}(x)}=1-\exp\left(-\frac{1}{(s-r-1)}\int_y^q D_2(x,t)\,dt\right), \tag{2.15}$$

where q is as defined in (2.12).

Corollary 2.2. It may be noted that at $\gamma_i \neq \gamma_j$ but $m_1 = \cdots = m_{n-1} = m > -1$.

$$\frac{G_{s-r}\left(\frac{\overline{F}(y)}{\overline{F}(x)}\Big|\,\gamma_{r+1}-\gamma_s+1,\ldots,\gamma_s-\gamma_s+1\right)}{a_s^{(r)}(s)}=\sum_{i=r+1}^{s}a_i{}^r(s)\left(\frac{\overline{F}(y)}{\overline{F}(x)}\right)^{\gamma_i-\gamma_s}=\left[1-\frac{\overline{F}(y)^{m+1}}{\overline{F}(x)^{m+1}}\right]^{s-r-1}$$

Thus (2.8) reduces to

$$\frac{\{\overline{F}(y)\}^{m+1}}{\{\overline{F}(x)\}^{m+1}}=1-\exp\left(-\frac{1}{(s-r-1)}\int_y^\beta D_2(x,t)\,dt\right), \quad m>-1. \tag{2.16}$$

as obtained by Ahsanullah *et al.* (2009).

Remark 2.2. With the convention $X(0,\, n,\, \tilde{m},\, k)=x=\alpha$, at $r=0$, Theorem 2.2 reduces to

$$G_s(\overline{F}(y)\,|\,\gamma_1-\gamma_s+1,\ldots,\gamma_s-\gamma_s+1)=\sum_{i=1}^{s}a_i(s)[\overline{F}(y)]^{\gamma_i-\gamma_s}=a_s(s)\exp\left[-\int_y^\beta D(t)\,dt\right],\ \text{if } \gamma_{1:s-1>\gamma_s} \tag{2.17}$$

and for $m_1 = \ldots = m_{n-1} = m > -1$

$$\overline{F}(y)^{m+1}=1-\exp\left[-\frac{1}{s-1}\int_y^\beta D(t)\,dt\right], \tag{2.18}$$

whereas for $\gamma_{r+1}=\ldots=\gamma_s$,

$$-\log(\overline{F}(y)=\exp\left[-\frac{1}{s-1}\int_y^q D(t)\,dt\right] \tag{2.19}$$

where q is as defined in (2.12) and

$$g_{j|s}(y) = E[h(X(j,n,\tilde{m},k)) \mid X(s,n,\tilde{m},k) = y],$$

and

$$D(y) = \frac{g'_{j|s}(y)}{[g_{j|s-1}(y) - g_{j|s}(y)]} = D_2(\alpha, y),$$

as obtained by Bieniek (2009), Khan and Khan (2011).

Corollary 2.3. Under the assumptions given in Corollary 2.1 and Corollary 2.2,

$$\overline{F}(x) = \left[\frac{e^{I_2}}{e^{I_1} + e^{I_2} - 1}\right]^{\frac{1}{m+1}}, \qquad m > -1 \tag{2.20}$$

and

$$\overline{F}(y) = \left[\frac{e^{I_2} - 1}{e^{I_1} + e^{I_2} - 1}\right]^{\frac{1}{m+1}}, \qquad m > -1 \tag{2.21}$$

where

$$I_1 = \int_\alpha^x A_1(t, y)\, dt, \quad I_2 = \int_y^\beta A_2(x, t)\, dt,$$

and

$$A_1(x, y) = \frac{D_1(x, y)}{(s - r - 1)}, \quad A_2(x, y) = \frac{D_2(x, y)}{(s - r - 1)}$$

Similarly for records, we have

$$\overline{F}(x) = \exp\left[-\frac{e^{I_1} - 1}{e^{I_1} + e^{I_2} - 1}\right], \tag{2.22}$$

and,

$$\overline{F}(y) = \exp\left[-\frac{e^{I_1}}{e^{I_1} + e^{I_2} - 1}\right]. \tag{2.23}$$

Acknowledgment

We are thankful to Referees and Professor Ahsanullah for their fruitful advice, help and suggestions.

References

Ahsanullah, M. and Raqab, M.Z. (2004): Characterizations of distributions by conditional expectations of generalized order statistics. *J. Appl. Statist. Sci.,* 13, 41-48.

Ahsanullah, M. and Beg, M.I. (2008): On characterizing distributions via regression on pairs of generalized order statistics. *Calcutta Statist. Assoc. Bull.* 60, 71-79.

Ahsanullah, M.; Khan, M.J.S. and Khan, A.H. (2009): On characterization of continuous distributions conditioned on a pair of non-adjacent generalized order statistics. *J. Staist. Theory Appl.* 8, 353-372.

Bieniek, M. (2007): On characterizations of distributions by regressions of adjacent generalized order statistics. *Metrika*, 66, 233-242.

Bieniek, M. (2009): A note on characterizations of distributions by regression of non-adjacent generalized order statistics. *Aust. N.Z.J. Stat.*, 51, 89-99.

Bieniek, M. and Szynal, D. (2003): Characterizations of distributions via linearity of regression of generalized order statistics. *Metrika*, 58, 259-271.

Cramer, E. (2002): Contribution to Generalized Order Statistics. *Ph. D. thesis*, University of Oldenburg.

Cramer, E. and Kamps, U. (2003): Marginal distributions of sequential and generalized order statistics. *Metrika*, 58, 293-310.

Cramer, E.; Kamps, U. and Keseling, C. (2004 a): Characterization via linear regression of ordered random variables: a unifying approach. *Comm. Statist. Theory Methods*, 33, 2885-2911.

Cramer, E.; Kamps, U. and Rychlik, T. (2004 b): Unimodality of uniform generalized order statistics with application to mean bound. *Ann. Inst. Statist. Math.*, 56, 183-192.

Kamps, U. (1995): *A Concept of Generalized Order Statistics*. Teubner, Stuttgart.

Keseling, C. (1999): Conditional distributions of generalized order statistics and some characterizations. *Metrika*, 49, 27-40.

Khan, A.H. and Alzaid, A.A. (2004): Characterization of distributions through linear regression of non-adjacent generalized order statistics. *J. Appl. Statist. Sci.*, 13, 123-136.

Khan, A.H. and Khan, M.J.S. (2011): Characterization of continuous distributions through generalized order statistics-an alternative approach. *Inter. J. Appl. Math. Statist.*, 21, 52-59.

Khan A.H.; Khan, R.U. and Yaqub, M. (2006): Characterization of continuous distributions through conditional expectation of function of generalized order statistics. *J. App. Prob. Statist.,* 1, 115-131.

Mathai, A.M. (1993): *A Handbook of Generalized Special Functions for Statistical and Physical Sciences*. Oxford Science Publications, New York.

Raqab, M.Z. and Abu-Lawi, L.N. (2004): Characterizations of continuous distributions based on expectations of generalized order statistics. *J. Appl. Statist. Sci.*, 13, 101-116.

In: Research in Applied Statistical Science
Editor: Mohammad Ahsanullah, pp. 249-259
ISBN: 978-1-63321-818-5

Chapter 21

EMPIRICAL BAYES METHOD ON CHANGEPOINTS ESTIMATION OF TUMOR GROWTH PROFILES IN XENOGRAFT EXPERIMENTS

Wei Ning*
Department of Mathematics and Statistics
Bowling Green State University, Bowling Green, OH, US

Abstract

In this paper, a method based on the empirical Bayes method is proposed to estimate the change point of a changepoint model, which is used to model the regrowth points of the tumor volume in xenograft experiment. Comparing with the other Bayes changepoint models, the advantage of using the proposed method is to estimate the parameters of the prior distributions through the marginal distribution of the data, instead of choosing values for the parameters vaguely in the prior distributions. The results show that the proposed method can discover the change pattern in the data successfully.

Keywords: Changepoint, Empirical Bayes Method, Prior Distribution, Maximum Likelihood Estimate

AMS Subject Classification: 62F15, 62F10, 62P10

1. Introduction

In translational cancer research, the tumor xenograft model is often used to model the effects of the therapies on controlling the volumes of the cancer cells. In xenograft experiment, immunocompromised mice are grafted with cancer cells and treated with anti-cancer therapies. During the experiment, the volumes of the cancer cells will be monitored before and after the treatment to investigate the effects of the therapies and possible tumor regrowth.

*E-mail address: wning@bgsu.edu (Corresponding author)

Tumor regression and regrowth is complicated and involves several biological processes. Tumor growth patterns can be quite different depending on the treatment. For example, the volumes of the untreated tumors may show the consistent growing pattern during the whole study period, while the the volumes of the tumors under the radiation treatment may regress for a period of time at the beginning and then subsequently regrow, that is, the growth of the volumes may follow a change pattern. It is commonly to use the time point until the tumor volume doubling as the endpoint in these studies to investigate such a pattern. However, it has disadvantages. First, some important biological aspects of a treatment effect may be missed due to ignoring the measurements taken after time-to-tumor doubling. The single estimate of doubling time does not address the biological mechanisms underlying different patterns of tumor growth.

Several methods have been proposed to choose an endpoint for tumor regrowth studies by comparing the tumor volumes at selected time points. For example, the Wilcoxon-Mann-Whitney test is used to compare the tumor volumes between treatments at a certain time point but it ignores the information at the other time points. Sauer et al.(2005) proposed to use Friedman repeated-measures ANOVA on ranks to compare the volumes between treatments at a given time point while considering the correlation of measurements on the same tumor in longitudinal data analysis. However, the censored data which is below the limit of quantitation is ignored. Tan (2002, 2005) and Fang (2004) developed a t-test via the EM algorithm and, Bayesian approaches for testing differences between two treatment regimens by analyzing longitudinal data and taking into account the censored data. However, their models can not be used to model the tumor growth pattern because their hypotheses are based on comparing either specific time points or a function of selected time points. Begg (1980) pointed out that the determination of the magnitude of the delay depends on the choice of the endpoint of the tumor size critically. Although these endpoints are useful to estimate the tumor growth delay, they neither capture all of the data nor address the different mechanisms underlying tumor growth. Recently, Zhao et al. (2011) proposed a Bayesian hierarchical changepoint approach to model the tumor xenograft volume profile data and estimate tumor growth profiles for 12 treatment groups and made comparisons between treatments. Same as the classical Bayesian method, they chose pre-specified prior distributions for the parameters. However, choices of the priors are somewhat subjective, which may not reflect the true distributions of parameters.

In this paper, we propose an estimation method based on the empirical Bayes method to estimate the regrowth point in a changepoint model for the selected tumors from the radiation treatment group. The paper is organized as follows. In section 2, we briefly introduce the change point problem and the empirical Bayes method. The details of the data from the xenograft experiment, the model and the estimated results will be given in section 3. Some discussion will be provided in section 4.

2. Change Point Problem and Empirical Bayes Method

2.1. Change Point Problem

In statistics, a single change point can be viewed as a place or time point such that the observations before and after that point follow different distributions. Multiple change points can be detected similarly. Change point problems have received a lot of attention since Page (1954, 1955) who introduced a simple process to detect a single change. Chernoff and Zacks (1964), Gardner (1969), Hawkins (1992), Sen and Srivastava (1975), and Worsley (1979) studied the testing and estimation of a change in the mean of a normal model. Hsu (1977), Inclán (1993), Chen and Gupta (1997) studied the change point problem for the variance. Worsley (1986) provided confidence regions and tests for a change-point in a sequence of exponential models. Kim and Siegmund (1989) studied the change point problem for the simple linear regression model by proposing a LRT test. Chen and Gupta (2000) discussed change point problems for various parametric models with different approaches. Asymptotic results of the LRT-based test can be found in Csörgő and Horváth (1997). Ning and Gupta (2009) considered the change point problem for the first four moments for the generalized lambda distribution and applied such a model to detect the variations of the copy numbers of DNA. Ning (2011) and Ning et al. (2011) proposed a nonparametric method for the detection of the change points with different types of changes in the mean.

2.2. Empirical Bayes Method

Bayesian analysis depends on a prior distribution for the model parameters. This prior can be parametric or nonparametric depending on unknown parameters which can in turn be drawn from second-stage prior. This sequence of parameters and priors result in a hierarchical model. However, the prior parameters from the final stage of such a hierarchical model have to be assumed known to stop the hierarchy. The empirical Bayes method uses the marginal distribution of the data to estimate the prior parameters from the last stage instead of making such assumption, and the proceeds as though the prior were known.

Morris (1983) proposed a parametric empirical Bayes approach by taking the density function of the prior distribution of the parameters as a parametric form. Robbins (1955, 1983) chose such a density function to be a nonparametric form to achieve more flexibility. Van Houwelingen (1979), Maritz and Lwin (1989) made further generalization and modernization. More details refer to Carlin and Louis (2000).

3. Detection of Change Points of Tumor Growth

3.1. Data Description

This data is taken from the study of the combination of a novel Chk1/2 targeted agent, AZD7762, with gemcitabine and radiation to treat pancreatic cancer (See Morgan et al, 2010). In this study, MiaPaCa-2 cells were injected subcutaneously into the flanks of mice (typically 6 or 7 per treatment group). Treatment with gemcitabine AZD7762, and/or radiation were initiated when the average tumor volume reached 100 mm^3. Tumor volume was monitored by measurement with calipers two times per week, until the tumors reached approximately 800 mm^3, the mouse died, or 120 days, whichever occurred first, and tumor volume was calculated.

We choose eight tumors from the treatment of radiation to illustrate our method. The tumor sizes were logarithmically with base 2 since the tumor growth rates are typically close to exponential both before and after the nadir.

3.2. Statistical Method

We denote $\mathbf{Y}_k = (y_{ik})$ as the $\log(\text{volume})$ of the kth tumor at the time t_i. In our data, we choose eight tumors from the radiation treatment, therefore $k = 1, \cdots, 8$. Since we use the same changepoint model to analyze eight tumors, for the simplicity, we drop the subindex k. For the given data of $\log(\text{volume})$, $\mathbf{Y} = (Y_1, Y_2, ..., Y_n)$, we assume that it follows a normal distribution with the mean μ_i and the constant variance σ^2, which is described by the following tumor-level piecewise linear model,

$$\begin{aligned} Y_i|\mu_i &\sim N(\mu_i, \sigma^2) \\ \mu_i &= a_i - b_{1i}(t_i - r_i)_{\_} + b_{2i}(t_i - r_i)_+, \end{aligned} \tag{1}$$

where $z_{\_} = \min(z, 0)$ and $z_+ = \max(z, 0)$. For each component of μ_i, we choose the priors as follows:

$$\begin{aligned} a_i &\sim N(\mu_a, \sigma_a^2), \\ b_{1i} &\sim N(\mu_{b1}, \sigma_{b1}^2), \\ b_{2i} &\sim N(\mu_{b2}, \sigma_{b2}^2), \\ r_i &\sim N(\mu_r, \sigma_r^2). \end{aligned}$$

When the animal receives the radiation treatment, the average size of the $\log(\text{volume})$ of the tumor will be decreased for a period of time at the beginning, and the average size of the volume will regrow after passing a certain time point. Our purpose is to estimate such a change point time.

To choose appropriate values for the means and variances for the prior distributions, we use the empirical method to find the maximum likelihood estimates (MLEs) from the marginal distribution of the data, instead of setting up the values personally. We rewrite the parameters as the population means plus an individual-specific random effect measured as a deviation from the mean, that is,

$$\begin{aligned} a_i &= \mu_a + \gamma_{0i}, \\ b_{1i} &= \mu_{b1} + \gamma_{1i}, \\ b_{2i} &= \mu_{b2} + \gamma_{2i}, \\ r_i &= \mu_r + \gamma_i. \end{aligned} \tag{2}$$

Therefore, in a vector notation, the population mean parameters are denoted by $\boldsymbol{\beta} = (\mu_a, \mu_{b1}, \mu_{b2}, \mu_r)^{\mathrm{T}}$. The random effects $\boldsymbol{\eta} = (\gamma_{0i}, \gamma_{1i}, \gamma_{2i}, \gamma_i)^{\mathrm{T}}$ are assumed independent, therefore, it follows a multivariate normal distribution with the priors distributions specified as

$$\boldsymbol{\eta} \sim MN((a_i, b_{1i}, b_{2i}, r_i)^{\mathrm{T}}, \boldsymbol{\Sigma}),$$

where $\boldsymbol{\Sigma} = \mathrm{diag}(\sigma_a^2, \sigma_{b1}^2, \sigma_{b2}^2, \sigma_r^2)$. With the notations above, we can express Y_i as

$$Y_i|\mu_i \sim N(\mu_i, \sigma^2) = N(f(\boldsymbol{\beta}, \boldsymbol{\eta}), \sigma^2). \tag{3}$$

It is easy for us to check the marginal distribution of $\boldsymbol{Y}$ is a nonlinear function of the parameters $(\boldsymbol{\beta}, \boldsymbol{\eta}, \sigma^2)$ from which we expect to obtain MLEs. Therefore, we consider the first order Taylor expansion of f at the expected values of the random effect, that is, $\boldsymbol{\eta} = 0$. We have

$$\mu_i = f(\boldsymbol{\beta}, \boldsymbol{\eta}) \approx f(\boldsymbol{\beta}, \mathbf{0}) + f'_{\boldsymbol{\eta}}(\boldsymbol{\beta}, \mathbf{0})\boldsymbol{\eta}, \tag{4}$$

where

$$\begin{aligned}\mu_i = f(\boldsymbol{\beta}, \boldsymbol{\eta}) &= \mu_a + \gamma_{0i} - (\mu_{b1} + \gamma_{1i})(t_i - (\mu_r + \gamma_i))_- \\ &+ (\mu_{b2} + \gamma_{2i})(t_i - (\mu_r + \gamma_i))_+.\end{aligned}$$

Thus, we obtain the approximation for the right side of (4) as

$$\begin{aligned}\mu_i = f(\boldsymbol{\beta}, \boldsymbol{\eta}) \approx &\mu_a - \mu_{b1}(t_i - \mu_r)_- + \mu_{b2}(t_i - \mu_r)_+ \\ &+ \gamma_{0i} - (t_i - \mu_r)_- \times \gamma_{1i} + (t_i - \mu_r)_+ \times \gamma_{2i} \\ &+ (\mu_{b1} I(t_i < \mu_r + \gamma_i) - \mu_{b2} I(t_i \geq \mu_r + \gamma_i)) \times \gamma_i.\end{aligned} \tag{5}$$

Since (5) is a linear equation of $\boldsymbol{\eta}$, it is easy to see that μ_i can be approximated by a normal distribution with the mean and variance as follows:

$$\begin{aligned}E(\mu_i) &= \mu_a - \mu_{b1}(t_i - \mu_r)_- + \mu_{b2}(t_i - \mu_r)_+, \\ Var(\mu_i) &= \sigma_a^2 + (\mu_{b1}(t_i - \mu_r)_-)^2 \sigma_{b1}^2 + (\mu_{b2}(t_i - \mu_r)_+)^2 \sigma_{b2}^2 \\ &+ (\mu_{b1} I(t_i < \mu_r + \gamma_i) - \mu_{b2} I(t_i \geq \mu_r + \gamma_i))^2 \sigma_r^2.\end{aligned}$$

Therefore, we rewrite our two-stage hierarchial model as

$$\begin{aligned}Y_i|\mu_i &\sim N(\mu_i, \sigma^2) \\ \mu_i &\sim N(E(\mu_i), Var(\mu_i)).\end{aligned} \tag{6}$$

Then the marginal distribution of y_i is

$$y_i \sim N(E(\mu_i), Var(\mu_i) + \sigma^2) \triangleq N(\mu_T, \sigma_T^2),$$

and therefore the marginal distribution of $\boldsymbol{y}$ is

$$\begin{aligned}m(\boldsymbol{y}) &= \prod_{i=1}^{n} m(y_i) = \prod_{i=1}^{n} \frac{1}{\sqrt{2\pi}\sigma_T} e^{-\frac{1}{2\sigma_T^2}(y_i - \mu_T)^2} \\ &= \left(\frac{1}{\sqrt{2\pi}\sigma_T}\right)^n e^{-\frac{1}{2\sigma_T^2}\sum(y_i - \mu_T)^2}.\end{aligned} \tag{7}$$

Therefore, the log-likelihood function is

$$\log m(\boldsymbol{y}) = -n \log 2\pi - \frac{n}{2} \log \sigma_T^2 - \frac{1}{2\sigma_T^2} \sum_{i=1}^{n} (y_i - \mu_T)^2 \tag{8}$$

Based on the above log-likelihood function, we can find the MLEs of the parameters $(\boldsymbol{\beta}, \boldsymbol{\eta}, \sigma^2)$.

From the equation (2), the estimated regrowth time point is

$$\hat{r}_i = \hat{\mu}_r + \hat{\gamma}_i. \tag{9}$$

Table 1. Estimates for Eight Tumors

	Tumor 1		Tumor 2		Tumor 3		Tumor 4	
Para	Est	SE	Est	SE	Est	SE	Est	SE
μ_a	4.659	0.163	2.804	0.094	8.042	0.153	3.706	0.171
μ_{b1}	0.004	0.0004	0.043	0.004	-0.017	0.001	0.028	0.009
μ_{b2}	0.003	0.0002	0.018	0.001	0.001	< 0.0001	0.010	0.003
μ_r	21.552	2.101	20.243	1.371	10.958	0.712	25.064	1.862
γ_{0i}	9.966	0.949	11.596	0.938	15.769	1.600	17.805	1.021
γ_{1i}	13.102	1.244	12.126	0.960	16.997	1.346	9.168	1.770
γ_{2i}	13.021	1.107	11.388	1.690	18.419	1.216	1.927	0.136
γ_i	6.481	0.477	5.080	0.336	19.594	0.657	14.135	0.407
σ^2	16.186	3.003	12.926	0.938	4.253	0.506	13.703	1.521
	Tumor 5		Tumor 6		Tumor 7		Tumor 8	
Para	Est	SE	Est	SE	Est	SE	Est	SE
μ_a	6.616	0.144	8.151	0.162	6.157	0.178	8.118	0.171
μ_{b1}	-0.003	0.001	-0.009	0.001	0.004	0.002	-0.003	0.0005
μ_{b2}	-4e-5	$< 1e-5$	0.003	0.002	0.001	0.0004	0.002	0.0001
μ_r	18.301	0.743	15.321	1.016	18.936	2.536	37.031	1.906
γ_{0i}	14.595	0.969	4.937	1.631	13.289	2.331	8.820	0.953
γ_{1i}	13.916	1.805	11.095	0.713	11.451	0.787	13.602	1.796
γ_{2i}	10.812	0.804	2.303	0.506	18.801	1.732	18.530	2.945
γ_i	14.483	1.048	13.298	1.530	11.594	1.468	1.180	0.120
σ^2	3.530	0.969	20.633	1.631	15.175	2.535	19.032	3.013

3.3. Results

We apply the method proposed in previous section to estimate the regrowth time point for the eight selected tumors treated with the radiation. Table 1 lists the estimates for the parameters with the standard errors for all eight tumors. From the results, the estimated regrowth time point for each individual tumor is 28.03, 25.32, 30.55, 39.20, 32.78, 28.62, 30.54 and 38.21 days respectively, which is the time when the tumor volume reaches the nadir. The average regrowth time point for the entire group is around 32 days. The results are matched with the ones obtained by Zhao et al. (2011).

In the Appendix, Figure 1 to Figure 4 are the scatterplots of treatment time versus $\log(\text{volume})$. The red line on each graph corresponds to the estimated regrowth time point, from which we can see the proposed method works well in this data. The estimated change-point is a biologically meaningful parameter, but we will not explore more details on this biological interpretation since this is not the focus of this paper.

4. Discussion

In this paper, we propose a simple procedure based on the empirical Bayes method to estimate the change points where the regrowth of the tumor occurs in the xenograft experiment. The advantage of the proposed method is that we estimate the parameters of the prior distributions in the hierarchical model through the maximum likelihood function of the marginal distribution of the data. Compared to pre-specified prior distributions with given values of the parameters, our prior distributions contain more information obtained from the data because the estimates of the parameters are derived from maximizing the marginal distribution of the data, therefore, will be more accurately describe the true distribution of the parameters.

Acknowledgments

The author would like to thank the Department of Radiation Oncology at University of Michigan for sharing the xenograft data. The author also would like to thank Dr. Lili Zhao for her valuable discussion and suggestion.

Appendix

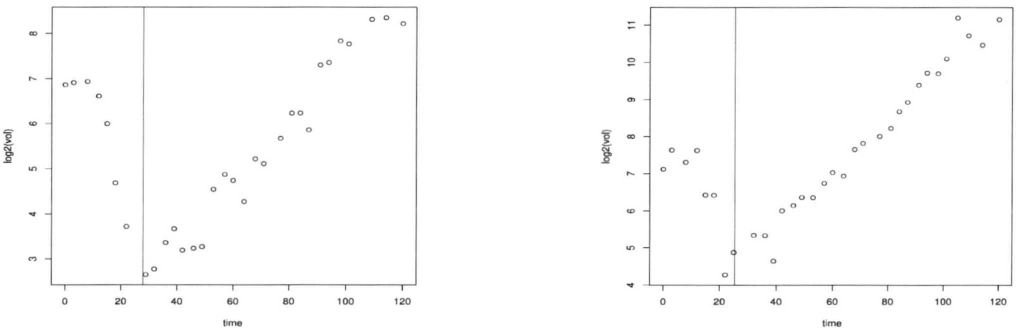

Figure 1. The two scatterplots, from left to right, top to bottom, represent the $\log(\text{volum})$ of the tumor 1 and tumor 2 versus the experiment time respectively. The estimated change point for Tumor 1 is $r_1 = 28.03$; The estimated change point for Tumor 2 is $r_2 = 25.32$.

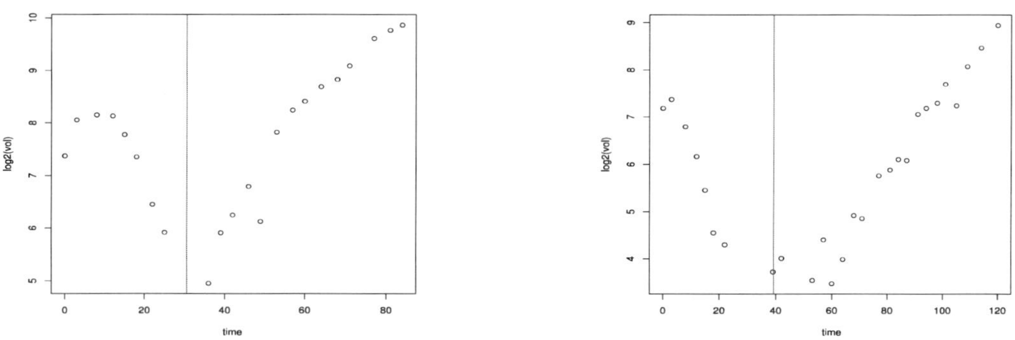

Figure 2. The two scatterplots, from left to right, top to bottom, represent the $\log(\text{volum})$ of the tumor 3 and tumor 4 versus the experiment time respectively. The estimated change point for Tumor 3 is $r_3 = 30.55$; The estimated change point for Tumor 4 is $r_4 = 39.20$.

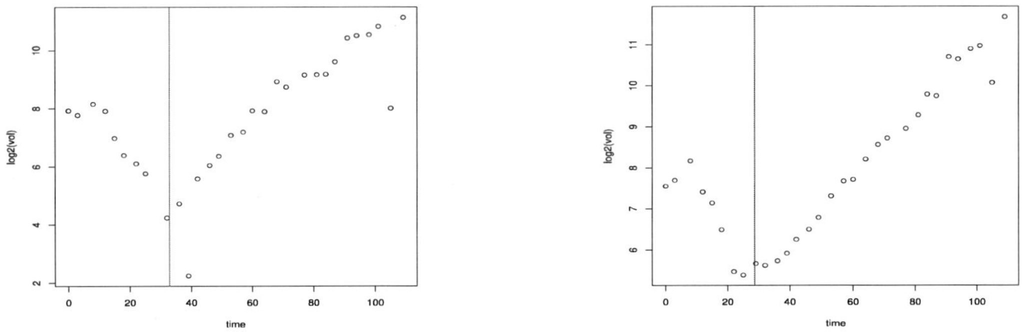

Figure 3. The two scatterplots, from left to right represent the $\log(\text{volum})$ of the tumor 5 and tumor 6 versus the experiment time respectively. The estimated change point for Tumor 5 is $r_5 = 32.78$; The estimated change point for Tumor 6 is $r_6 = 28.62$.

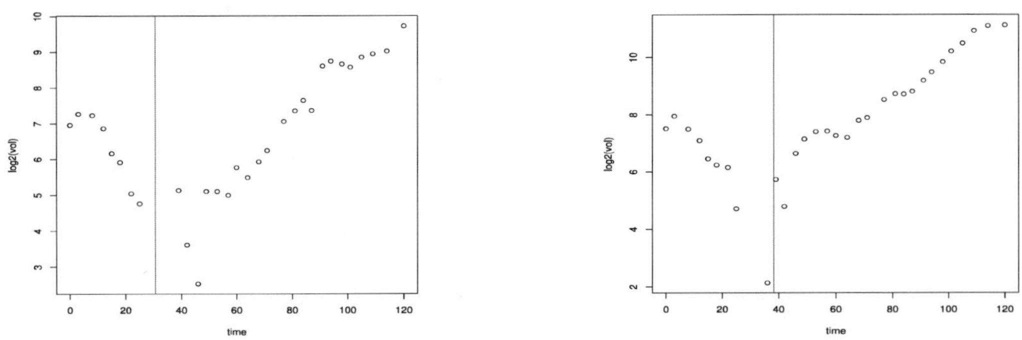

Figure 4. The two scatterplots, from left to right, represent the $\log(\text{volum})$ of the tumor 7 and tumor 8 versus the experiment time respectively. The estimated change point for Tumor 7 is $r_7 = 30.54$; The estimated change point for Tumor 8 is $r_8 = 38.21$.

References

Begg, A.C. (1980). Analysis of growth delay data: potential pitfalls. *British Journal of Cancer,* 42, 93C97.

Bellera, C.A., HANLEY, J.A., Joseph, L., Albertsen, P.C. (2008). Hierarchical change-point models for biochemical markers illustrated by tracking postradiotherapy prostate-specific antigen series in men with prostate cancer. *Ann Epidemiol,* 18, 270-82.

Carlin, B.P. and Louis, T.A. (2000). *Bayes and Empirical Bayes Methods for Data Analysis,* London: Champman & Gall/CRC.

Chen, J. and Gupta, A. K. (1997). Testing and Locating Variance Changepoints With Application to Stock Prices. *Journal of American Statistical Association,* 92, 739-747.

Chen, J. and Gupta, A. K. (2000). *Parametric Statistical Change Point Analysis.* Boston: Birkhäuser.

Chernoff, H. and Zacks, S. (1964). Estimating the Current Mean of a Normal Distribution Which is Subject to Changes in Time. *Annals of Mathematical Statistics,* 35, 999-1018.

Csörgő, M., Horváth, L. (1997). *Limit Theorems in Change-Point Analysis.* John Wiley& Sons: New York.

Fang, H.B., Tian, G.L., Tan, M. (2004). Hierarchical models for tumour xenograft experiments in drug development. *Journal of Biopharmaceutical statistics,* 14, 931-945.

Gelman, A., Carlin, J.B., Stern, H.S., Rubin, D.B. (1996). *Bayesian Data Analysis.* New York: Chapman & Hall/CRC.

Gardner, L. A. (1969). On Detecting Change in the Mean of Normal Variates. *Annals of Mathematical Statistics,* 40, 116-126.

Hall, C.B., Ying, J., Kuo, L., Lipton, R.B. (2003). Bayesian and profile likelihood change-point methods for modeling cognitive function over time. *Computational Statistics and Data Analysis,* 42, 91-109.

Hawkins, D. M. (1992). Detecting Shifts in Functions of Multivaraite Location and Co-variance Parameters. *Journal of Statistical Planning and Inference,* 33, 233-244.

Hewitt, H.B., Blake, E.R. (1968). The growth of transplanted tumors in pre-irradiated sites. *British Journal of Cancer,* 22, 808.

Hsu, D. A. (1977). Tests for Variance Shifts at an Unknown Time Point. *Applied Statistics,* 26, 179-184.

Inclán, C. (1993). Detection of Multiple Changes of Variance Using Posterior Odds. *Journal of Business and Economic Statistics,* 11, 189-300.

Kim, H. J. and Siegmund, D. (1989). The Likelihood Ratio Test for a Change-point in Simple Linear Regression. *Biometrika,* 76, 409-423.

Kiuchi Kiuchi, A.S., Hartigan, J.A., Holford, T.R., Rubinstein, P., Stevens, C.E. (1995). Change Points in the Series of T4 Counts Prior to AIDS. *Biometrics,* 51, 236-48.

Maritz, J.S. and Lwin, T. (1989). *Empirical Bayes Methods,* 2nd ed. London: Chapman and Hall.

Morgan, M.A., Parsels, L.A., Zhao, L., et al. (2010). Radiosensitizaion by the Chk1/2 inhibitor, AZD7762 involves abrogation of the G2 checkpoint and inhibition of homologous recombination repair. *Cancer Research,* 70, 4972-4981.

Morris, C.N. (1983). Parametric empirical Bayes Inference: Theory and applications. *Journal of American Statistical Association,* 78, 47-65.

Ning, W. (2011). Empirical Likelihood Ratio Test for a Mean Change Point Model with a Linear Trend Followed by an Abrupt Change. *Journal of Applied Statistics.* DOI:10.1080/02664763.2011.628647.

Ning, W. and Gupta, A. K.,(2009). Change point analysis for generalized lambda distributions. *Communication in Statistics-Simulations and Computation.* 38, 1789-1802.

Ning, W., Pailden, P. and Gupta, A.K. (2011). The Empirical Likelihood Ratio Test for the Epidemic Change Point Model. *Journal of Data Science,* 10(2012), 107-127.

Page, E.S. (1954). Continue inspection schemes. *Biometrika,* 41, 100-235.

Page, E.S. (1955). A test for a chance in a parameter occurring at an unknown point. *Biometrika,* 42, 523-527.

Robbins, H. (1955). An empirical Bayes approach to statistics. In *Proceedings 3rd Berkeley Symposium on Mathematical Statistics and Probability,* 1, Berkeley, CA: Univ. of California Press, 157-164.

Robbins, H. (1983). Some thoughts on empirical Bayes estimation. *Annals of Statistics,* 1, 713-723.

Sauer, L.A., Dauchy, R.T., Blask, D.E., Krause, J.A., Davidson, L.K., Dauchy, E.M. (2005) Eicosapentaenoic Acid Suppresses Cell Proliferation in MCF-7 Human Breast Cancer Xenografts in Nude Rats via a Pertussis ToxinCSensitive Signal Transduction Pathway. *The American Society for Nutritional Sciences,* 135, 2124-2129.

Shusterman, S., Grupp, S.A., Barr, R., Carpentieri, D., Zhao, H., Maris, J.M. (2001) Angiogenesis Inhibitor TNP-470 Effectively Inhibits Human Neuroblastoma Xenograft Growth, Especially in the Setting of Subclinical Disease. *Clinical Cancer Research,* 7, 977-84.

Slate, E., Lark, L. (2001) Using PSA to detect prostate cancer onset: an application of Bayesian retrospective and prospective changepoint identification. *Journal of Educational and Behavioral Statistics,* 26, 443-468.

Spiegelhaler, D., Thomas, A., Best, N., Gilks, W. (2000) Winbugs: Bayesian inference using Gibbs sampling, version 1.3. Cambridge, MA: MRC biostatistics Unit.

Tan, M., Fang, H.B., Tian, G.L., Houghton, P.J. (2002) Small sample inference for incomplete longitudianl data with trunction and cersoring in tumour xenograft models. *Biometrics,* 58, 612-620.

Tan, M., Fang, H.B., Tian, G.L., Houghton, P.J. (2005) Repeated measures models with constrained parameters for incomplete data in tumour xenograft experiments. *Statistics in Medicine,* 24, 109-119.

Trott, K.R., Kummermehr, J. (1982) Split dose recovery of a mouse tumour and its stroma during fractionated irradiation. *British Journal of Radiology,* 55, 841-846.

van Houwelingen, J.C. (1977). Monotonizing empirical Bayes estimators for a class of discrete distributions with monotone likelihood ratio. *Statistica Neerlandica,* 31, 95-104.

Zhao, L.L., Morgan, M.A., Parsels, L.A., Maybaum, J., Lawrence, T.S., Normolle, D. (2011). Bayesian hierarchical changepoint methods in modeling the tumor growth profiles in xenograft experiments. *Clinical Cancer Research,* 17(5), 1-7.

INDEX

A

ACF, 77, 183, 184, 185, 187, 188, 189, 190, 191, 193, 195, 197
ADA, 77
aging process, 201
agriculture, 133
AIDS, 257
alcoholism, 124
algorithm, vii, 11, 13, 15, 17, 250
alternative hypothesis, 16
ANOVA, 250
anti-cancer, 249
antigen, 256
Asia, 113, 115, 116, 124

B

base, 30, 251
Bayes estimates, viii, 45, 56
behaviors, ix, 183
bias, ix, 137, 140, 141, 163, 166, 169, 183, 187, 190, 191, 192, 193, 194, 196
biological processes, 250
births, 24, 113
bounds, 25, 154
Brazil, 199, 206

C

Cairo, 45, 133
cancer, ix, 119, 124, 125, 129, 131, 249
cancer cells, 249
candidates, 71
carcinogenicity, 204
carcinoma, 119, 131
category a, 106, 111
CDC, 77
chemotherapy, 124
cirrhosis, 124
classes, viii, 93, 94, 212, 213, 216, 217
coding, 111
coefficient of variation, 153, 158, 180
cognitive function, 257
compilation, 24
complexity, 61, 63, 69, 70
computation, 8, 15, 28, 123, 124, 148, 150, 187, 189
computer, 28, 61, 62
computing, 105, 150, 166, 196
conditional mean, 144, 145, 152
conditioning, 104
configuration, 12
construction, 75, 76, 96
continuous data, 144
continuous random variable, 173, 174, 175
contour, 239
convention, 243, 245
correlation, 29, 250
correlations, 183, 188
cost, 12, 133, 134
cumulative distribution function, 46, 135
cycles, 144

D

data analysis, 116, 120, 250
data availability, 33
data set, x, 24, 81, 104, 113, 116, 123, 124, 125
deaths, 24
decay, 188, 189
decomposition, 26, 27, 64, 103, 104, 105, 112, 114, 115
degenerate, 6, 7
demographic change, 24
demographic data, 24
demography, 35, 199
derivatives, 14, 15, 140, 202
destruction, 125, 128
detection, 251
developed countries, vii, 23, 29
developing countries, vii, 23, 24, 29, 30
deviation, 16, 30, 109, 119, 126, 252
discrete random variable, 143, 144, 145, 147, 148, 149, 150
diseases, 124

distribution, viii, ix, x, 2, 3, 4, 6, 8, 9, 12, 14, 15, 37, 38, 40, 41, 43, 45, 46, 47, 51, 52, 53, 54, 55, 57, 58, 83, 84, 85, 86, 87, 88, 89, 90, 91, 108, 121, 122, 123, 130, 131, 133, 134, 135, 141, 142, 143, 145, 146, 147, 148, 149, 150, 151, 152, 153, 154, 155, 156, 159, 160, 161, 162, 163, 165, 166, 173, 174, 175, 177, 179, 180, 181, 196, 202, 203, 204, 219, 220, 221, 222, 223, 224, 225, 226, 229, 230, 231, 233, 235, 238, 251, 253, 254, 255
distribution function, ix, 2, 37, 145, 155, 160, 162, 220, 231, 233, 238
DNA, 251
DOI, 258
drugs, 11

E

economics, 174
Egypt, 45, 133, 173
embolization, 124
engineering, 174
environmental issues, 142
environmental stress, 46
equipment, 144, 200
Europe, 113, 114, 115, 116, 124
exponential distributions, vii, 37, 221

F

families, ix, 43, 143, 144, 147, 151, 152, 154
fertility, 24, 36
Filipino, 125
financial, 90
flexibility, 202, 251
forecasting, 36
formula, 12, 51, 224
France, 37

G

geometry, 75, 78
Germany, vii, 23, 24, 28, 29, 30, 31, 33, 34, 36, 103, 117, 235
goodness-of-fit tests, vii, 9, 10
grading, 91
graph, 35, 106, 128, 254
grouping, 16
growth, 45, 135, 250, 256, 257

H

hazards, x, 199, 200, 201, 203
hemolytic uremic syndrome, 45
hepatocellular carcinoma, ix, 119, 121, 124, 125, 128, 129, 130
heterogeneity, viii, 59, 61
hypothesis, 1, 2, 8, 9, 12, 103, 152, 203, 204
hypothesis test, 152, 203, 204

I

IBD, 212, 214
identification, 196, 258
identity, 235
immunocompromised, 249
immunodeficiency, 200
incidence, 60, 124, 130, 211, 212, 213, 214, 215, 216
income, 113, 114, 115, 116
independent variable, 29
India, vii, 23, 24, 28, 29, 30, 31, 33, 82, 83, 90, 93, 103, 209, 219, 237
individuals, 124, 200, 205
industry, 144, 200
inequality, 66, 67, 68, 69
infant mortality, 114, 115
inflammation, 124
inflation, 12, 104
information matrix, 15, 52, 60, 61, 136, 157, 203
inhibitor, 257
integration, 14, 126, 225, 229, 239
intervals, 55
intervention, 196
Iran, 143
Iraq, 35
irradiation, 259
issues, 45, 104, 120, 167, 185, 187
iteration, 15
iterative solution, 18

J

Jordan, 57, 155, 183, 197

L

lead, 86, 170
leukemia, 45
life tables, vii, 23, 207
lifetime, ix, 47, 130, 143, 144, 153, 200
light, viii, 29, 59, 78
linear function, 160, 167
linear model, 12, 21, 116, 120, 202, 203, 252
liver, ix, 119, 121, 124, 125, 129
liver cancer, 124, 129
liver transplant, 124, 125
logistic regression, vii, 11, 12, 13, 21, 135, 142
longitudinal study, 15

M

magnitude, 184, 250
manipulation, 70
manufacturing, 199
marginal distribution, x, 249, 251, 252, 253, 255
marketing, 94
marriage, 24
mass, viii, 83, 84, 85, 143, 149
matrix, 13, 14, 15, 16, 17, 18, 21, 26, 27, 54, 60, 78, 103, 104, 105, 106, 107, 108, 109, 110, 111, 112, 136, 181, 211, 212, 213, 214, 216
measurement(s), viii, 12, 21, 93, 101, 102, 133, 134, 250
median, 43, 128, 129, 142, 187
medical, 94, 199
metastatic cancer, 124
methodology, vii, x, 11, 12, 13, 18, 20, 24, 199, 204
Miami, 119
mice, 249, 251
migration, 24, 120
modelling, 200
models, vii, viii, x, 11, 12, 17, 20, 103, 104, 107, 125, 130, 135, 154, 174, 180, 183, 184, 185, 186, 187, 188, 189, 194, 196, 200, 206, 207, 208, 238, 249, 250, 251, 256, 257, 258
modernization, 251
modifications, 94
Monte Carlo method, vii, 1, 2
Monte Carlo simulation, viii, 45, 46
mortality, vii, 23, 24, 29, 36, 113, 114, 115, 116
multidimensional, 21
multiple interpretations, 174

N

National Institute of Mental Health, 17, 20
National Institutes of Health, 131
Nepal, vii, 23, 24, 28, 29, 30, 31, 32, 33, 34, 35, 36
New England, 130
New Zealand, 43
non-linear equations, 166, 167, 168
normal distribution, 2, 6, 8, 9, 10, 13, 45, 54, 135, 203, 204, 252
North America, 124
null, vii, 1, 8, 211
null hypothesis, vii, 1
numerical computations, 55, 166

O

oil, 113, 114, 115, 116
one dimension, 169
optimization, 160
ordered random variables, 247
orthogonal sets, viii, 75

P

Pakistan, 90, 91, 131, 235
pancreatic cancer, 251
parallel, 75, 78, 194
parameter estimates, 15
parameter estimation, 13, 187
Pareto, 47, 173, 174, 221, 235
parity, 24
percentile, 160
periodicity, 185, 189, 192
Philadelphia, 207
placebo, 11, 15, 16, 17, 94, 101
playing, 120
pneumoconiosis, 45
pneumonia, 200
policy makers, vii, 23
poor performance, 29
population, 24, 29, 30, 125, 134, 142, 158, 163, 230, 240, 243, 252
predictor variables, 26
probability, viii, 13, 15, 16, 25, 26, 28, 36, 39, 46, 58, 83, 84, 85, 91, 110, 121, 125, 135, 137, 143, 149, 153, 156, 173, 174, 180, 181, 220, 238, 239
probability density function, 46, 121, 135, 137, 156, 238
probability distribution, 91, 135, 137, 153, 173, 180, 220, 238
probability theory, 174
profit, 114
project, 24, 90
proportionality, 46, 201
proposition, 187
prostate cancer, 256, 258

Q

queuing theory, 84

R

radiation, 124, 250, 251, 252, 254, 255
radiation therapy, 124
random errors, 59
real numbers, 223
recombination, 257
recovery, 259
recurrence, viii, x, 83, 84, 86, 87, 199, 200, 205, 219, 221, 225, 226, 228, 230, 233, 235
regression, vii, viii, x, 1, 9, 11, 12, 13, 17, 20, 21, 23, 26, 103, 104, 105, 106, 107, 112, 114, 116, 117, 135, 142, 162, 196, 199, 200, 201, 206, 208, 238, 247, 250
regression analysis, 103, 104, 112, 116, 200, 201
regression method, 26

regression model, vii, viii, 1, 11, 12, 13, 17, 20, 21, 23, 103, 104, 106, 116, 117, 142, 196, 206, 208
regrowth, x, 249, 250, 253, 254
reliability, 46, 143, 144, 153, 154, 180, 199
repetitions, 56
requirements, vii, 11, 12
resection, 125, 128
residual error, 12
response, ix, 13, 15, 16, 119, 122, 123, 126
revenue, 24
risk, 55, 56, 57, 201, 202
river basins, 197
river flows, 194
root(s), 105, 106, 210, 212, 214, 215, 216

S

sample mean, 186
sample variance, 111, 159
Saudi Arabia, 45, 75
scaling, 105, 106
schizophrenia, 17
sensitivity, ix, 28, 109, 119, 128
shape, ix, 46, 119, 121, 126, 129, 133, 135, 137, 141, 156, 158, 160, 165, 166, 167, 196, 204
simple linear regression, 251
simple random sampling, 135, 142
simulation, vii, viii, ix, 11, 13, 17, 18, 21, 45, 46, 55, 108, 109, 111, 116, 155, 156, 166, 183, 187, 191, 197
simulations, ix, 133, 190
skewness, 9, 46
software, 17, 90, 104, 105, 123, 124, 125, 126, 129
solution, 16, 26, 158, 163
South America, 124
stability, 106, 110
standard deviation, ix, 30, 120, 122, 125, 126, 128, 129, 131, 205
standard error, 13, 16, 107, 112, 115, 195, 205, 254
state(s), 33, 61, 104, 114, 115
statistical inference, 84, 120
statistics, vii, x, 2, 8, 9, 37, 38, 43, 58, 108, 109, 125, 131, 133, 142, 144, 153, 160, 162, 163, 174, 180, 203, 204, 219, 220, 221, 224, 225, 226, 227, 230, 235, 237, 238, 247, 250, 257, 258
stress, viii, 45, 46, 200
stress-strength function, viii, 45
structure, 2, 10, 84, 187
subgroups, 127, 128
substitutions, 66
survival, ix, x, 23, 25, 26, 28, 36, 45, 119, 120, 124, 125, 126, 127, 128, 129, 133, 135, 142, 143, 155, 156, 174, 199, 200, 201, 207, 208
survivorship function, vii, 23, 26
Switzerland, 103
symmetry, 70
syndrome, 200
synthesis, 197

T

techniques, 156
test statistic, 8, 18, 19, 29
testing, vii, 1, 2, 8, 9, 12, 13, 16, 17, 18, 19, 20, 103, 120, 130, 250, 251
therapy, 120
time lags, ix, 183, 187, 189, 190, 192, 195
time series, ix, 183, 184, 185, 186, 187, 194, 195, 196, 197
time variables, vii, 23
transformation, 2, 5, 26, 27, 51, 53, 56, 161, 166, 238
transformations, vii, 1, 3, 6, 7, 135
translation, 43
transmission, 24
treatment, vii, viii, 1, 11, 12, 15, 16, 17, 59, 60, 61, 63, 64, 69, 78, 79, 94, 99, 101, 102, 120, 125, 126, 128, 129, 153, 200, 205, 209, 249, 250, 251, 252, 254
tumor(s), x, 124, 125, 128, 200, 204, 205, 249, 250, 251, 252, 254, 255, 256, 257, 259
tumor growth, 250, 251, 259
type I error rate, vii, 11, 12, 13, 16, 18, 19, 20

U

United States (USA), 82, 124, 130, 234
urban, 24, 36
urban areas, 24

V

variables, vii, viii, 1, 37, 45, 46, 55, 56, 84, 103, 104, 105, 106, 107, 109, 110, 111, 112, 113, 114, 116, 143, 144, 152, 153, 154, 166, 238
variance-covariance matrix, 105
variations, 251
vector, x, 13, 15, 60, 104, 105, 199, 200, 201, 204, 252

X

X-axis, 191, 192, 193, 195